Data-Driven Decision Making in Entrepreneurship

Tools for Maximizing Human Capital

Editors

Nikki Blacksmith
Co-founder & CEO of Blackhawke Behavior Science
Washington, DC, USA

Maureen E. McCusker
Co-Founder, Blackhawke Behavior Science
Senior Manager, Talent Analytics, Marriott International
Richmond, VA, USA

CRC Press
Taylor & Francis Group
Boca Raton London New York

CRC Press is an imprint of the
Taylor & Francis Group, an **informa** business
A SCIENCE PUBLISHERS BOOK

Cover illustration courtesy of Dane Wagner and Jonathan Boulos

First edition published 2024
by CRC Press
2385 NW Executive Center Drive, Suite 320, Boca Raton FL 33431

and by CRC Press
4 Park Square, Milton Park, Abingdon, Oxon, OX14 4RN

CRC Press is an imprint of Taylor & Francis Group, LLC

Library of Congress Cataloging-in-Publication Data (applied for)

ISBN: 978-1-032-05277-9 (hbk)
ISBN: 978-1-032-05278-6 (pbk)
ISBN: 978-1-003-19685-3 (ebk)

DOI: 10.1201/9781003196853

Typeset in Times New Roman
by Radiant Productions

Dedication

To my parents, for their unwavering love and support.

Nikki Blacksmith, Ph.D.

To all the entrepreneurs relentlessly fighting the systemic challenges to their success, and to my business partner and co-founder, for her strength and support.

Maureen E. McCusker, Ph.D.

Preface

When I was a graduate student in Industrial-Organizational (I-O) Psychology, I did not receive any formal training on entrepreneurship. So, when I began consulting for a venture capital firm, I read as much as possible about the science and practice of entrepreneurship and venture capital. That's when I ran across a couple of statistics that blew my mind, but not in a good way: (1) about 97% of venture capital in the United States is awarded to male founders and primarily white men with ties to Silicon Valley, and (2) only a tiny fraction (about 10%) of startups succeed. With the accumulated scientific knowledge regarding organizational performance across fields (e.g., management, psychology, economics), those are inexcusable metrics. As a Latina and as a woman, I took them quite personally. As an I-O psychologist, I was disheartened and felt culpable.

If an established organization were to hire 97% men, it would constitute a violation of labor law: a blatant disregard of the Civil Rights Act of 1964. Well, successful startups eventually expand beyond 15 people, necessitating compliance with Federal labor laws, including the Civil Rights Act of 1964. Therefore, prioritizing diversity as a fundamental goal for startups should be a legal imperative. Plus, diversity fuels the success of both the startup and the investor because it positively impacts the bottom line. So why are we seeing procedural injustice? Why aren't startups led by founders from marginalized groups getting funded?

Also, why is the failure rate so high? Founders can access many free tools and extensive information in the public forum. Numerous communities exist to support founders, including co-working spaces, incubators, accelerators, and venture studios. Venture capitalists have extensive knowledge and experience building companies. So why, then, are the success rates of venture-backed startups so meager? How are venture capitalists selecting startups to fund? How are venture investment decisions made? So many questions.

In seeking answers, I devoured any information I could find on how venture capital decisions are made. It quickly became apparent: we face familiar problems disguised by an unfamiliar context. First, the venture capital industry, characterized by a predominantly male composition, prioritizes founders with 'warm introductions,' contributing to the disproportionate funding of male founders. In I-O terms, unconscious biases contaminate the selection decision. Second, when conducting due diligence on the founders, reference calls and unstructured interviews are the most common methods, which we know (from I-O research) are some of the least fruitful methods to assess human capital.

Early versions of selection methods and processes in the 20th century were no different. Then, I-O psychologists introduced psychometric tests and mechanical decision aids to increase the accuracy and fairness of employment decisions. As of 2019, more than 75% of Fortune 500 companies used psychometric testing during recruitment and selection because they work. As demonstrated by the widespread use of standardized selection methods in large corporations, I-O psychologists have made a substantial impact.

In the 20th century, I-Os also advocated for policy change surrounding employment practices. I-Os were appointed to the Advisory Committee on Testing and Selection to assist with developing the Uniform Guidelines on Employee Selection Procedures, a framework designed to help organizations comply with Federal law prohibiting discrimination. Our field has undoubtedly fueled positive change. However, the beneficiaries primarily are established corporations. The neglect of startups in I-O science has contributed to the current state of the entrepreneurial ecosystem: it is exclusive and performing poorly. The absence of I-O psychology interventions at the earliest stages of a company is unfortunate and, frankly, irresponsible.

I take responsibility for that neglect and aim to catalyze change and embed psychological science and practice into the entrepreneurship ecosystem. We need a healthy entrepreneurship ecosystem now more than ever. As a global society, we look to startups and organizations to develop innovative solutions that will help us address our foremost challenges. I could explain more about the apocalyptic nature of our problems, but I'll spare you the rant. The point is: humanity is suffering and in danger. We can no longer engage in extravagant investing; it is unacceptable that 9 out of 10 startups fail. We do not have the financial or planetary resources to waste, and time is ticking. These problems will not magically disappear or dissipate. We only have one option, to take action now!

Psychology and related fields can play a significant role in addressing global plights vis-à-vis organizational improvement. We should invest our time, research, and grants into promoting a healthy and socially just entrepreneurial ecosystem. We've seen our practices work with large organizations, and startups are no more than newly conceived organizations. We've made a difference before, and we can do it again.

That is why my co-editor, Maureen McCusker, and I poured our spirit and energy into this book, intentionally intertwining science and practice. Each chapter illustrates how applying psychological science in the entrepreneurial ecosystem can help improve the startup success rate and increase diversity. Concurrently, the chapters present gaps in the scientific literature and call for future research. We also designed this volume to arouse advocacy for transforming labor law and policy. So to you, our reader, thank you for taking the time to learn and join the cause. We hope this book inspires you to take action—whether advancing our scientific understanding of startups, applying I-O tools, or advocating for change. We need both science and practice to create a healthy, innovative, and productive entrepreneurship ecosystem.

Nikki Blacksmith

Acknowledgements

It is our belief that (1) science ought to be applied for the betterment of society, and (2) applied practice grows stronger with a reliance on science. Accordingly, we are grateful for all the scientists and practitioners who enable the functioning of our society. However, it is also our belief that the hardest work lies in translating science into practice and translating practice into scientific research questions. It is from these translations that unthinkably innovative breakthroughs emerge, large-scale societal changes form, and society's most pressing and difficult problems are addressed. Those who lead and effectively execute these translation efforts are few and far between prodigies, making an impact, and driving change. We are honored, humbled, and profoundly grateful that some of the top translators in the organizational and entrepreneurial sciences agreed to contribute to this volume. We acknowledge and thank the authors of this volume, from whom we have learned, grown, and been inspired by their shared passion and enthusiasm for leveraging human capital and data to transform the entrepreneurship ecosystem for the better. We thank them for their courage to drive innovative work that challenges the status quo for the advancement of *all people* in the entrepreneurial ecosystem. We have concluded our work on this volume feeling as though we have built a community of translators, innovators, change-makers, colleagues, and friends. For all our readers, we welcome you to join us.

Contents

About the Contributors

Allison Piper Kimball, MBA, M.S.

Allison Piper Kimball is an investor, mentor, and board advisor to early-stage companies. A former energy industry executive, Allison worked closely with myriad entrepreneurs at all stages of company growth and development, from building a startup to taking companies in their early years from startup to sale and IPO. Allison is the managing partner of Wave 27 Ventures, which invests in angel and early-stage companies. She serves on the board and leadership team of the Houston Angel Network, on the board of directors of two private early-stage companies, and on the limited partner advisory committee of a venture capital fund. Allison is a volunteer judge and mentor to university and accelerator business plan and startup competitions and is a Tulane University "Innovator In Residence," providing mentoring to students and area small business startups. Allison holds an MBA in Finance and Information Systems from Tulane University, a Master of Science in Public Health from The University of North Carolina at Chapel Hill, and a Bachelor of Science from Clarkson University in New York. Allison is a doctoral student at Business School Lausanne in Lausanne, Switzerland, focusing on entrepreneurship.

Anthony Roberson, M.A.

Anthony Roberson holds his Bachelor of Arts degree in Psychology from Albright College and a Master's Degree from Penn State University in Industrial-Organizational Psychology. With experience on consulting projects for multiple Fortune 500 companies on developing, implementing, and assessing creativity initiatives and working various applied projects through his academic programs, Anthony has developed skills in leadership assessment and development, diversity, equity, and inclusion (DEI), professional communication, and program coordination. He has also developed technical skills from working on projects such as developing a technical report for a DEI assessment tool, analyzing client data, and developing slide decks for client presentations. Currently a Ph.D. student at the University of Nebraska, Omaha, and a Consulting Assistant at Mattingly Solutions, Anthony aspires to help organizations select, develop, and support a diverse array of leaders and create an inclusive and safe culture where every employee can thrive.

Christine Yin Man Fong, Ph.D.

Christine Yin Man Fong is an Assistant Professor at the Tilburg University. Her research focuses on the impact of job design on the well-being and performance of employees. Her work has been published in *Career Development International, Applied Psychology: An International Review,* and *Sustainability.*

Dane Luke Wagner, M.S.

With over a decade of experience in data analytics and UX research, Dane has contributed to industry research on topics such as loss prevention, behavior modification, employee satisfaction, and memory/cognitive processes. He currently works for Blackhawke Behavior Science where he is a Co-Founder and Chief Product Officer. Dane's research passions lie in the integration of modern technology to standardize qualitative research, bridging the science practitioner gap through applied technologies education, and improving employee experience models for smaller organizations and businesses. He holds a Master's degree in Industrial-Organizational Psychology and is still active in the College of Psychology as well as the College of Technology and Computer Science at Florida International University.

Holly Holladay-Sandidge

Holly Holladay-Sandidge is a doctoral student in Organizational Science at the University of North Carolina at Charlotte. Her research interests include entrepreneurship, organizational strategy, and strategic HR. Currently, she is most interested in the relationship between founder and organizational identity, as well as ownership of work. Prior to her doctoral studies, Holly owned and operated her own small business, taught, and worked as a research associate at Mississippi State University. She has consulted for numerous businesses and organizations on topics related to organizational management, marketing, and starting new ventures.

Ian M. Hughes, M.A.

Ian M. Hughes is a doctoral candidate studying Industrial-Organizational Psychology at Bowling Green State University. His research, focused on the intersection of social exchanges and interactions with employee well-being, has been featured in outlets such as the *Journal of Business and Psychology*, *Occupational Health Science*, and more. Beyond this stream of research, he also has an interest in developing psychometrically sound measures to assess various phenomena. He plans on pursuing a job in academia as a professor, where he can spend the rest of his days teaching, researching, and mentoring young minds. Finally, he enjoys spending time with his cat.

Jennifer Pelt Wisdom, Ph.D. MH, ABPP

Jennifer Pelt Wisdom, Ph.D. MPH ABPP, is a licensed clinical psychologist and board-certified business and organizational psychologist who specializes in research and training associated with human capital development, organizational change, and leadership through her company, Wisdom Consulting. She has worked with complex health care, government, and educational environments for more than 25 years, including serving active duty in the U.S. military, working with non-profit service delivery programs, and as faculty in higher education. She is the author of more than 85 peer-reviewed scientific papers and the popular press *Millennials' Guides* series, which includes *Millennials' Guide to Work*; *Millennials' Guide to Management and Leadership*; *Millennials' Guide to Diversity, Equity, and Inclusion; Millennials' Guide to Workplace Politics*; and *Generation Z's Guide to Work*. She is the 2022–2023 President of the Society for Psychologists in Leadership. She earned a Ph.D. in Clinical Psychology from The George Washington University in Washington, DC, and a Master of Public Health in Epidemiology and Biostatistics from Oregon Health & Science University. She is an intrepid adventurer based in New York City and Portland, Oregon. She can be reached at www.leadwithwisdom.com.

Jerome Katz, Ph.D.

Jerome Katz (Ph.D., University of Michigan) is an internationally recognized expert on entrepreneurship and entrepreneurship education. He taught at Wharton before moving to Saint Louis University where he has been over a third of a century, today as the Brockhaus Chair of Entrepreneurship and Director of Saint Louis University's nationally ranked entrepreneurship program. Jerome is the lead author of *Entrepreneurial Small Business* (McGraw-Hill) one of the top three texts globally in entrepreneurship and small business, which like this chapter melds his personal experiences with the latest in research-based entrepreneurship pedagogy. He is also the founder of the Emerald series *Advances in Entrepreneurship, Firm Emergence and Growth,* and the Sage series "Entrepreneurship and the Management of Growing Enterprises." He serves on nine editorial boards and has received the Foundational Paper Award of the AOM Entrepreneurship Division as well as having eight articles republished in compendia of classic entrepreneurship papers. He has been quoted in more than 200 articles in the major magazines in business and entrepreneurship, as well as the major American newspapers. He has had a lifelong interest in developing peoples' potential, serving as co-chair of the Academy of Management's Mentoring Committee, creator of the AOM Entrepreneurship Division's Late-Career Consortium, and as a Longenecker Fellow of USASBE. He has served as a former Vice-President for Research of ICSB, and a former Governor and Entrepreneurship Division Chair of the Academy of Management. Much of his entrepreneurship education materials are available at his website http://eweb.slu.edu/.

Jessica Hart, MBA

Jessica is a consultant at Boston Consulting Group where she is passionate about improving the way our society works through both the private sector, government, and non-profit organizations. She graduated from Harvard Business School in the MBA class of 2021 where she conducted the research behind this chapter and established the HBS Impact Fund which seeks to expand access to affordable, flexible capital for Black, Brown, and Immigrant entrepreneurs. Jessica lives in London and holds a BEng in Electronic Engineering from the University of Manchester.

Karthik Varada, Ph.D.

Karthik Varada is a Project Researcher at the University of Tokyo, studying ESG and climate-tech investments. Previously, Karthik was the founding member of Asia Women Impact Fund, a $100m impact investing fund at the Sasakawa Peace Foundation. Karthik has advised several venture capital funds and startups on impact investments and impact management. Karthik holds a Ph.D. from the University of Tokyo, focused on ESG and gender-lens investing. He holds a master's degree from the University of Tokyo and a bachelor's in engineering from the Indian Institute of Technology, Hyderabad.

Kelly Diouf

Kelly is a Research Analyst for Blackhawke Behavior Science, where she conducts research on selection, motivation, coaching, and team development. She is gaining an expertise in psychometric testing. As an aspiring I-O psychologist, Kelly is dedicated to helping entrepreneurs succeed by translating industrial-organizational science to a wider audience, and by creating products designed to develop leadership skills and team cohesion. She holds a B.A. in Psychology from American University and is taking a gap year to accumulate knowledge and experience in the industrial-organizational psychology field.

Kelsie Colley, M.A.

Kelsie (she/her) is an Industrial Organizational Psychologist, ABD. As a People Research Consultant at Zoom and the global lead of Neurodiversity at Zoom Affinity Group, she's on a relentless quest to bring the scientist-practitioner model to life. Kelsie leverages her passion for research, advising, and inclusive practices to empower organizations through data-driven decision making, DEI, Employee Listening, People Analytics, and Leadership Consulting. She's authored peer-reviewed papers, book chapters, and white papers in various fields, from selection and game-based cognitive ability testing to autism, DEI, and data analysis. As an autistic professional with ADHD, Kelsie knows that representation and employee voice are catalysts for success. Her research delves into the intricate connections between technology, strategy, culture, and diversity equity and inclusion, all with the aim of translating research into actionable practices that support employee flourishing. Kelsie holds a Master's degree in Industrial/Organizational Psychology, is near the final stages of her PhD journey, and proudly contributes her insights to the Society for Industrial/Organizational Psychology (SIOP).

Lindsey M. Freier, M.A.

Lindsey M. Freier is a doctoral candidate studying Industrial-Organizational Psychology at Bowling Green State University. Her research focuses on predictors of employee well-being, particularly those related to the social environment and culture of organizations, creativity and innovation, and judgement and decision making. Her research has been published in journals such as the *International Journal of Selection and Assessment* and *Occupational Health Science*. She particularly enjoys combining these three, often disparate, streams of research, and is passionate about applying her research to positively impact both employees and organizations.

Neil Morelli, Ph.D.

Neil Morelli, Ph.D., is the Chief I-O Psychologist at Codility, a technical hiring platform for companies such as Microsoft, Tesla, and Paypal to evaluate the engineering skills of their job candidates. At Codility, Dr. Morelli specializes in validating and creating effective hiring processes through technology-enabled talent assessment. Before joining Codility, Dr. Morelli was the VP of Product at HighMatch, an Atlanta-based pre-hire assessment software company, and the Head of Product and Selection Science for The Cole Group, a Bay Area executive search firm serving high-growth startups and technology companies. Dr. Morelli has also taught graduate-level Talent Acquisition at the University of Georgia and authored peer-reviewed publications in the *International Journal of Selection and Assessment*, the *Journal of Business and Psychology, The Handbook of Employee Commitment*, and *The Cambridge Handbook of Technology and Employee Behavior*.

Nilima Ajaikumar

Nilima Ajaikumar is a People and Organization consultant with broad experience supporting organizations in the private, public, and non-profit sectors on organizational strategy and design, HR transformation, talent, and change management initiatives. She leads research and people analytics at Deeper Signals and is currently pursuing doctoral studies in Organizational Psychology at Columbia University. Previously she worked in organizational consulting roles at PwC and Brookfield Properties.

Reece Akhtar, Ph.D.

Dr. Reece Akhtar is a co-founder and CEO at Deeper Signals. He is an organizational psychologist and data scientist specializing in applied personality assessment and computational psychometrics. As a lecturer at NYU and UCL, he has published scientific articles on personality and machine learning, talent management, and leadership. Previously he led product innovation at RHR International and Hogan Assessments Systems.

Rosalyn Sandoval, Ph.D.

Dr. Rosalyn Sandoval is an assistant professor of management and entrepreneurship in the College of Business at California State University, Monterey Bay. She is also an equity and entrepreneurship consultant for a new venture called Work4Change. Rosalyn is an expert on human behavior, organizations, entrepreneurship, and inequality. She believes in finding creative and innovative solutions to workplace issues that benefit the well-being of stakeholders and organizations. Rosalyn's scholarly interests broadly focus on diversity, equity, inclusion, and justice within entrepreneurship. Specifically, she focuses on the role that founders and leaders of new ventures play in shaping gender and racial inequality within their ventures. Before getting her Ph.D. in Organizational Science from the University of North Carolina in Charlotte, she had successfully started and led several new ventures, where she gained a valuable practical understanding of entrepreneurship and leadership. Now, as a researcher studying entrepreneurship, she has a chance to combine the practical with the theoretical and apply these findings to the benefit of entrepreneurs and new ventures. She has published several articles and book chapters in top journals such as the *Journal of Business and Psychology, Gender and Society*, and *Journal of Business Venturing*. When not working, she enjoys her full house of pets, time with family, being outdoors, and listening to audiobooks while putting together puzzles.

Samantha Dubrow, Ph.D.

Dr. Samantha Dubrow is Human-Centered Engineering Researcher at The MITRE Corporation. At MITRE, Samantha conducts applied research and development in Industrial-Organizational Psychology, teamwork and leadership, human factors, user experience, human-machine teaming, and multiteam system collaboration management. She helps teams and multiteam systems across a variety of government agencies utilize technology to improve their teamwork processes and job performance. Samantha holds a Ph.D. in Industrial-Organizational Psychology from George Mason University and has published articles in a variety of journals including *The Leadership Quarterly*, *Annual Review of Organizational Psychology and Organizational Behavior*, *Personnel Psychology*, *Human Performance*, *Journal of Managerial Psychology*, and in the proceedings of I/ITSEC. She has also presented at a variety of conferences, including the Society of Industrial-Organizational Psychology (SIOP), the Interdisciplinary Network for Group Research (INGRoup), and the International Meeting for Simulation in Healthcare (IMSH).

Sarah Resnick

Sarah Resnick is a second-year doctoral student in the Leadership and Organization Science department at Binghamton University School of Management. Prior to this, she earned her B.S. in Marketing from Case Western Reserve University and worked for many years in live event production. Sarah's research interests include networks and network contagion, employee activism, virtual teams, and diversity. She is co-organizing a panel symposium at the upcoming Academy of Management Conference.

Sertrice Grice, M.S.

Sertrice Grice, M.S. (she/her), is Co-Owner and Chief Consulting Officer of Mattingly Solutions, a woman-owned diversity, equity, and inclusion (DEI) consulting firm. Co-author of *Inclusalytics: How Diversity, Equity, and Inclusion Leaders Use Data to Drive Their Work*, Sertrice is passionate about helping organizations use metrics to create data-driven DEI strategies that drive meaningful change. Some of her clients include, DICK's Sporting Goods, Baltimore Tracks, and the National Association of Women Business Owners (NAWBO). Sertrice serves clients through DEI executive advising, measurement consulting, and building/delivering inclusive behavioral change solutions. She is also a DEI thought leader, a highly acclaimed public speaker, a leader in national organizations including the Society for Industrial and Organizational Psychology, and Blacks in I-O, and volunteers for her sorority Alpha Sigma Alpha. Sertrice received her B.S. in Psychology from Emporia State University and her M.S. in Industrial-Organizational Psychology from Radford University. She is currently located in Raleigh, North Carolina, and spends most of her free time cuddling her dog, Winston.

Tom Schmittzehe, MC/MPA, MBA

Tom is the founder and Managing Partner of Moonshot Ventures, a venture capital and impact investing firm operating in Southeast Asia, which in 2020 launched the Indonesia Women Empowerment Fund (IWEF), the first venture fund dedicated to Gender Lens Investing and supporting women startup founders in Indonesia. Tom has spent the last 25 years working with startups, including as an investor, advisor, and founder—primarily within the emerging sector of impact investing and in Asia. He previously held regional leadership roles with Ashoka, a global pioneer of impact investing, and he co-founded two successful impact investing firms in Asia. He has a particular interest in social innovation and has helped launch several new approaches in impact investing, particularly related to blended finance. He advises development agencies and multilaterals on social innovation and impact investing, such as the UNDP and USAID. Tom holds degrees from Oxford University, the Wharton School of the University of Pennsylvania, and the Harvard University John F. Kennedy School of Government (HKS).

Tomas Rosales, MSc

Tomas is currently the Vice President of Growth at ZUMI, a Kenya-based startup that focuses on empowering informal retailers. Previously, he worked at growth equity investor Verdane Capital. Whilst there, he designed and implemented the strategy for a €300 million impact fund, one of the first fully commercial funds to adopt the impact linked carry model. Prior to Verdane, Tomas worked at Boston Consulting Group with digital business models and social impact. He also spent time at the development agency UNOPS in Myanmar. Tomas graduated with an MBA from Harvard Business School, where he and Jessica conducted the research behind this chapter. He also wrote cases on social impact organizations. Tomas holds an MSc in Applied Economics and Finance from Copenhagen Business School.

Victoria Mattingly, Ph.D.

Dr. Victoria Mattingly (she/her) is CEO and Founder of Mattingly Solutions, a woman-owned, diversity, equity, and inclusion (DEI) consulting firm that uses data and science to improve the human experience at work. She earned her Ph.D. in industrial-organizational psychology from Colorado State University and has been providing DEI and learning & development consulting services to businesses and non-profits for over 10 years, including Amazon, Intel, and Duracell. Her research and practice focus on using psychometrics and behavior change science to co-create and deliver more evidence-based workplace solutions to clients who desire to make work better for people from *all* backgrounds and identities. She's a workplace inclusion and allyship scientist-practitioner, publishing peer-reviewed articles and presenting her work at academic conferences while also writing popular press publications to better reach her end users (business and people leaders), a sought-after business school instructor, and an esteemed DEI public speaker and thought leader. She is also co-author of the book *Inclusalytics: How Diversity, Equity, and Inclusion Leaders Use Data to Drive Their Work.*

Yik Kiu Leung, Ph.D.

Yik Kiu Leung is an Assistant Professor at the Jheronimus Academy of Data Science and Tilburg University. His research combines insights from psychology with data science methods to understand the well-being and mental health of entrepreneurs. His current research focuses on the link between mental health symptoms (e.g., ADHD and narcissism) and entrepreneurial outcomes (e.g., intention, success, and well-being). His research has been published in the *Journal of Business Venturing Insights* and the *Journal of Small Business Management* as well as featured in different media outlets.

Part I

Human Capital Assessment and Development

Chapter 1

Introduction

Nikki Blacksmith[1,*] and *Maureen E McCusker*[2]

> *"Scientists tend to resist interdisciplinary inquiries into their own territory. In many instances, such parochialism is founded on the fear that intrusion from other disciplines would compete unfairly for limited financial resources and thus diminish their own opportunity for research."*
>
> —Hannes Alfvén, 1986

The Power of Interdisciplinary Inquiry for Startups

Interdisciplinary inquiry incites innovation and ingenuity. History shows us this phenomenon countless times, well before the Nobel Prize-winning Physicist quoted above called out scientists' resistance and fear in doing so. Today, we have countless scientific fields (i.e., neurology, clinical psychology, family and marriage psychology, social psychology, organizational psychology, sports psychology, management, and entrepreneurialism) all working within our respective disciplines, yet all with similar aims to better understand and predict human behavior. Scientists and practitioners consistently make incredible advancements within their fields that can be directly or indirectly translated or applied in a separate discipline, but they are instead retained in their own disciplines' territory.

The purpose of this book is thus to inspire rather than resist the integration of the organizational sciences, especially industrial-organizational (I-O) psychology, and entrepreneurship research. Entrepreneurship is typically taught and studied within business schools in universities where courses in psychometrics and test development are rare. I-O psychology is housed in the psychology department

[1] American University, Blackhawke Behavior Science, 1709 21st St NW #52, Washington, DC 20009.
[2] American University, Blackhawke Behavior Science, 3325 Stuart Ave, Richmond, VA 23221.
Email: Maureenemccusker@gmail.com
* Corresponding author: nikkiblacksmith@gmail.com

where courses in entrepreneurship are nearly nonexistent. We are unaware of any I-O psychology programs that offer courses in entrepreneurship. The separation of these two disciplines is disappointing given the opportunities for cross-fertilization. While there have been attempts to integrate the broader field of psychology into entrepreneurship research (e.g., Frese 2009, Frese and Gielnik 2014), their impact has been limited and challenged (Davidsson 2016). The studies coming from "the psychology of entrepreneurship" research primarily focused on five broad areas: careers perspective, individual differences, health and well-being, cognition and behavior, and entrepreneurial leadership (Gorgievski and Stephan 2016). Most research examining the role of the person in entrepreneur took a trait-based approach, akin to the antiquated trait theories of leadership from the late 19th and early 20th century, suggesting entrepreneurs are born, not made (Davidsson 2016). While these approaches made critical advancements in the field, they are reductionist in that they overlook the role of the context (e.g., organizational culture, regional ecosystem) and situation (e.g., task characteristics, team cohesion).

Why I-O Psychology for Startups?

While the psychological field has much to offer entrepreneurship research, this volume focuses on I-O psychology because of its focus on *human capital* at multiple levels of analysis (e.g., individual, team, organization). The I-O psychology field has greatly advanced over the last century (Kozlowski et al. 2017). The I-O field, in particular, has made great advancements in organizational research methods and measurement tools to understand the complexities of human behavior in organizations (Cortina et al. 2017). Those in the I-O field have deep expertise in measurement error, validity, meta-analysis, rater agreement and aggregation, test theories and development, mediating and moderating effects, multi-level effects, and many other sophisticated methodological approaches that have enabled great advancements in understanding humans and collectives in organizations. I-O psychology embraces a multi-level approach to understanding organizations where individuals are nested in teams that are nested in multi-team systems, which are nested in the organization, which is nested in an external ecosystem. As such, we argue that entrepreneurship should be studied as a complex, adaptive system that involves all levels of an organization (i.e., individuals, teams, organizations, and ecosystems), dynamically interacting over time. Through this approach, we can gain a better understanding of how human capital impacts startup performance.

From Product to People

"The Lean Startup" by Eric Ries was published in 2011 and revolutionized the way founders and leaders approached entrepreneurship (Blank 2013, Gray 2021). It was perceived as a "holy book" and ignited a movement in the entrepreneurial ecosystem as well as in entrepreneurship education and scholarship (Gray 2021). Ries (2011) proposed a "build-measure-learn" approach to entrepreneurship. In other words, founders build a minimal viable product (MVP) and put it in the hands of the customer as soon as possible, and measure the feedback. From there, the founder iterates the product several times based on what they learned from the customer feedback. The predominant model before the lean startup model was to develop a product and release it when it was perfect (Gray 2021). Ries (2011) argued that perfecting a product before testing and validating led to higher risk, costly redesign, and a long design cycle.

The lean startup model was not without critics. For instance, Ben Horowitz from the infamous Andreessen Horowitz venture capital firm argued that the lean startup model missed the mark because it did not focus on aspects such as winning the market and maintaining financial stability (Koss 2021). Other strategies for building a startup include a focus on building intellectual property or disrupting a market (Gans et al. 2018). Moreover, most education and information about building a successful startup focuses on product development, marketing, and financial management.

What is lacking glaringly in all these strategies and proposed approaches to building startups is a focus on human capital. Without people, a startup cannot exist. People design the products. People ideate the strategy. People fundraise. People sell. The people execute. People are the foundation and spirit of every startup, and to be successful, a team must be united and work towards a common goal. Human resource problems, conversely, put the startup at risk for failure.

We argue that human capital should be a primary, rather than a secondary or tertiary, focus for startups. Without the right people all the other components are moot because it is the people who generate ideas, make decisions, and execute all of the business functions. People power all startup activities from generating the vision to closing sales to building products. People, therefore, are a startup's greatest asset. However, when not managed well, people become a startup's biggest liability.

As many of the chapters in this volume show, the primary reasons startups fail arguably stem from human capital. Approximately 65% percent of VC-backed startups fail because they did not build a qualified team at the start (Main 2021). In addition, team and teamwork challenges were mentioned by 23% of startups as a factor in their failure in a study by CB Insights. Lastly, people are ultimately the root cause of other reasons for failure such as lack of product-market fit or poor go-to-market strategy.

Ramifications of sidelining human capital management affect more than just the founders, employees, and investors. The entrepreneurial ecosystem is anything but diverse (see Chapter 9). The lack of diversity — at the macro-economic level — is a social justice issue. For startups, it leads to performance misses. Research has demonstrated that diversity can affect innovation, discrimination, communication, performance, and team success. Yet, 72% of startups are founded by an all-male team (Silicon Valley Bank 2020). Only 28% have at least one woman co-founder.

This Volume

This book is devoted to the research and data-driven practice of managing human capital in startups. By integrating insights, methods, and tools from the organizational and psychological sciences into the science of entrepreneurialism, the chapters in this volume grow the burgeoning and interdisciplinary science of startup success. Collectively, the chapters herein focus on the people as the agents underlying the functioning at various levels of the entrepreneurial ecosystem (i.e., the individual entrepreneur and the individual job, the entrepreneurial team and social/environmental context, and the collective startup and other human-based agents in the entrepreneurial ecosystem).

Part I of this volume focuses on methods of assessing and developing the human capital of entrepreneurs, entrepreneurial teams, and startup workers. *Chapter 2: Work Analysis-based Job Descriptions: The Secret to Finding the Right Startup Talent at the Right Time* by **Neil Morelli** begins by detailing work analysis, an indispensable method of understanding work in order to predicting performance. In *Chapter 3: Identifying and Measuring Entrepreneurial Talent in the Age of Artificial Intelligence*, **Nilima Ajaikumar** and **Reece Akhtar** bring to light predictive individual differences that can be the focus of future research studying person-based antecedents of entrepreneurial performance. *Chapter 4: Professional Human Capital Development for Startup Founders and Workers* by **Jennifer Wisdom** introduces learning and development into the entrepreneurship process. **Samantha Dubrow**, **Sarah Resnick**, and **Anwesha Choudhury** then provide a solid foundation for understanding entrepreneurial teams in *Chapter 5: Selection and Training for Teamwork: Implications for Diverse, Virtual, and Human-Machine Teams*. **Nikki Blacksmith, Kelly Diouf, and Maureen McCusker** conclude Part I with an explanation of how we can adapt and apply personnel selection research — a pinnacle of I-O psychology—to the venture capital decision making process in Chapter 6: *Human Capital Due Diligence: Leveraging Psychometric Testing for Wiser Investment Decisions.* Specifically, they discuss how human capital due diligence processes can benefit from psychometric testing.

Part II of this volume focused on the social-environmental and situational factors that are critical to understanding entrepreneurship performance and well-being. In *Chapter 7: Opportunity or Threat? Entrepreneurs' Well-Being and Performance in the Data-Driven Era*, **Yik Kiu Leung** and **Christine Yin** Man Fong explain that information and communication technologies are inseparable from entrepreneurial work which can offer positive and negative consequences. **Victoria Mattingly**, **Sertrice Grice**, **Kelsie Colley**, and **Anthony Roberson** provide insights on how to take a data-driven approach to study diversity, inclusion, and equity at the team and organizational level in *Chapter 8: Using Data to Build More Diverse, Equitable, And Inclusive Startups.* **Dane Luke Wagner** then discusses the growing amount of data in startups and how it can be mined to aid in human capital management in *Chapter 9: An Introduction to the Utilization and Application of Text Analysis.* **Lindsey Freier** and **Ian Hughes** bring to light the social and human complexities of the culture in entrepreneurial settings in *Chapter 10: Promoting Well-Being and Innovation in Startups: The Role of the Social Environment.* In *Chapter 11: Understanding the Basics of Startup Development Organizations*, **Allison Piper Kimball** reminds us that startups do not exist in isolation and talks about the growing role of startup development organizations in startup success.

Part III focuses on measuring organizational-level performance. In C*hapter 12: Cultures of Evaluation: Leveraging Academia for Due Diligence in Angel Investments*, **Jerome Katz** provides an insider view of how angel investors select the startups in which they invest. **Rosalyn Sandoval** and **Holly Holladay-Sandidge** argue that understanding startup performance is more than just financials in *Chapter 12: More Than Money: Considering Nonfinancial Measures of Organizational Performance in Startups*. Similarly, in *Chapter 13: Incentivizing Investors to Make Impactful Investments: Introducing a Model for Impact-Linked Carry*, **Jessica Hart**, **Karthik Varada**, **Tom Schmittzehe**, and **Tomas Rosales** share emerging methods for measuring environmental, social, and governance impacts.

References

Blank, S. (2013). Why the lean start-up changes everything. Harvard Business Review. Retrieved from https://hbr.org/2013/05/why-the-lean-start-up-changes-everything.

Cortina, J.M., Aguinis, H. and DeShon, R.P. (2017). Twilight of dawn or of evening? A century of research methods in the Journal of Applied Psychology. *Journal of Applied Psychology*, 102(3): 274.

Davidsson, P. (2016). A "business researcher" view on opportunities for psychology in entrepreneurship research. *Applied Psychology*, 65(3): 628–636.

Frese, M. (2009). Towards a psychology of entrepreneurship: An action theory perspective. *Foundations and Trends in Entrepreneurship*, 5(6): 437–496. doi:10.1561/0300000028.

Frese, M. and Gielnik, M.M. (2014). The psychology of entrepreneurship. *Annual Review of Organizational Psychology and Organizational Behavior*, 1: 413–438.

Gans, J., Scott, E.L. and Stern, S. (2018). Strategy for startups. Harvard Business Review. Retrieved from https://hbr.org/2018/05/strategy-for-start-ups#strategy-for-startups.

Gorgievski, M.J. and Stephan, U. (2016). Advancing the psychology of entrepreneurship: A review of the psychological literature and an introduction. *Applied Psychology*, 65(3): 437–468.

Gray, R. (2021). 10 years since 'The Lean Startup': A product developer's perspective. Forbes. Retrieved from https://www.forbes.com/sites/forbesbusinesscouncil/2021/02/17/10-years-since-the-lean-startup-a-product-developers-perspective/?sh=302801ad7d8f.

Koss, H. (2021). The Tech World Fell for Lean Startup. Was That a Mistake? Built In. Retrieved from https://builtin.com/founders-entrepreneurship/lean-startup.

Kozlowski, S.W., Chen, G. and Salas, E. (2017). One hundred years of the Journal of Applied Psychology: Background, evolution, and scientific trends. *Journal of Applied Psychology*, 102(3): 237.

Main, K. (2021). Sixty-five percent of VC-backed startups fail because they don't ask these two questions. Inc. Retrieved from https://www.inc.com/kelly-main/65-percent-of-vc-backed-startups-fail-because-they-dont-ask-these-2-questions.html.

Reis, E. (2011). The lean startup. *New York: Crown Business*, 27: 2016–2020.

Silicon Valley Bank. (2020). 2020 Women in US Technology Leadership Report. Retrieved from: https://www.svb.com/globalassets/library/uploadedfiles/content/trends_and_insights/reports/women-in-us-technology-leadership-2020-silicon-valley-bank.pdf.

Society for Industrial-Organizational Psychology. (2018). *Principles for the Validation and Use of Personnel Selection Procedures*. Bowling Green, OH: Society for Industrial-Organizational Psychology.

Chapter 2

Work Analysis-based Job Descriptions

The Secret to Finding the Right Startup Talent at the Right Time

Neil Morelli

Finding, hiring, and keeping the right team are critical win conditions for startups. In an analysis of more than 110 startup failures, CBInsights (2021a) found that 1 in 10 startups fail because they do not have the right team—one of the top 12 reasons overall. A 2020 McKinsey survey of nearly 500 global companies found that "transforming the talent strategy" resulted in the most value to top and bottom lines—more than 'tangible' transformations such as updating vendor management (Dhasarathy et al. 2021).

Academic studies have backed up these industry findings. Research has found that effective staffing and personnel development practices are consistent predictors of success, often falling under the "strategic human capital management (HCM)" category. The relationship between strategic HCM and company performance holds across industries and companies of all sizes (Alagaraja 2012, Katou 2009, Sung and Choi 2014).

Strategic HCM practices often include building more robust hiring systems, investing in training dollars, or overhauling employee recognition and feedback programs. Investing in strategic HCM ultimately increases revenue, productivity, and growth by increasing or improving a team's collective effort, innovation, and retention. In other words, successful talent in a startup initiates a virtuous chain reaction: talent supplies the competency, attitude, and culture that accomplishes the strategy, differentiating the startup from competitors, and successful differentiation culminates in competitive advantages and financial performance (Huselid et al. 2005, Ployhart et al. 2018).

Chief I-O Psychologist, Codility, 107 Cheapside, 9th Floor, London, UK.
Email: neil.morelli@codility.com

But forget the corporate-sounding adage to "transform the talent strategy." There is a simple yet effective secret to impactful HR processes and people decisions (i.e., knowing who to hire, promote, train, or exit)—the humble yet misunderstood job description.

This chapter offers a step-by-step guide for building better job descriptions using an evidence-based approach called work analysis, the systematic method for understanding work, and the human capabilities or characteristics it requires. Work analysis provides the bedrock data for the science and practice of industrial-organizational (I-O) psychology. Now, entrepreneurs can learn this data gathering methodology to empower evidence-based decisions that serve every area of their business as they scale.

Why Work Analysis-Powered Job Descriptions Matter to Startup Business

After learning about work analysis, entrepreneurs may ask: "Isn't the effort to analyze work in my business a luxury only large companies can afford?" Or think: "Work analysis sounds like time away from more important things like developing products, securing funding, and building my brand!" Yes, larger businesses are usually the ones to practice work analysis. And yes, work analysis can take time. But, failing to understand work (or workers) is often the root cause behind inefficient and costly people processes that erode a startup's effectiveness when engineering, fundraising, or marketing. If left unchecked, lackluster people practices can cause ripple effects that negatively impact a business for years. Here are a few examples of what can go wrong when startup work goes unanalyzed.

Costly Mishires

Failing to understand a job's purpose and capability requirements underlies a common hiring mistake: choosing candidate profiles based on how similar they are to anecdotal examples of individuals who have done the job before. These examples are usually former colleagues and bosses, leaders at competitors or notable businesses, or famous people who have been successful in a similar role. Although looking for people who have 'been there and done that' at notable brands might offer some signal, overly relying on profiles or pattern-matching backgrounds without first understanding the work often nets decisions based on irrelevant factors, personal biases, outdated assumptions, and happenstance.

And there's more. Profile-based decisions are more likely to focus on the wrong predictors–reputation, likeability, and inflated or irrelevant experiences — they are also more likely to have misaligned expectations. Role expectations theory provides a helpful explanation for this problem. Work psychologists describe work as a social exchange between two parties with shared expectations about a company's needs and a worker's obligations (Dierdorff and Morgeson 2007). Expectations need to be communicated well for the social exchange to work effectively. In other words, there needs to be a consensus between employer and employee on what should be done in a job. One study found that agreement between an employer and employee on role expectations was lower when personal traits defined a job (e.g., one might say, "a candidate should be like Dave: smart, funny, and a numbers guy"). Lower consensus resulted in poorer employee performance and attitudes like job satisfaction and commitment (Dierdorff and Morgeson 2007).

If leaders misunderstand the work their startup needs it can lead to mishires with real costs. First, many startups choose to outsource hiring and invest significant funds into external recruiters that source and attract candidates that are not a good fit. A mishire compounds costs due to backfill hiring expenses, lost compensation, severance, productivity disruption, and opportunity costs. The consulting firm, GhSmart, estimated that "the average hiring mistake" costs firms 15 times an employee's base salary, and an executive mishire costs 27 times the base salary (Smart and Street 2008, Smart 2012). Without correctly identifying the traits and abilities required for work, companies end up with biased hiring practices that ultimately miss great talent and waste precious resources.

Reactive Workforce Planning

Iterating and learning quickly through trial-and-error is standard startup practice, but how entrepreneurs set and execute their talent strategy is one exception. An effective talent strategy finds the "talent quality sweet spot," where enough high-quality individuals are recruited and hired to achieve its performance objectives, but not too many where resources are wasted on acquiring and retaining them (Ployhart et al. 2018). Therefore, an effective talent strategy understands how work is changing to best anticipate future business needs.

Consider how the 'data scientist' role has changed as an illustrative example. Data scientists help firms make scientific or data-driven decisions to run more effectively. Today's data scientists collect and analyze datasets and suggest hypotheses and actions. However, this was not always the case. The label *data scientists* evolved from the term "data miner," popularly used from the 1960s to the 1980s, to describe people who collected, analyzed, and suggested actions from data.

Although the label "data miner" was still in use through the 1990s, computing power and data storage capacity increased significantly as cloud services became more available in the mid-2000s (Ben-David 2020). More data and interconnected systems offered more sophisticated approaches and tools for modeling and visualizing data, ushering in the role's "data scientist" label. As the role evolves into engineering and maintaining infrastructure for automated systems, machine learning-focused software engineers and machine learning researchers are now growing in demand (Fig. 1). This is one example of how a job title can take on new meanings and assumptions as technology changes. Therefore, assuming what work is required in the role may miss an opportunity to hire the right person for the present and fail to understand what will be needed in the future.

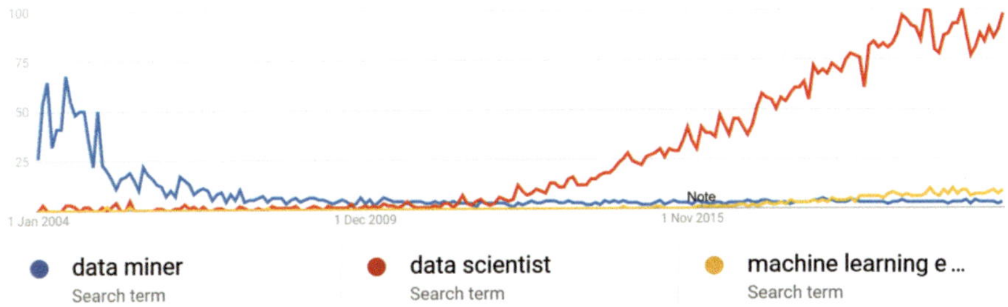

Figure 1. Google trends data on the use of data scientist-related job titles.

Systemic, Unconscious Biases

Finally, a cursory understanding of the organization's work leads to job descriptions that are fertile soil for unconscious and implicit bias. Job descriptions are often a company's first impression with potential job candidates. Descriptions that misrepresent the work performed or add irrelevant requirements can scare diverse, qualified candidates away and reflect poorly on the startup's employment brand. In other words, without an in-depth understanding of the job's mission, tasks, and capability requirements, job descriptions suffer from an informational vacuum usually filled with biased and fluffy language that sounds best to the author.

For example, a 2011 study (Gaucher et al.) discovered job advertisements that used more masculine-perceieved, such as "dominance" or "competitive," versus more feminine-percieved, such as "support" or "understand," received more male candidates. In addition, women perceived more masculinely worded job advertisements as *less* appealing than ads with gender-neutral wording (e.g., referring to a company's "excellence" rather than "market dominance"). Similar results were found in a 2018 study (Hentschel et al.) that surveyed more than 150 women applying for a German entrepreneurial training program. Women who were given a program advertisement that used the masculine linguistic form of the word "entrepreneur" in German were less interested in the training program and more likely to say they would not be a good fit.

The words in job advertisements and descriptions matter if a new venture wants to promote diversity and avoid legal concerns when hiring a team. The right words are easier to find when focusing on the job's critical and required capabilities and requirements rather than the implicitly discriminatory language of what "sounds good."

If work analysis holds the key to generating clear job expectations based on today's characteristics and requirements for more effective job descriptions, what's involved in performing a work analysis, and what's feasible for startup leaders?

How to Conduct a Feasible Work Analysis

Traditionally called job analysis, I-O psychologists define work analysis as systematically gathering data about how, where, with whom, with what, and why work is performed. Thus, work analysis is a practical, evidence-based method for finding and communicating "the essential nature of the job" to those who are not job experts (Morgeson et al. 2020).

Work analysts gather and summarize qualitative and quantitative data by reviewing documentation, interviewing experts, observing the work being performed, and administering surveys. With this classical definition in mind, it could be that "work analysis" sounds too complex and time-consuming to be useful for a startup leader. However, there are ways to condense work analysis into four steps that are more feasible for even the busiest entrepreneur.

Figure 2. A feasible work analysis process.

Qualify the Job for Analysis Using Business Strategy

Suppose an entrepreneur hires a frontend engineer ably suited to the job who successfully performs her tasks. But rather than frontend features, the product needs improved backend infrastructure for better scalability and integrations. Would performing the 'front-end engineer' job successfully matter much to helping the product scale and work with the wider technology ecosystem? Probably not. Of course, this example is simplistic and easily avoidable. Nevertheless, it illustrates what goes wrong when startup leaders fixate on finding the "right person" for a job without asking whether a job successfully performed is relevant to business objectives.

In other words, a "qualified" job is one that, if successfully performed, achieves strategic business goals (Tett and Burnett 2003). From the opposite perspective, an "unqualified" job can never be "successfully performed," even with the perfect person in the role, because job performance is out-of-sync with the business strategy. Thus, successful work analysis begins with qualifying whether the job performs the work a startup needs. So, how does one "qualify the job?" By considering the business strategy.

Fortunately, considering the business strategy should take the least amount of homework as most business leaders, especially entrepreneurs, have some business strategy in mind. But should the strategy not exist or is not well-articulated, here are a few questions to consider:

1. How does the business make money? What is its mission statement, and how does it plan to accomplish its mission (Martin 2005)?
2. What are the company's shared values? (e.g., innovation, culture, competition; Ployhart et al. 2006).
3. What separates the company from competitors? What are its competitive advantages (i.e., assets that are valuable, rare, and not easily inimitable)?
4. What are the primary objectives over the next 12 months that will help the company differentiate and provide value?

Next, entrepreneurs should ask themselves this question: Would the "perfect" person for this job contribute to the business's objectives? Answering "yes" to this question means that the job is worth analyzing (Ployhart et al. 2018). Answering "no" means that even if the work is understood perfectly and the "right" person is found, this person has a low probability of success because the job's mission is out of sync with what the business needs.

Good timing is a big part of a startup's success or failure. Similarly, the first step in a work analysis is to confirm that now is the right time for this job to exist. In addition, contextualizing the job's purpose within broader goals helps create a mission statement that defines success in the role. Once success is defined, the right talent requirements can be identified.

Form a Council of Job Experts

Entrepreneurs need to know a little about a lot. But it's unlikely that they will be familiar with every job in the business, let alone an expert. Therefore, once a job has been qualified for analysis, one must identify and organize a small group, or "job council," of subject matter experts (SMEs) that understand the target job (Harvey et al. 2007). A job council should have at least three people although groups as large as 8 to 10 can help avoid biases that come with smaller, more homogenous groups.

Although a job's functional leaders, such as sales, product, engineering, or marketing, should be added to a council when analyzing individual contributor roles, many fast-growing or early-stage startups have yet to hire functional leaders. In this case, entrepreneurs must be creative when sourcing job council members. Board members knowledgeable about the job, former managers or leaders in their network, advisors, investors, or consultants can also serve on a job council. For individual contributor-level jobs, consider paying job incumbents (either contractors or employees at other companies) for a day or two of their time to serve on a job council. For leadership roles, executive recruiters are knowledgeable experts. Regardless of position or title, entrepreneurs should recruit council members by answering the following questions: Have they directly performed, managed, or observed this job firsthand? Could this person clearly and accurately describe what this job does on a day-to-day basis?

Entrepreneurs forming a council should remind SMEs that although being on a 'job council' may sound like a significant time investment, participation often only requires a few hours in individual conversations or during a live meeting for each job. As the startup scales, standing job councils with rotating employee membership are advisable. Job council membership can become a prestigious station, as members serve as visible employee representatives of their job families, offering recognition and voice to successful, tenured employees.

Identify Job Duties (or Outcomes)

Once a job council is identified, each member (individually or corporately) should be asked about an average day in the life of the job. The goal is to generate around ten work task statements describing the role's observable actions. For many jobs, these statements include a verb (an action word), an object (on what or whom the action is performed), and a qualifier (a description of how, why, or where the action is performed; Harvey et al. 2007). Stated as a question, the job council should answer the following (Morgeson et al. 2020):

- What tasks are performed in this job?
- How does this job complete tasks?
- Why does the job complete these tasks?

An example backend developer's job duty might be: "Designs and develops scalable software architectures to protect and provide software scalability."

Job duty statements can describe the essential nature of existing jobs in behavioral terms. However, behavioral descriptions can be difficult for new jobs (e.g., cryptocurrency engineer), executive-level positions, and modern "jobs" quickly becoming more dynamic and unobservable (Chait and Stross 2021, Dierdorff and Morgeson 2007).

Thus, ask the council to focus on outcomes rather than actions for newly created or senior jobs. In other words, the job council should list the goals or objectives that, if accomplished, progress the business toward its mission. For additional guidance on creating job-level outcomes, see Smart and Street's (2008) approach to building executive recruiting scorecards or the objective and key result methodology pioneered by Google (Castro 2021).

List Required Capabilities and Link Them to The Duties or Outcomes

Next, the job council should consider each duty or outcome statement and think about the capabilities, or competencies, required to perform it. This process is what psychologists call the 'inferential leap,' moving from what a person must do to what a person must be capable of doing. This leap is required to properly identify and measure the capabilities that matter when making people decisions such as who to hire, train, or promote.

Capabilities, also called competencies, are the knowledge, skills, abilities, and other human characteristics (KSAOs) needed for effective performance (Campion et al. 2011). Admittedly, identifying capabilities can be more complicated than it sounds. To help, the job council can consider these questions (Ployhart et al. 2018):

- What competencies are currently vital for effective performance in the role?
- How do these competencies align with your business strategy?
- Which competencies can be mapped onto specific strategic goals?
- Which competencies differentiate the business from competitors?

But, what if no one on the council is familiar enough with the job to answer these questions? Or the entrepreneur has trouble coming up with the correct labels and definitions for competencies?

Fortunately, several free, high-quality resources exist for choosing competencies:

- The Occupational Network (O*Net): A database created and maintained by the U.S. Department of Labor covers work activities and competencies for nearly 1,000 jobs.
- Office of Personnel Management MOSAIC Competencies: 325 competencies and definitions resulting from a United States government survey of over 200 jobs.
- Thought leader articles: Whitepapers, popular press articles, and blogs by trusted thought leaders often list KSAOs needed for performance.
- Job descriptions: Capabilities and qualifications from other companies' job descriptions. Be careful not to over-index on these, as many aren't developed using work analysis.
- Training courses and content: Curricula and course content may list necessary knowledge, technical skills, problem-solving capabilities, and interpersonal skills.

When developing the characteristics and capabilities list, entrepreneurs should go beyond just the people they know in these roles. Focusing on people who have successfully performed similar jobs often provides a biased sample of the capabilities required to complete the tasks or outcomes in a unique work environment. Not accounting for context can lead to people who performed work needed in the past, not the work needed today (Kristof-Brown et al. 2005).

Finally, once these activities and requirements have been identified the 'job council' can link them back to the work duties or outcomes. Connecting competencies to work duties/outcomes creates a tally of the most critical competencies.

With smaller job councils, this can be done through a meeting to provide binary links through consensus. With larger job councils, or if a live session is impractical,

Table 1. Example work activity to competency linkage for a junior mobile developer.

How important is this competency for effectively performing the junior mobile developer job?

0 = NA
1 = Not Important
2 = Somewhat important
3 = Important
4 = Very Important
5 = Extremely Important

Work Activity / Competency	Swift	Objective-oriented programming	Asynchronous operations	Design Patterns	UX Design Principles	Documentation	Debugging	Collaborative Problem-Solving	Teamwork	Independence / Autonomy	Adaptability	Dependability	Initiative
Develop frontend mobile interfaces using Swift.													
Create new features using objective oriented or reactive programming languages.													
Build and modify asynchoronous operations with the backend.													
Follow MVVM design patterns to to correct errors.													
Modify software to maintain compability with new hardware or improve performance as per user requirements or specifications.													
Improve user interface based on user feedback and analytic research.													
Document created or modified code for collaboration with other development teams.													
Consult with managers and senior engineers to clarify program intent, identify problems, and suggest changes.													
Confer with systems analysts, engineers, programmers and others to obtain information on project limitations and capabilities, performance requirements and interfaces.													
Total													

each council member can link them using an importance scale (see Table 1 for an example). Averages can then be computed for the competencies that apply to the most duties or outcomes. Table 1 provides a template for linking work activities to competencies, using a junior mobile developer as an example.

How to Compose the Job Description

A good job description includes the job's mission, duties or outcomes, requirements, and attractive information about the company and the work environment. A job description with these elements is effective because it helps create "predictive hypotheses," or testable expectations about which worker capabilities predict future work performance. These expectations build the foundation for choosing the best methods to find people who have the necessary capabilities (Sanchez and Levine 2012).

Work analysis provides the critical components of an effective job description: where the job sits in the organization and how it accomplishes its top-level goals (i.e., the "Mission Statement or Purpose Summary" job description section). The individual work activities and tasks (i.e., the "Job Duties" section) or successful outcomes. And a list of the human competencies and capabilities required to perform the job duties or achieve the outcomes in the work environment (i.e., the "Minimum and Preferred Qualifications" section). In some cases, describing the mission and duties (or outcomes) would be enough to communicate the job's essential nature to non-experts. Still, minimum and preferred qualifications are often needed to set the capability requirements for employees with a high chance of successfully performing the job (Buster et al. 2005). Thus, many job descriptions should often include minimum and preferred qualifications. In the section below, each component is described in more detail with best practices.

1) About [Company Name]

   Add the company's mission and vision, the work environment, culture, and what makes it different from other companies in its industry or segment.

Writing about the company is an opportunity to add attractive marketing copy to the job description. Research shows that successful hires are more than a good fit with the job; they are also a good fit with the organization's cultural values and working styles (Kristof-Brown et al. 2005). This section should communicate why the company is the right one to work for in evocative and creative language. However, one should avoid letting profile-type descriptions or marketing lacquered job characteristics (e.g., "rockstar," "ninja") seep into this section. Not only are these descriptors in poor taste, they also undermine the actual qualifications derived from the specific qualifications derived from the analysis. Balanced amounts of realistic and attractive information yield better candidates with higher chances of success once on the job (Pavur 2010).

2) Job Mission

   Add how this job will help support or achieve the business's strategic objectives. Try to succinctly capture why this job exists and why the position is open now.

Now, review notes from the job qualification work analysis step. How does the job serve the business strategy? This information may feel unnecessary to some, but equifinality, or the phenomenon that jobs performed in different ways lead to the same results, means that new hires often negotiate their roles over time and innovate how they perform their work (Dierdorff and Morgeson 2007). Thus, regardless of how well the duties, outcomes, and characteristics are identified, the job mission statement will provide the conceptual definition of success. In other words, this statement is critical as it clearly and succinctly communicates to the future job holder why this job matters to the business. All other decisions can stay oriented with this north star in mind.

3) Duties or Outcomes

   Add a filtered list of critical job duties or outcomes.

This section is where additional time and effort spent in the work analysis will pay off. "Filtered" and "critical" are the key terms in this section's description. Consider how an outsider might interpret the job duty or outcome statements:

- Will they be familiar with bespoke or niche tools and systems?
- Will they recognize all acronyms?
- Are metrics understood industry-wide?

Filter the duties by how critical they are based on the competencies required. Perhaps some of the competencies needed to perform these tasks aren't necessary on day one. Or, maybe some duties aren't related to more than one competency, making it an infrequent and less valuable assignment to communicate. In this section, less is more. The critical goal is to share the essential nature of the job with outsiders either via job duties (if it is a more individual contributor, entry-level, or familiar role) or outcomes (if it is an executive, senior, specialized, or novel role).

4) Qualifications

   Add the critical competencies to the qualifications list. Remember there is a difference between "minimum qualifications" and "preferred qualifications." This section should communicate how the company will evaluate candidates for the role and what capabilities are expected for success.

Next, there should be a curated list of qualifications. As a reminder, competencies or capabilities are KSAOs people need to perform a job. For example,

- What topics, ideas, concepts, processes, methods, rules, or frameworks should a successful candidate understand?
- What people skills, thinking skills, tools, systems, coding languages, or technologies should a successful employee be able to demonstrate or use?
- What characteristics, attitudes, working styles, and motivations does a successful employee likely have? (TIP: The answers should be things that have been linked to the duties or outcomes on your list provided by the job council.)

What about experience? Of course, experience is often the quickest way of evaluating if a candidate has the competencies and capabilities necessary for the role. But pre-hire work experience qualifications are often arbitrarily set and are weaker predictors of future work performance than many assume (Van Iddekinge et al. 2019). Research has found that beyond the first year or two of experience, experience matters little in predicting work performance and may be a liability for some executive-level roles (Hamori and Koyuncu 2014). While it may be impossible to remove experience qualification requirements from the job description, entrepreneurs need to think critically about how much experience is essential and how those experiences reflect the critical competencies.

The Benefits of Better Job Descriptions: The Technical Interview Case Study

As a real-world example of this method's power, consider how much easier recruiting gets with a well-formulated job description in hand. It can help identify online communities where active candidates are likely to spend time. It can guide founders to the right recruiter who will best source and attract passive candidates. And it can provide a guide for a technical interview that evaluates the skills and competencies that matter, rather than pattern-match superficial signals such as work histories or

YOUR LOGO

Job Title:	Junior Mobile Developer	**Schedule:**	Full Time
Job Category:	Engineering	**Position Type:**	Salaried
Location:	HQ Based or Remote	**Reports To:**	Director of Engineering

About Your Business

We believe what we do is life-changing and we love creating experiences of a lifetime for our customers. Our business is a early-stage software company ready to change the world. We love what we do, and we give it all we've got – on the job and off. When our customers use our product it's not just software we're providing. It's an experience. Our team is always ready for a challenge and goes beyond in everything they do. If you have a genuine drive to improve on all fronts, we invite you to join us.

Job Mission

The Junior Mobile Developer helps achieve important engineering functions that contribute to several strategic objectives. Success in this role means being contributing to key mobile application projects, spearheading process and procedure improvements, and keeping an eye on code quality, maintainability, and readability. A successful junior mobile developer will help our business achieve our adoption and profitability metrics over the next year, helping us raise our next funding round at an attractive valuation.

Job Duties

- Develop frontend mobile interfaces using Swift.
- Create new features using objective oriented or reactive programming languages.
- Build and modify asynchoronous operations with the backend.
- Follow MVVM design patterns to to correct errors.
- Modify software to maintain compability with new hardware or improve performance as per user requirements or specifications.
- Improve user interface based on user feedback and analytic research.
- Document created or modified code for collaboration with other development teams.
- Consult with managers and senior engineers to clarify program intent, identify problems, and suggest changes.
- Confer with systems analysts, engineers, programmers and others to obtain information on project limitations and capabilities, performance requirements and interfaces.

Minimum Qualifications

- 3 years of computer science education or 1 year of experience building mobile applications (e.g., writing, debugging, and documenting code in mobile development language such as Swift, Android, or Kotlin)

Preferred Qualifications

- **Technology languages**: Swift
- **Objective-oriented Programming**: Understands and applies issues related to pointer/reference semantics and memory leaks (e.g. sharing, object ownership, garbage collection, NULL pointer/reference, cyclic structures, memory allocation, copy-on-write).
- **UX Design Principles:** Understands and applies responsive design principles for both desktop and mobile websites or applications.
- **Collaborative Problem Solving:** The ability to detect problems, recognize important information, and link various data; to trace potential causes and look for relevant details.
- **Adaptability:** The ability to remain fully functional by adapting to changing circumstances.
- **Dependability:** Sets high-quality standards and strives for continuous improvement and quality assurance.

Your Company Job Description: Junior Mobile Developer
Last Revision Date: Today's Data by Your Name

Figure 3. Example job description for a junior mobile developer.

educational pedigrees. This chapter closes by expanding on the last example. This is a case study of how a work analysis-based job description informs an integral part of the hiring process, the technical interview.

The unstructured and ad hoc technical interview is notorious for its dogmatic acceptance by technical hiring managers for recruiting developers. But what if recruiters and hiring managers could improve the effectiveness of technical interviews while also accounting for a new job type? Take hiring the "machine learning engineer" discussed previously as an example.

In this example, the business is a startup that makes pitch decks more searchable for investors and analysts to explore (CBInsights 2021b). The minimum viable product has relied on humans reviewing pitch decks, manually entering data, and building dashboards. The startup begins to gain traction with sales and realizes that reaching its revenue goals requires automating the manual work of scraping, wrangling, cleaning, and analyzing data. Although the leadership team has heard that other startups are having problems hiring "machine learning engineers," they determine this job is critical to achieving next year's strategic goals and objectives. What's more, they realize success in the position means more than preparing and analyzing data. It also means setting the future product strategy for the data aggregation and search tool.

To learn about the duties and required capabilities of the role, the founding team forms a council of job experts. The council understands how the machine learning (ML) engineer role has evolved over the past 18 months and includes a former colleague, an investor who has ML-based companies in their portfolio, an academic, and a contract ML engineer. As the job is still being shaped, and the full-scale product has yet to go to market, the council decides to list successful outcomes for the job and the capabilities required to achieve them. Two 60-minute calls and a working lunch get the council to a finalized set of job outcomes mapped to capabilities.

These inputs serve as the information they need to craft a job description. The description sells the benefits of working at the startup and the company's mission, vision, and working styles while stating why the role exists, what outcomes define success, and the required capabilities.

Usually, the technical interview would involve the founder or CTO (if the founder does not have a technical background) asking brainteasers, an experience-type question, and an open-ended question to identify the "smart" people who can "talk the talk." Typically, hiring decisions relate to the candidate's likeability and reputation. But this time, the hiring process starts with a job description based on work analysis. The role is qualified as being critical to the business and day-to-day capabilities include writing production-ready code that sources, prepares, and models data, working with application programming interfaces (APIs) and building robust software architecture.

On top of technical proficiency, this role needs to lead and manage others. Capabilities such as building a team and creating a longer-term product strategy related to search capabilities in the application are also critical. With these capabilities in mind, the startup's leaders decide to separate a repetitive, generic interview into two parts: a technical interview that measures job knowledge via a work sample (e.g., a pair programming session or job talk) and a behavioral interview (i.e., "tell me about a time when" questions) that measure leadership and team-building skills. Finally, all candidates are asked an open-ended question about their values to see if they align with the company values to assess culture fit.

A better interview process captures more consistent and precise signals that predict future behavior and outcomes. A better interview also creates a better candidate experience and saves the decision-makers time gathering data and collaborating. And when the startup makes a hire—the new employee's job description provides a clearly stated mandate with activities that can be molded to the job's real-world contours.

Conclusion

Imagine the competitive advantages of qualifying whether the strategic roles are appropriate for the company-level objectives, communicating a well-articulated purpose and mission to company outsiders, crisply defining job duties or outcomes, and translating a shortlist of capability requirements into a consistent, structured interview process. Think of a job that is needed in the next 3–6 months; how would the requirements identified through the work analysis differ from what one might have suspected? How do work analysis-based job descriptions and interview processes overcome the limitations of a 'profile-centric' process? How does this effort prepare the startup for future success as the job changes? What would happen if startup leaders knew more clearly that their new hires would achieve the outcomes or duties they identify and contribute to their business by fulfilling the job's mission?

Startups must move fast. Market conditions, competitors, customer needs, and everything else change quickly. Although this chapter provided an approach to creating job descriptions that is flexible and agile, it may still feel onerous and slow. If that's true, consider one final anecdote about the costs of failing to choose the right person.

Brian Acton, the co-founder of WhatsApp, applied to work at Facebook in 2009. It's hard to say how or why, but Facebook rejected Brian's job application. Five years later, Facebook acquired WhatsApp for about 20 billion dollars, making it one of the most expensive software acquisitions (Anders 2014) at the time. While it is unclear if hiring Acton would have led to Facebook building WhatsApp internally, saving the acquisition costs, it is probably safe to assume that Facebook's hiring process missed a high potential, high-impact candidate that would have offered an outsized benefit to the company. This mistake likely cost Facebook billions of dollars. One out of every ten startups falls short of their growth, funding, or revenue objectives due to a mismatched team. Investing time in collecting data to secure the right talent is a small yet crucial step towards attaining these goals and beyond.

References

Alagaraja, M. (2012). National human resources development in practice: An interview with M.V. Subbiah. *Human Resource Development International*, 15(4): 515–524. http://dx.doi.org/10.1080/13678868.2012.690971.

Anders, G. (2014, February 19). *He Wanted a Job; Facebook Said 'no'—In a $3 Billion Mistake.* Forbes. https://www.forbes.com/sites/georgeanders/2014/02/19/he-wanted-a-job-facebook-said-no-in-a-3-billion-mistake/?sh=62c6fada5244.

Ben-David, J. (2020, January 22). *The History of Cloud Computing: Two Decades in Review (Part 1).* Turbonomic. https://blog.turbonomic.com/cloud-computing-two-decades-in-review.

Buster, M.A., Roth, P.L. and Bobko, P. (2005). A process for content validation of education and experienced-based minimum qualifications: An approach resulting in federal court approval. *Personnel Psychology*, 58(3): 771–799. https://doi.org/10.1111/j.1744-6570.2005.00618.x.

Campion, M.A., Fink, A.A., Ruggeberg, B.J., Carr, L., Phillips, G.M. and Odman, R.B. (2011). Doing competencies well: Best practices in competency modeling. *Personnel Psychology*, 64(1): 225–262. https://doi.org/10.1111/j.1744-6570.2010.01207.x.

Castro, F. (2021). *The Beginner's Guide to OKR.* Felipe Castro. https://felipecastro.com/en/okr/what-is-okr/.

CBInsights. (2021a, August 3). *The Top 12 Reasons Startups Fail.* https://www.cbinsights.com/research/startup-failure-reasons-top/.

CBInsights. (2021b, January 20). *Higher and Higher.* https://us1.campaign-archive.com/?u=0c60818e26ecdbe423a10ad2f&id=7ea4cd0935&e=9ade47dbbc&.

Chait, D. and Stross, J. (2021). *Talent Makers: How the Best Organizations Win Through Structured and Inclusive Hiring.* John Wiley & Sons.

Dhasarathy, A., Frazier, D., Khan, N. and Steagall, K. (2021, March 11). *Seven Lessons on How Technology Transformations can Deliver Value.* McKinsey Digital. https://www.mckinsey.com/business-functions/mckinsey-digital/our-insights/seven-lessons-on-how-technology-transformations-can-deliver-value.

Dierdorff, E. and Morgeson, F. (2007). Consensus in work role requirements: The influence of discrete occupational context on role expectations. *Journal of Applied Psychology*, 92(5): 1228–1241. https://doi.org/10.1037/0021-9010.92.5.1228.

Gaucher, D., Friesen, J. and Kay, A.C. (2011). Evidence that gendered wording in job advertisements exists and sustains gender inequality. *Journal of Personality and Social Psychology*, 101(1): 109–128. https://doi.org/10.1037/a0022530.

Harvey, J.L., Anderson, L.E., Baranowski, L.E. and Morath, R.A. (2007). Job analysis: Gathering job specific information. pp. 57–95. *In*: D.L. Whetzel and G.R. Wheaton (eds.). *Applied Measurement: Industrial Psychology in Human Resources Management.* Mahwah, NJ: Lawrence Erlbaum Associates.

Hentschel, T., Horvath, L.K., Peus, C. and Sczesny, S. (2018). Kick-starting female careers: Attracting women to entrepreneurship programs. *Journal of Personnel Psychology*, 17(4): 193–203. https://doi.org/10.1027/1866-5888/a000209.

Homori, M. and Koyuncu, B. (2014). Experience matters? The impact of prior CEO experience on firm performance. *Human Resource Management,* 54(1): 23–44. https://doi.org/10.1002/hrm.21617.

Huselid, M., Becker, B.E. and Beatty, R.W. (2005). *The Workforce Scorecard: Managing Human Capital to Execute Strategy.* Harvard Business Press.

Katou, A.A. (2009). The impact of human resource development on organizational performance: Test of a causal model. *Journal of Behavioral and Applied Management*, 10(3): 335–356. https://www.proquest.com/scholarly-journals/impact-human-resource-development-on/docview/1151704161/se-2?accountid=10901.

Kristof-Brown, A.L., Zimmerman, R.D. and Johnson, E.C. (2005). Consequences of individuals' fit at work: A meta-analysis of person-job, person-organization, person-group, and person-supervisor fit. *Personnel Psychology*, 58(2): 281–342. https://doi.org/10.1111/j.1744-6570.2005.00672.x.

Martin, S.L. (2005). General thoughts on needs analysis and evaluation. *The Industrial-Organizational Psychologist*, 42(4): 91–94.

Morgeson, F.P., Brannick, M.T. and Levine, E.L. (2020). *Job and Work Analysis: Methods, Research, and Applications for Human Resource Management.* SAGE Publications, Inc. p. 186.

Pavur, Jr., E. (2010). Use job descriptions to support leadership. *The Psychologist Manager Journal*, 13: 119–122.

Ployhart, R.E., Schneider, B. and Schmitt, N. (2006). *Staffing Organizations: Contemporary Practice and Theory* (3rd ed.). Lawrence Erlbaum Associates Publishers.

Ployhart, R.E., Weekley, J.A. and Dalzell, J. (2018). *Talent Without Borders: Global Talent Acquisition for Competitive Advantage.* Oxford University Press.

Sanchez, J.I. and Levine, E.L. (2012). The rise and fall of job analysis and the future of work analysis. *Annual Review of Psychology*, 63: 397–425. https://doi.org/10.1146/annurev-psych-120710-100401.

Smart, G. and Street, R. (2008). *Who.* Ballantine Books.

Smart, B. (2012). *Topgrading* (3rd ed.). Penguin Books.

Sung, S.Y. and Choi, J.N. (2014). Multiple dimensions of human resource development and organizational performance. *Journal of Organizational Behavior*, 35(6): 851–870. https://doi.org/10.1002/job.1933.

Tett, R.P. and Burnett, D.D. (2003). A personality trait-based interactionist model of job performance. *Journal of Applied Psychology*, 88(3): 500–517. https://doi.org/10.1037/0021-9010.88.3.500.

Van Iddekinge, C.H., Arnold, J.D., Frieder, R.E. and Roth, P.L. (2019). A meta-analysis of the criterion-related validity of prehire work experience. *Personnel Psychology*, 72(4): 571–598. https://doi.org/10.1111/peps.12335.

Chapter 3

Identifying and Measuring Entrepreneurial Talent in the Age of Artificial Intelligence

Nilima Ajaikumar[1] and *Reece Akhtar*[2,*]

In academia and business, entrepreneurship has received considerable interest given its allure of autonomy, innovation, and ability to produce considerable amounts of wealth and value (Hisrich et al. 2007). The startup is the new 'garage rock band' with its promises of fame and fortune. Yet, this analogy is sobered by the fact that many startups fail to grow or become profitable businesses (Shane 2008). The question of which entrepreneurial ventures do go on to achieve success and growth becomes of critical interest. As entrepreneurship is a key driver of economic and social progress, knowing the predictors of entrepreneurial success yields important implications for the way institutions support such activity (Kuratko 2016).

The academic study of entrepreneurship draws researchers from a variety of disciplines and is studied across multiple levels of analysis — reflecting the significant impact that it has on society. Yet, all entrepreneurial activity begins with an individual choosing to take advantage of an opportunity to create economic, technological, or social value. From this perspective, the field of Industrial-Organizational (I-O) psychology can offer significant insights for institutions that are looking to invest, educate or support promising entrepreneurs, or enterprises that aspire to maintain their competitive advantage and nurture "*intrapreneurial*" activity among its employees. Once *what* psychological characteristics are predictive of entrepreneurial success is known, the next challenge becomes understanding *how* to measure such talents. Psychometric tools are widely used to achieve this, however,

[1] 06 W 97 St., Apt #13H, New York, NY, 10025, 929 334 8485.
[2] 430 W 34th St #16E, New York, NY, 10001.
Email: na2841@tc.columbia.edu
* Corresponding author: reece@deepersignals.com

modern data science practices and technologies powered by artificial intelligence (AI) algorithms offer new ways to identify entrepreneurial talent with greater scale, efficiency, and accuracy.

This chapter is designed to provide practitioners with a clear understanding of the psychological literature on entrepreneurial talent and achievement, with the aspiration that it will improve the effectiveness of their investment and talent management decisions. This chapter aims to address three questions:

What are the psychological predictors of entrepreneurial success?
Can entrepreneurial talent be reliably identified and measured?
Can AI technology improve the identification of entrepreneurial talent?

What is "Entrepreneurial Talent"?

While the role entrepreneurs have played in driving economic growth globally and transforming the world is well recognized, what it means to be entrepreneurial is less well understood. Traditionally, entrepreneurship is associated with the context of establishing or owning a company (Shane 2008). The nature of this perspective however prescribes a categorical view of entrepreneurship. If a person owns or has founded a business, they are considered to be an entrepreneur. If not, then they are not an entrepreneur. Much of this viewpoint could be attributed to what people have come to understand of entrepreneurship from the successful founders or business owners that are publicized widely and frequently. The names of world-renowned tech entrepreneurs of Silicon Valley are quickly referenced in relation to entrepreneurship. But how about the founders and business owners who were not as successful? Or those whose startups did not take off? If an individual's venture was unsuccessful, would they be considered to not be entrepreneurial? Surely not. Especially given the number of failures some of the most recognized entrepreneurs had to overcome before they found success. Defining entrepreneurship in this manner is therefore criticized to be too narrow and restrictive in context (McKenzie et al. 2007).

This then raises the question of how do we better understand entrepreneurship.

A broader perspective of entrepreneurship is to refer to it in relation to the efforts associated with bringing about innovation, growth, and value creation (Shane and Venkataraman 2000). With this definition, entrepreneurship has not been deemed a characteristic confined to those who own businesses alone. The aforementioned entrepreneurial efforts could occur within an established organization and are more specifically referred to as *intrapreneurship* or *corporate entrepreneurship* (Antoncic and Hisrich 2001). It could also be observed in innovative efforts taken to develop new technologies. This form of entrepreneurship is termed *technological entrepreneurship* (Venkataraman 2004). Further, entrepreneurial activities could also be noted in activities that focus on the betterment of social interests. This is referred to as *social entrepreneurship* (Mair and Martí 2006). These various perspectives shed light on how entrepreneurship may not focus on the sole outcome of creating a business, though it could result in it (McKenzie et al. 2007).

At the crux of these broader definitions of entrepreneurship is the understanding that entrepreneurship involves engaging in activities that involve recognizing and exploiting opportunities, innovating, and creating value (Kuratko 2007). What is key to this perspective is that it views entrepreneurial activity to be closely related to an individual's personality (Kuratko 2007, McKenzie et al. 2007). Entrepreneurial talent involves the ability of an individual to anticipate future circumstances, identify opportunities that others have yet to spot, connect the dots, and bring in insights from various domains to create value (Ahmetoglu et al. 2011).

Taking this view shifts the focus on entrepreneurship entirely. When trying to identify entrepreneurial talent, the question that takes center stage is no longer whether or not an individual owns a business, rather, who are the individuals that demonstrate certain traits that are characteristic of entrepreneurial talent? These individuals could be found across various realms of society, industries and organizations, even in the most unseeming roles. For example, an academic could be considered to be entrepreneurial. Both individual contributors and senior executives within an organization could demonstrate entrepreneurial talent. Entrepreneurs could be spotted everywhere.

If that is the case, then how do we identify them? What is it that distinguishes these individuals?

The Science that Underpins Personality Research

With this broader perspective of entrepreneurial talent, the science of individual differences has become a key area in research related to entrepreneurship (Brandstätter 2011). Studies have shifted focus from the underlying motivational factors that contribute to entrepreneurial activity to the 'trait approach' (Rauch and Frese 2007). With this approach, the emphasis is on identifying the specific traits and psychological attributes that are characteristic of entrepreneurs (Zhao and Siebert 2006).

Before delving into the specific traits associated with entrepreneurial talent, this chapter will deviate briefly to first explore the science that forms the foundation of personality research. The study of the psychological constructs associated with personality and individual differences spans over a hundred years. Broadly, it holds the view that an individual's orientation in thinking, feeling, and behaving could be accurately measured. In psychology, the specific field of study that investigates the use of measures to assess psychological characteristics is referred to as psychometrics.

Psychometric assessments are tools that help objectively assess the extent to which an individual tends to possess or exhibit certain traits. Psychometric assessments typically take the form of surveys, though technology has introduced more innovative ways of obtaining this self-reported data. A key advantage of using psychometric assessments is that they are more objective than relying on an individual's intuitive judgment. Psychometric assessments thus help overcome human biases and provide a more data-driven approach to assessing personality traits.

When using psychometric assessments, to measure talent, it is also important to understand how best to evaluate its quality. The reliability and validity of an assessment are key concepts to understand in this process and are used by academics and practitioners to ensure they are employing psychometric assessments that do what they are supposed to do. Reliability refers to the degree to which an assessment measures a particular characteristic consistently and provides the same results every time it is administered. For example, an individual taking a psychometric assessment for entrepreneurial talent should expect to obtain the same results every time they complete the assessment. If not, it is hard to be confident in the accuracy of the assessment. Validity refers to whether the tool is assessing what it claims to measure. So a psychometric assessment of entrepreneurial talent should measure entrepreneurial talent and not some other personality trait.

Understanding the reliability and validity of a psychometric measure is therefore very important when evaluating appropriate psychometric assessments to employ. Moreover, psychometric assessments undergo rigorous statistical testing to obtain acceptable evidence that demonstrates their reliability and validity. As a result, several scientific studies over the years have shown that using psychometric assessments is a far more accurate method to measure talent than human judgment (Chamorro-Premuzic 2017, Kuncel et al. 2010).

Can Entrepreneurial Talent be Accurately Measured?

A framework that is central to personality research and psychometric assessments is the Five Factor Model (FFM) (McCrae and Costa 1987). This framework views personality based on the five dimensions: openness, conscientiousness, extroversion, agreeableness, and emotional stability (also referred to as neuroticism). There is extensive research that demonstrates the various degrees to which each of these distinct traits could predict important outcomes within the workplace. Conscientiousness, for example, is significantly correlated with job performance ($r = 0.22$; Barrick and Mount 1991).

The FFM, also commonly referred to as the Big Five framework, has been central to research on the personality profile of entrepreneurs (Costa and McCrae 1985b). Zhao and Siebert (2006) conducted a meta-analysis that demonstrated that entrepreneurs tend to demonstrate higher levels of openness and conscientiousness, and lower levels of agreeableness and neuroticism, in comparison to managers within an organization. Outcomes from this study also suggested that entrepreneurs tend to be creative, systematic, and prepared to defy social norms. While this depiction aligns with the general stereotypes associated with entrepreneurs, what this study also helps understand is that entrepreneurial talent could be distinguished by these broad personality traits, rather than the traditional view of equating it with business ownership (Shane 2008). Seeking these broader traits in individuals allows researchers to identify entrepreneurial talent across various professional and social contexts.

Further to the FFM, a meta-analytic study conducted by Rauch and Frese (2007) identified narrow traits including the need for autonomy, self-achievement, stress tolerance, innovativeness, self-confidence, and proactive personality to be better predictors of entrepreneurial outcomes (average $r = .25$). In addition, a study conducted by Stewart and Roth (2001) identified that entrepreneurs tend to have a propensity for taking on greater risk. This finding should come as no surprise given the level of risk that is typically involved in the pursuit of innovative activities, especially when pursuing novel opportunities and navigating unchartered territories.

The Measure of Entrepreneurial Tendencies and Abilities (META; Ahmetoglu et al. 2011) is another psychometric measure that defines entrepreneurial behavior through four narrow traits: *vision, creativity, opportunism, and proactivity. Vision* refers to an individual's tendency to have ambitious goals, continuously strive for improvement, and have a general desire to make a difference in the world. Those around them often find these individuals to be inspirational, though some may also find their plans to be unrealistic. *Creativity* describes an individual's tendency to be imaginative, open-minded, and think outside the box. Highly creative Individuals tend to not conform to established boundaries and are always looking for new ways of doing things. *Opportunism* focuses on an individual's tendency to identify business opportunities around them that may not be evident to others. These individuals tend to be highly aware of business trends and more optimistic than others about the prospects of business ventures. *Proactivity* concerns the tendency of individuals to work quickly to achieve goals efficiently. Proactive individuals tend to be dominant, fearless, and willing to lead people and projects.

What is different about META when compared to the other psychometric assessments that measure entrepreneurial talent is that it assesses entrepreneurial traits by assessing the extent to which individuals demonstrate the broader entrepreneurial behaviors described earlier—recognizing and exploiting opportunities, innovation, and value creation (Shane and Venkataraman 2000). Studies have demonstrated META to predict a range of entrepreneurial outcomes (Ahmetoglu et al. 2011, 2018, Akhtar et al. 2013, Almeida et al. 2013, Leutner et al. 2014). These include behaviors associated with various forms of entrepreneurship (corporate, invention, social), developing prototypes, seeking funding for innovations, and overcoming organizational problems. META is also identified to be a significant predictor of employee engagement, suggesting that employees with higher levels of entrepreneurial talent, also tend to be more energetic and dedicated at work (Ahmetoglu 2015). Higher levels of employee engagement are known to contribute to several other organizational outcomes including higher productivity and financial outcomes (Schaufeli and Bakker 2004). Further, individuals who score high on the META assessment are reported to have significantly higher plans to start their businesses. A word of caution on this finding is that individuals who tend to score higher on having startup plans also tend to have higher intentions to quit an organization. Thus, in some ways, high levels of entrepreneurial talent within an organization could also pan out to have both favorable and unfavorable outcomes (Ahmetoglu 2015). Adequate retention measures will need to be in place to ensure such individuals are not quick to quit to seek self-employment. What is most important to note is that META is a powerful measure that is considered to predict entrepreneurial success over and above other assessments that define entrepreneurship in ways described earlier (Ahmetoglu et al. 2011, Akhtar et al. 2013).

For practical purposes, the results of these studies provide a few key takeaways. First and foremost, it demonstrates the relationship between personality traits and entrepreneurial talent, enabling the identification of individuals who have these tendencies across a broader range of contexts. Second, entrepreneurial talent can be measured. Psychometric tools, such as FFM measures and the META inventory, help not just identify these individuals but also serve as useful data points to predict successful entrepreneurial outcomes. Further, these tools could help develop entrepreneurial talent within employees working within the corporate, social, and creative realms, through employing targeted training interventions (Tiernay and Farmer 2011). Research indicates that organizations with entrepreneurial talent stand to gain a competitive advantage in the markets in which they operate (Lumpkin 2007). Assessing and investing in the development of entrepreneurial talent along with sustaining an organizational culture that is supportive of entrepreneurial achievement could be a valuable source of leverage for organizations operating in competitive industries. Finally, startups and other new ventures could also consider using psychometric assessments to assess for fit as they consider the employee makeup of their organization, to minimize risk and increase the chances of recruiting individuals that are best suited to perform under the unique demands of these environments.

Can AI Improve the Identification of Entrepreneurial talent?

While psychometric assessments are established tools, advancements in technology and digital services present significant opportunities to change the way investors and leaders identify and invest in entrepreneurial talent. If each person is generating a sea of data each day, through their digital footprint, revealing accurate and objective signals into their behavior, personality, skills, and values, having to physically take an assessment, begins to seem very antiqued. Putting together the fact that there is a high degree of entrepreneurial failure, and that current recruitment practices have limitations (i.e., resumes are often embellished (Turczynski 2021), interviews tend to be biased (Derous et al. 2016), and there is an inequality in access to funding opportunities (Bjerk 2008, Clayton et al. 2014)), a strong case could be made for mining a would-be entrepreneur's online behavior for digital talent signals to overcome these issues. This would usher in more effective and equitable practices around investment and talent management decisions. For example, organizations looking to hire or promote entrepreneurial employees could leverage data generated from tools such as Slack to measure the extent to which individuals are creative, proactive, and opportunistic. Similarly, financial institutions that are looking to serve unbanked entrepreneurs with micro-loans could leverage their digital footprints from social media platforms to build accurate psychometric properties to confirm their trustworthiness and integrity.

The application of AI to mine digital footprints or sources of novel data that reveal signals of entrepreneurial talent is still in its infancy. However, these technologies are increasingly being proven through scientific research and are already being applied to other areas of talent identification and selection including coaching, development, and culture/engagement. Accordingly, researchers can review these innovations to understand their efficacy and eventual impact on the entrepreneurial domain.

The way institutions assess talent has largely remained unchanged in the last 50 years. Institutions still rely on cumbersome methods such as the interview or resume, while the main innovation that has happened to scientific tools like psychometric assessments is to go from paper and pencil to online formats. These methodologies are limited not only due to their inability to scale and lack of sophistication, but also because they have simply not kept up with the changing nature of work and current ways of operating. Given that the premise of assessment practices is to make predictions about how one will behave in the future, the age of AI creates an exciting opportunity to improve current methods by incorporating large amounts of diverse and objective data points about individuals to build highly accurate profiles than can shape hiring decisions in seconds.

AI-powered talent algorithms use more and varied data than traditional talent methods. They can take advantage of the millions of digital records people generate each day. For example, the average person now produces over 146,880 megabytes of data a day — a statistic that increases each year (Andre 2021). Similarly, sensor-packed devices are continuously capturing digital records of the decisions people make, the content they consume and the people with whom they interact. These two sources of data alone provide all the input needed to bring about a new era of talent assessment. The use of AI tools can overcome the current limitations in assessing entrepreneurial talent in four ways:

1. Digital platforms and devices can objectively measure behavior, removing the need for biased human evaluations (Javed and Brishti 2020).
2. AI algorithms can be optimized to maximize the prediction between digital records and indicators of entrepreneurial success (Javed and Brishti 2020).
3. AI's ability to scale and reach a greater number of individuals from different backgrounds, can lower the barrier to entry to underrepresented populations (Javed and Brishti 2020).
4. AI algorithms can provide greater transparency into how data is used and weighted to reach a hiring or investment decision, thereby producing fairer and more ethical practices (Javed and Brishti 2020).

The previous section discussed how one's personality is predictive of entrepreneurial outcomes. Characteristics such as creativity, opportunism, and extraversion can be reliably measured using psychometric tools and used to build accurate evaluations of one's potential to be a successful entrepreneur. Can such characteristics be measured by AI and digital footprints? In a series of studies, a team from Cambridge University demonstrated computer algorithms could be trained to interpret the pattern of "likes" on Facebook and produce accurate classifications of users' personality traits that were previously identified to predict entrepreneurial success (Kosinski et al. 2013, Leutner et al. 2014). Others have replicated this work, extending the findings to relationships between the images people post, their language, and even dark side personality traits (interpersonally dysfunctional traits that negatively impact work performance and shape entrepreneurial activity; Akhtar et al. 2013, 2018, Liu et al. 2016, Ortigosa et al. 2014).

In a follow-up study, the same team of Cambridge researchers sought to compare how accurate their social media algorithms were compared to one's colleagues, friends, and families (Youyou et al. 2015a). They not only found that one's digital footprints provide more accurate evaluations of one's personality than their colleagues and family members, but the amount of data needed to achieve such accuracy was shockingly small. For example, with only 10 Facebook likes, their algorithm could surpass the behavioral ratings supplied by close colleagues. With just 150 likes, the evaluations surpassed that of one's closest family members. Furthermore, their algorithm could also predict an individual's career values better than human raters — critical for seeking out an entrepreneurial career (Almeida et al. 2013).

Summarizing the scientific literature investigating the accuracy of predicting psychological characteristics from digital footprints, Azucar et al. (2018) conducted a meta-analysis of 14 separate studies containing different types of social media records from a variety of platforms to estimate the true prediction of one's personality. Through this process, they arrived at a few conclusions. First, the correlation between digital footprints acquired through social media, and personality ranges between .29 and .40. This moderate effect size is what is commonly found when studying the relationship between personality and entrepreneurial achievement (Leutner et al. 2014). Second, the accuracy of predictions increases when more than one type of digital footprint is used. This finding is intuitive and is analogous with well-designed assessment procedures whereby an investor or recruiter will gather evaluations from a variety of sources using a variety of methods before reaching a decision.

From a practical perspective, it is easy to see how these findings could be applied to the entrepreneurial domain. Rather than request an individual to complete a long and expensive psychometric assessment as part of a loan application or hiring process, a secure and private web application would allow applicants to donate their digital footprints, allowing an algorithm to objectively evaluate their talents in microseconds. Such a process would be cheap, require no time investment from the applicant, and allow the time-pressed and underrepresented to better participate in the process. The adoption of such tools would have the added benefit of increasing practitioners' use of scientific concepts and constructs in their talent practices, moving away from gut instinct or intuitive evaluations of one's potential, and enabling organizations to start to leverage 50 years of research that has demonstrated the importance of the personality in the pursuit of entrepreneurial success. In addition, researchers will be rewarded with high volumes of high-quality behavioral data that can be used to test new hypotheses about how entrepreneurial opportunities are identified and exploitative. If it is possible to gather accurate data across large populations, the viability of testing macro-level hypotheses significantly increases.

While the future of talent identification is bright and filled with potential, developers and users of such tools must consider their ethical implications. AI-powered algorithms have the potential to become a "weapon of math destruction" when they are scalable, opaque, and used in high stakes settings. Ethical or responsible algorithms must first seek the consent of the user ("what data am I asking for?"), be fully transparent ("how does the algorithm work and how is it weighting the requested data?"), and should complement, not replace human decision-making. Furthermore, the development of such algorithms would benefit from following EEOC assessment guidelines, which state clear requirements regarding a tool's reliability, validity, and fairness (for a primer on how AI algorithms work and how to use them ethically, see Leutner et al. 2022).

Conclusion

This chapter attempted to provide practitioners with a brief introduction to the science of entrepreneurial talent and how it can be readily identified. Much of the world of work is riddled with subjectivity, anecdotes, and intuition — this leads to overconfidence, inefficiency, and bias, that holds back people, communities, and economies. Entrepreneurship has a tremendous ability to generate progress, prosperity, and wellbeing. It is therefore critical that promising entrepreneurs are identified, supported, and developed. Science, data, and technology are the antidotes to the current challenges holding back the entrepreneurship process; addressing these challenges is key to unlocking future value.

What could practitioners do to help realize this promise?

1. **Look for the right traits:** The reviewed evidence suggests that creativity, opportunism, proactivity, and vision are the most consistent psychological predictors of entrepreneurial success. Avoid captivating, yet ineffective, characteristics such as overconfidence and narcissism as they do not correlate with competence and often lead to unfavorable predicaments (Chamorro-Premuzic 2019).
2. **Trust data, not intuition:** Scientific assessments that are reliable and predictive should be included as part of the evaluation process in assessing entrepreneurial talent. This ensures that all individuals are accurately assessed against the same criteria, not intuitions or biases that lead to ineffective and potentially harmful decisions.
3. **Integrate AI into the processes:** Data-driven algorithms tend to outperform the evaluations of intuitive raters and are powerful to help calibrate and guide high-stakes decisions. Integrating novel sources of data, and new methodologies can increase the visibility and access of entrepreneurs across communities and social strata (Chalmers et al. 2020, Chamorro-Premuzic 2017, Youyou et al. 2015b).

References

Ahmetoglu, G., Leutner, F. and Chamorro-Premuzic, T. (2011). Economics: Understanding the relationship between individual differences in trait emotional intelligence and entrepreneurship. *Personality and Individual Differences*, 51(8): 1028–1033.

Ahmetoglu, G. (2015). *The Entrepreneurial Personality: A New Framework and Construct for Entrepreneurship Research and Practice.* Goldsmiths, University of London.

Ahmetoglu, G., Akhtar, R., Tsivrikos, D. and Chamorro-Premuzic, T. (2018). The entrepreneurial organization: The effects of organizational culture on innovation output. *Consulting Psychology Journal*, 70(4): 318–338. https://doi.org/10.1037/CPB0000121.

Akhtar, R., Ahmetoglu, G. and Chamorro-Premuzic, T. (2013). Greed is good? Assessing the relationship between entrepreneurship and subclinical psychopathy. *Personality and Individual Differences*, 54(3): 420–425. https://doi.org/10.1016/j.paid.2012.10.013.

Akhtar, R., Winsborough, D., Ort, U., Johnson, A. and Chamorro-Premuzic, T. (2018). Detecting the dark side of personality using social media status updates. *Personality and Individual Differences*, 132: 90–97. https://doi.org/10.1016/J.PAID.2018.05.026.

Almeida, P.I., Ahmetoglu, G. and Chamorro-Premuzic, T. (2013). Who wants to be an entrepreneur? The relationship between vocational interests and individual differences in entrepreneurship. *Journal of Career Assessment*, 22(1): 102–112.

Andre, L. (2021). *53 Important Statistics about How much Data is Created Every Day*. https://financesonline.com/how-much-data-is-created-every-day/.

Antoncic, B. and Hisrich, R.D. (2001). Intrapreneurship: Construct refinement and cross-cultural validation. *Journal of Business Venturing*, 16(5): 495–527.

Azucar, D., Marengo, D. and Settanni, M. (2018). Predicting the Big 5 personality traits from digital footprints on social media: A meta-analysis. *Personality and Individual Differences*, 124: 150–159. https://doi.org/10.1016/j.paid.2017.12.018.

Barrick, M.R. and Mount, M.K. (1991). The big five personality dimensions and job performance: A meta-analysis. *Personnel Psychology*, 44(1): 1–26.

Bjerk, D. (2008). Glass ceilings or sticky floors? Statistical discrimination in a dynamic model of hiring and promotion. *Economic Journal*, 118(530): 961–982. https://doi.org/10.1111/j.1468-0297.2008.02157.x.

Brandstätter, H. (2011). Personality aspects of entrepreneurship: A look at five meta-analyses. *Personality and Individual Differences*, 51(3): 222–230.

Chalmers, D., MacKenzie, N.G. and Carter, S. (2020). Artificial intelligence and entrepreneurship: implications for venture creation in the fourth industrial revolution. *Entrepreneurship Theory and Practice*, 45(5): 1028–1053. https://doi.org/10.1177/1042258720934581.

Chamorro-Premuzic, T. (2017). *The Talent Delusion: Why Data, not Intuition, is the Key to Unlocking Human Potential*. London, UK: Little, Brown Book Group.

Chamorro-Premuzic, T. (2019). *Why do so Many Incompetent Men become Leaders? (and what to do about it)*. Harvard Business Review Press.

Clayton, N., Williams, M. and Howell, A. (2014). *Unequal Opportunity: How Jobs are Changing in Cities*. London: Centre for Cities.

Costa, P.T., Jr. and McCrae, R.R. (1985b). *The NEO Personality Inventory Manual*. Odessa, FL: Psychological Assessment Resources.

Derous, E., Buijsrogge, A., Roulin, N. and Duyck, W. (2016). Why your stigma isn't hired: A dual-process framework of interview bias. *Human Resource Management Review*, 26(2): 90–111. https://doi.org/10.1016/j.hrmr.2015.09.006.

Hisrich, R., Langan-Fox, J. and Grant, S. (2007). Entrepreneurship research and practice: A call to action for psychology. *American Psychologist*, 62(6): 575–589. https://doi.org/10.1037/0003-066X.62.6.575.

Javed, A. and Brishti, J.K. (2020). *The Viability of AI-based Recruitment: A Systematic Literature Review*. Umea University.

Kosinski, M., Stillwell, D. and Graepel, T. (2013). Private traits and attributes are predictable from digital records of human behavior. *Proceedings of the National Academy of Sciences of the United States of America*, 110(15): 5802–5805. https://doi.org/10.1073/pnas.1218772110.

Kuncel, N.R., Ones, D.S. and Sackett, P.R. (2010). Individual differences as predictors of work, educational, and broad life outcomes. *Personality and Individual Differences*, 49(4): 331–336.

Kuratko, D.F. (2007). Entrepreneurial leadership in the 21st century: Guest editor's perspective. *Journal of Leadership & Organizational Studies*, 13(4): 1–11.

Kuratko, D.F. (2016). Entrepreneurial Leadership in the 21st Century: Guest Editor's Perspective. *Journal of Leadership & Organizational Studies*, 13(4): 1–11. https://doi.org/10.1177/10717919070130040201.

Leutner, F., Ahmetoglu, G., Akhtar, R. and Chamorro-Premuzic, T. (2014). The relationship between the entrepreneurial personality and the Big Five personality traits. *Personality and Individual Differences*, 63: 58–63. https://doi.org/10.1016/j.paid.2014.01.042.

Leutner, F., Akhtar, R. and Chamorro-Premuzic, T. (2022). *The Future of Recruitment: Using the New Science of Talent Analytics to Get Your Hiring Right*. Emerald Group Publishing.

Liu, L., Preoţiuc-Pietro, D., Samani, Z.R., Moghaddam, M.E. and Ungar, L. (2016). Analyzing personality through social media profile picture choice. *Proceedings of the 10th International Conference on Web and Social Media*, 211–220.

Lumpkin, G.T. (2007). Intrapreneurship and innovation. pp. 237–263. *In*: J.R. Baum, M. Frese and R. Baron (eds.). *The Psychology of Entrepreneurship*. Mahwah, New Jersey: Lawrence Erlbaum Associates.

Mair, J. and Marti, I. (2006). Social entrepreneurship research: A source of explanation, prediction, and delight. *Journal of World Business*, 41(1): 36–44.

McKenzie, B., Ugbah, S.D. and Smothers, N. (2007). "Who is entrepreneur? Is it still the wrong question? *Academy of Entrepreneurship Journal*, 13: 23–43.

McCrae, R.R. and Costa, P.T. (1987). Validation of the five-factor model of personality across instruments and observers. *Journal of Personality and Social Psychology*, 52(1): 81.

Ortigosa, A., Carro, R.M. and Quiroga, J.I. (2014). Predicting user personality by mining social interactions in Facebook. *Journal of Computer and System Sciences*, 80(1): 57–71. https://doi.org/10.1016/J.JCSS.2013.03.008.

Rauch, A. and Frese, M. (2007). Let's put the person back into entrepreneurship research: A meta-analysis on the relationship between business owners' personality traits, business creation, and success. *European Journal of Work and Organizational Psychology*, 16(4): 353–385.

Schaufeli, W.B. and Bakker, A.B. (2004). Job demands, job resources, and their relationship with burnout and engagement: A multi-sample study. *Journal of Organizational Behavior: The International Journal of Industrial, Occupational and Organizational Psychology and Behavior*, 25(3): 293–315.

Shane, S. (2008). *The Illusions of Entrepreneurship: The Costly Myths That Entrepreneurs, Investors, and Policy Makers Live By*. New Haven: Yale University Press.

Shane, S. and Venkataraman, S. (2000). The promise of entrepreneurship as a field of research. *Academy of Management Journal*, 25: 217–226.

Stewart Jr, W.H. and Roth, P.L. (2001). Risk propensity differences between entrepreneurs and managers: A meta-analytic review. *Journal of Applied Psychology*, 86(1): 145.

Tierney, P. and Farmer, S.M. (2011). Creative self-efficacy development and creative performance over time. *Journal of Applied Psychology*, 96(2): 277.

Turczynski, B. (2021). *Lying on a Resume (2020 study)*. Retrieved from: https://resumelab.com/resume/lying.

Venkataraman, S. (2004). Regional transformation through technological entrepreneurship. *Journal of Business Venturing*, 19(1): 153–167.

Youyou, W., Kosinski, M. and Stillwell, D. (2015a). Computer-based personality judgments are more accurate than those made by humans. *Proceedings of the National Academy of Sciences*, 112(4): 1036–1040. https://doi.org/10.1073/PNAS.1418680112.

Youyou, W., Kosinski, M. and Stillwell, D. (2015b). Computer-based personality judgments are more accurate than those made by humans. *Proceedings of the National Academy of Sciences of the United States of America*, 112(4): 1036–1040. https://doi.org/10.1073/pnas.1418680112.

Zhao, H. and Seibert, S.E. (2006). The big five personality dimensions and entrepreneurial status: A meta-analytical review. *Journal of Applied Psychology*, 91(2): 259.

Chapter 4

Professional Human Capital Development for Startup Founders and Workers

Jennifer P Wisdom

Entrepreneurs and individuals leading startups have multiple responsibilities in a fast-paced environment with shifting priorities. For many startups, professional human capital development programs — activities provided or supported by the organization to support continuing education and career training for its employees and leaders — take a back seat to the more urgent issues of securing funding, developing products, and generating sales. Many startups do not look to these professional development activities for value creation within the organization; rather they view them as a way to control costs or maintain legal requirements (Bendickson et al. 2017). Human capital development is not emphasized despite findings that startups led by leaders with greater educational training and startups with higher levels of training have a higher likelihood of survival and growth (Fuller-Love 2006, Prommer et al. 2020). For startups' long-term growth and valuation, investment in professional development can contribute to positive outcomes for the organization and its employees.

Professional development in startups occurs within the context of multiple growth stages (Gaponova and Korshunov 2018, Maltz and Sokol 2018, Kimball, this volume). Typically, in the formation stage, startups begin with a founder and a small set of colleagues allowing them to clarify expectations and resolve disagreements quickly within a shared culture. As the startup expands into the validation stage, communication, and norms for relating may become fractured as specialized teams form and onboarding practices across the startup may not support shared core values, expectations, and communication styles. The growth stage is marked by a shift of the startup to clarify and institute formal talent management and human resources

Wisdom Consulting, 140 W 69th St #45C, New York, NY, 10023, 504-484-5679.
Email: jpwisdom@gmail.com

practices. Across these stages (i.e., formation, validation, growth), the leader who oversees these processes rarely has human resources expertise and much broader responsibilities than people-related concerns. Further, leaders' perspectives toward these human resource processes may be as merely administrative and efficiency oriented as opposed to viewing them as an opportunity to create organizational culture, develop staff, and improve the company's infrastructure (Maltz and Sokol 2018). Startup strategies to address professional development can vary during these different growth stages, depending on the perspective of the people-focused leader, the goals of the startup, and the available funding and timeline for work.

This chapter describes the role and practice of professional development in startups. The chapter begins with a description of the strategic contributions of professional development programs to organizations and discusses how these programs can uniquely benefit startups. Next, it summarizes factors associated with companies' engagement with training at different stages of organizational development, including the impact of professional development training on organizational success. Finally, the chapter concludes with a discussion of models for professional development that are applicable to startups, including how to balance priorities for resource allocation and maximize organizational benefits.

Strategic Contributions of Professional Development Programs to Startups

The primary growth mechanism of businesses is human capital, which resides in the individual workers and in the joint relationships that workers create (Nahapiet and Ghoshal 1998). Human capital flourishes in high-performance work systems. These high-performance work systems include human resource processes, policies and practices for staffing, self-management teams, decentralized decision making, training, flexible work assignments, communication, and compensation (Evans and Davis 2005). As the managing entity of high-performance work systems, human resource offices facilitate effective professional development as well as selection, compensation, and performance management of workers (Bendickson et al. 2017). Therefore, within a smooth-running organization, the business has a high-performance work system, which includes human resource practices that oversee professional development to grow human capital.

Companies that use high-performance work systems tend to have better workers: they recruit workers of high quality who fit both the organization and job; these workers receive training and therefore have higher skills, are more likely to stay in the organization, have higher levels of commitment and satisfaction; and are more likely to be engaged with organizational goals (Bendickson et al. 2017, Gong et al. 2010, Pfeffer 2007). Professional development activities are demonstrated to increase understanding and performance in teachers (Thurlings and den Brok 2017), human services professionals (Lauer et al. 2013), physicians (Forsetlund et al. 2021), and nurses (Bell et al. 2021). In this regard, it is in startups' interest to ensure their workers are well-trained and have a high level of professional proficiency within these high-performance work systems.

Professional development in the context of startups includes two primary training targets: leaders and workers. Several authors suggest that startup professional development programs should target all employees, specifically because startup workers tend to have to be "multitaskers" covering several roles due to initial lean staffing, learning as they go, and making fast decisions in highly uncertain environments (Fuller-Love 2006, Gaponova and Korshunov 2018).

Professional Development for Startup Leaders

Startup leaders often engage in entrepreneurial training via startup development organizations such as accelerators, incubators, venture studios, and co-working communities (see Kimball in this volume). These fixed-term, cohort-based startup educational and mentorship programs are offered through private companies, higher education institutions, and U.S. federal agencies such as the National Science Foundation (Cohen et al. 2019, Garciá-González and Ramírez-Montoya 2021, Krysiewicz and Isakowitz 2019). Other undergraduate, graduate, and certificate entrepreneurship programs are offered through schools of engineering, biomedical and business administration. Entrepreneurship training generally includes topics such as defining customer segments and product-market fit, understanding customer needs, creating value propositions, clarifying the product market, financing, marketing, developing business models, and more (e.g., Krysiewicz and Isakowitz 2019). Research linking the engagement of startup leaders in these programs and successful startup outcomes is mixed, primarily because of heterogeneity in program aims, small sample sizes of studies, and lack of uniform definitions (Cohen et al. 2019, Kimball 2022).

Research demonstrates founders have continuous needs for different kinds of skills. These include technical skills (e.g., spreadsheets, data collection/analysis, computer programming, search engine optimization, content management, and project management; Bradford 2016); leadership skills (e.g., communication, strategy, prioritization, financial expertise, networking); and psychological aspects of entrepreneurship (e.g., goal setting, entrepreneurial identity, task strategy and planning, feedback seeking, and learning; Frese 2009). Additional reports suggest that although entrepreneurs emphasize strategic thinking skills as more critical than financial skills (Council for Excellence in Management and Leadership 2002), management skills to create efficiency contribute most to a firm's success (Collins 2001).

Methods for professional development activities for founders can include (a) formal training and education, either in person or online; (b) experiential learning, including developmental job assignments and feedback; (c) developmental relationships such as mentoring and knowledge exchange programs, and (d) self-managed learning, which is individual self-assessment of personal capabilities and self-directed initiatives to improve skills (Abbott and Dahmus 1992, Prommer et al. 2020). Some founders continue to receive support during the startup from mentor relationships formed during pre-startup accelerator or entrepreneurship programs.

Despite the proliferation of training for aspiring entrepreneurs in the formation stage of companies and the documented need for startup founder professional development, no reports on the prevalence of founder professional development during the validation and growth stages of a startup were found. This need-availability gap suggests opportunities for research into leader professional development during the startup's early years. Researchers should first identify the prevalence of professional development, specifically during validation and growth stages, then clarify critical components of founder professional development associated with startup success during different startup stages.

Professional Development for Startup Workers

Startup workers' professional development skills include specific technical skills as well as leadership and management, developing management systems and techniques, team building, planning, delegation, and financial management (Fuller-Love 2006). A study of university affiliated entrepreneur trainees and established entrepreneurs in Russia indicated that about half of the respondents suggested training would be useful for all workers in a startup on negotiation, project management, and company planning and management decision making (Gaponova and Korshunov 2018). Between 20–30% of respondents also suggested training for all staff in conflict management, human resources management, personal effectiveness, fundamentals of entrepreneurship, crisis management, and cross-cultural communication (Gaponova and Korshunov 2018).

These skills may be offered internally through human resources offices (once established), contracted out to external trainers/mentors, or most likely, skipped altogether. Workers who begin with the company in the early startup stages are more likely to become managers over time; training to enhance management skills and cultural adaptations is particularly important for them.

Organization-Wide Development

In addition to the individual benefits to startup leaders and workers of receiving professional development, specific organization-wide benefits may be of particular interest to startups and emerging organizations. Professional development for startup founders and workers can have a positive impact on organization sustainability and profitability, establish the company as a learning organization, offer legitimacy for having established human resources systems, increase startup performance, reduce the likelihood of startup failure, and contribute to a competitive advantage if the company can offer to employees what other companies cannot (Bendickson et al. 2017, Fuller-Love 2006, Senge 1994, Wright and McMahan 1992).

Professional Development Program Models

Numerous researchers have recommended specific content for professional development activities that would benefit startup founders and workers (Bradford 2016, Frese 2009, Fuller-Love 2006, Gaponova and Korshunov 2018, Morris et al. 2013). Table 1 integrates these authors' recommendations for training and provides additional suggestions regarding human capital professional development topics across stages of startup development.

Table 1. Content and for startup founders and staff human capital development during three stages of company formation.

	Founders		Staff	
	Content	**Methods**	**Content**	**Methods**
Formation	*Technical skills* • Spreadsheets • Data collection/ analysis • Computer programming • Search engine optimization • Content management *Leadership skills* • Communication • Strategy • Prioritization • Financial expertise • Conflict management • Networking *Psychological aspects of entrepreneurship* • Goal setting • Stress management • Entrepreneurial identity • Task strategy and planning • Feedback seeking • Self-efficacy	• Competitions • Game programs • Seminars • Coaching	*Technical skills* • Spreadsheets • Data collection/ analysis • Computer programming • Search engine optimization • Content management *Management skills* • Project management • Team building • Planning • Delegation • Conflict management • Negotiation • Financial management • Team building *Psychological aspects of entrepreneurship* • Goal setting • Stress management • Task strategy and planning • Problem solving • Self-efficacy	• Team building activities • Seminars • Socialization • Direct coaching by founders
Validation	All of the above plus: • Corporate values and culture • Learning styles • Cross-cultural communication • Process improvement/ efficiency • Resource leveraging • Risk management	• Short-term programs Coaching from investors and partners • Self-learning through business literature	All of the above plus: • Corporate values and culture • Human resources management • Personal effectiveness • Crisis management • Cross-cultural communication • Process improvement/ efficiency • Resource leveraging • Risk management	• In-house training

Table 1 contd. ...

...Table 1 contd.

	Founders		Staff	
	Content	**Methods**	**Content**	**Methods**
Growth	All of the above plus: • Advanced leadership competencies	• International/ foreign internship programs • Long term programs or master's programs in business administration Consultations with senior experts	All of the above plus: • Advanced management competencies • Organizational dynamics	• Comprehensive training program

No reports of professional human capital development programs specifically for startup founders within companies in validation or growth stages were found. There are many reasons startups do not offer professional development programs including: (a) few senior managers have formal management training, which leads to a lack of training culture; (b) founders, rather than specialized human resources leaders, may make decisions about training, often discounting training entirely; (c) founders may not have time or resources to develop formal systems for professional development or to choose appropriate training; (d) founders may view prospective trainers as having insufficient experience or expertise to adequately train their team; (e) many startups have a crisis-oriented approach to work and do not consider other ways of managing day-to-day activities; (f) many management training programs are not applicable to startups (e.g., they focus on functional models of separate specialized departments that have not yet been formed); (g) founders may fear training programs may show either their or their business's weakness; (h) business growth and profits may be a lower priority than founder autonomy and independence (Fuller-Love 2006, Perren 1999, Stanworth and Gray 1991, Westhead and Storey 1996).

Prommer et al. (2020) acknowledged the dearth of professional development in startups and asked industry experts for their predictions on the future of leadership development activities in startups. They predicted most startups will engage in leadership programs over the next five to 10 years, aiming to develop the leadership abilities of all employees, not only founders. They also predicted the primary leadership development methods will be informal, unplanned, and involve online experiential learning and developmental relationships, especially coaching (Prommer et al. 2020).

The Future of Professional Development in Startups

The status of professional development in startups indicates that there is a high need for professional development for both founders and workers. Further, startups led by leaders with greater educational training and startups with higher levels of training have a higher likelihood of survival and growth, and there is substantial interest in startups providing ongoing professional development to both founders and workers (Fuller-Love 2006, Prommer et al. 2020). In practice, however, there is little evidence of professional development programs in startups; there are many barriers to startups offering such programs, and most startups consider training an expense rather than an investment in their human capital or the future of the business.

Although this chapter provides a summary of findings related to professional development in startups, there is much work to be done. Researchers and evaluators should clarify terminology to increase the precision and applicability of findings. Terminology that would benefit from specificity includes what programs are called (e.g., "professional development programs," "leadership development," continuing education, or founder and worker training), what training topics are covered (e.g., management, team building, communication), and outcome measures (e.g., business profitability, startup survival, valuation). Clarification of terminology is the first step to reducing variation in the quality and consistency of available training.

Next, human resources professionals, organizational psychologists, and other industry leaders should conduct thorough surveys of professional development activities in startups to identify active components of training that are most beneficial to startups. For example, merely offering professional development may result in employee interest and retention because of the startup's commitment to its workers and growth more than the specific content of the training. Developing a strong evidence base that identifies what ongoing training is most beneficial, and cost-effective ways to provide it during all stages of a startup will serve the startups, their founders, and their workers.

Finally, it would be useful to reduce the variation in the quality and content of training by clarifying content standards for startup founders' and workers' professional development. It would also be helpful to provide standards for trainers to ensure they are operating with a clear understanding of startup culture, desired outcomes of training, and effective training methods best serve startups, their founders, and their workers.

Practical Recommendations for Professional Development Programs

Current startup "accelerator" and entrepreneurial training for aspiring founders should include information on both (a) the role of management, leadership, and efficiency in startup success and (b) the importance of ongoing professional development throughout the lifecycle of a startup. Coaching and mentorship through these programs could also emphasize this value. To make decisions about offering professional development in startups, there are significant decision points:

- *Who is the target audience for professional development?* Both founders and workers would benefit from professional development training. Leaders may wish to consider each group's needs separately and identify prospective cost-benefit analyses for providing development to each group.

- *What kind of content is desired?* Professional development training can address technical skills, leadership skills, or psychological aspects of entrepreneurship. Content can be guided by needs at different stages of startup development and should focus at least initially on aspects of the startup that are slowing innovation or success, however defined (Gaponova and Korshunov 2018). For each training, objectives should be established, and outcomes measured.
- *Should the training be provided internally or by external contractors?* Internal programs can be much more targeted to individual startup cultures but require significant organizational resources and expertise on staff to create programming and ensure training access, and track training outcomes. Early startups may not have this human resources expertise available on staff. Externally contracted programs will be less targeted toward a startup's specific culture but could take less time for startup staff to coordinate and administer. University programs may have more consistent quality than non-academic providers but may not provide programs of appropriately targeted intensity.
- *What training mechanisms should be used to deliver professional development?* As indicated above, options may include in person or online formal training and education, experiential learning, developmental relationships (e.g., mentoring, coaching), and self-managed learning (Abbott and Dahmus 1992, Prommer et al. 2020).
- *What is the budget for professional development?* Costs should be considered both in terms of the cost of providing the training (e.g., instructor time, materials) and in terms of the cost of worker absence from the workplace. Ideally, professional development budgets are closely tied to the expected return on investment.
- *How should professional development be incentivized?* Leaders can reinforce the importance of the startup as a learning organization, include professional development goals in staff performance metrics, or provide financial incentives for participation.
- *How should the startup adjust professional development programs with the growth of the startup and its employees*? Leaders should track the aspects of business performance that led to the implementation of professional development programs. They should adjust the target audience, content, delivery methods, and assessment based on observed results of programming.
- *How should the startup track and measure professional human capital development outcomes?* Professional development outcome measurement includes five levels. Four are measured on an individual worker level (Tian et al. 2007): worker's satisfaction with training (level 1), worker's changes in knowledge and attitudes (level 2), change in worker behavior and practice (level 3), and improved work outcomes (level 4). A fifth level is organizational outcomes (e.g., higher worker retention, organizational performance, company valuation or survival, and process efficiency). Startups should aim to assess all five levels utilizing pre-post tests (to address levels 1 and 2), connecting worker performance reviews and coaching training (levels 3 and 4), and organization-wide metrics such as employee retention (level 5). Professional external

consultants can assist startups in setting up a proper system of evaluation to ensure maximum impact and efficiency of professional development training.

Conclusion

Professional human capital development programs demonstrate promise to improve the effectiveness of startup workers and the outcomes of startup endeavors. Startup leaders would be well served to create professional development plans tailored to the needs of their organization, which can contribute to higher worker satisfaction, retention, and performance.

References

Abbott, J. and Dahmus, S. (1992). Assessing the appropriateness of self-managed learning. *Journal of Management Development*, 11(1): 50–60.

Bell, D.F., Pestka, E. and Forsyth, D. (2021). Outcome evaluation: Does continuing education make a difference? *The Journal of Continuing Education in Nursing*, 38(4).

Bendickson, J., Muldoon, J., Ligouri, E. and Midgett, C. (2017). High performance work systems: A necessity for startups. *Journal of Small Business Strategy*, 27(2): 1–12.

Bradford, L. (2016). 9 Technical Skills to Learn as an Aspiring Startup Founder. *Forbes*, June 27, 2016. Accessed at: https://www.forbes.com/sites/laurencebradford/2016/06/27/9-digitaltechnical-skills-to-learn-as-an-aspiring-startup-founder/?sh=46283c8723a3.

Cohen, S., Fehder, D.C., Hochberg, Y.V. and Murray, F. (2019). The design of startup accelerators. *Research Policy*, 28(7): 1781–1797.

Collins, J.C. (2001). Good to Great: Why Some Companies Make the Leap... and Others Don't. Harper Business.

Council for Excellence in Management and Leadership. (2002). *Joining Entrepreneurs in their World: Improving Entrepreneurship, Management and Leadership in UK SMEs.* Council for Excellence in Management and Leadership.

Evans, W.R. and Davis, W.D. (2005). High-performance work systems and organizational performance: The mediating role of internal social structure. *Journal of Management*, 31: 758–775.

Forsetlund, L., O'Brien, M.A., Forsén, L., Mwai, L., Reinar, L.M., Okwen, M.P., Horsley, T. and Rose, C.J. (2021). Continuing education meetings and workshops: Effects on professional practice and healthcare outcomes. *Cochrane Database of Systematic Reviews*. Available at: https://doi.org/10.1002/14651858.CD003030.pub3.

Frese, M. (2009). Towards a psychology of entrepreneurship: An action theory perspective. *Foundation and Trends in Entrepreneurship*, 5(6): 437–496.

Fuller-Love, N. (2006). Management development in small firms. *International Journal of Management Reviews*, 8(3): 175–190.

Gaponova, O.S. and Korshunov, I.A. (2018). Deploying a corporate learning system at the innovative startup. *Russian Education & Society*, 60(4): 289–314.

García-González, A. and Ramírez-Montoya, M.S. (2021). Social entrepreneurship education: Changemaker training at the university. *Higher Education, Skills and Work-Based Learning*, 11(5): 1236–1251.

Gong, Y., Chang, S. and Cheung, S.Y. (2010). High performance work system and collective OCB: A collective social exchange perspective. *Human Resource Management Journal*, 20(2): 119–137.

Krysiewicz, A. and Isakowitz, T. (2019). I-Corps as a training tool for new technology development. In *Academic Entrepreneurship for Medical and Health Sciences*. https://doi.org/10.21428/b2e239dc.d88e0110.

Lauer, P.A., Christopher, D.E., Firpo-Triplett, R. and Buchting, F. (2013). The impact of short-term professional development participant outcome: A review of the literature. *Professional Development in Education*, 40(2): 207–227.

Maltz, M. and Sokol, M. (2018). Startup lessons for everyone. *People & Strategy Journal*, 54–56. Available at: https://www.shrm.org/executive/resources/people-strategy-journal/Summer2018/Pages/default.aspx.

Morris, M.H., Webb, J.W., Fu, J. and Singhal, S. (2013). A competency-based perspective on entrepreneurship education: Conceptual and empirical insights. *Journal of Small Business Management*, 51(3): 352–369.

Nahapiet, J. and Ghoshal, S. (1998). Social capital, intellectual capital, and the organizational advantage. *Academy of Management Review*, 23(2): 242–266.

Perren, L. (1999). Factors in the growth of microenterprises (Part 1: Developing a framework). *Journal of Small Business and Enterprise Development*, 6(4): 366–385.

Pfeffer, J. and Veiga, J.F. (1999). Putting people first for organizational success. *The Academy of Management Executive*, 13(2): 37–48.

Prommer, L., Tiberius, V. and Kraus, S. (2020). Exploring the future of startup leadership development. *Journal of Business Venturing Insights*, 14: e00200.

Senge, P. (1994). *The Fifth Discipline*. Doubleday.

Stanworth, J. and Gray, C. (1991). *Bolton 20 Years On The Small Firm in the 1990s.* Small Business Research Trust, Paul Chapman.

Thurlings, M. and den Brok, P. (2017). Learning outcomes of teacher professional development activities: A meta-study. *Educational Review*, 69(5): 554–576.

Tian, J., Atkinson, N.L., Portnoy, B. and Gold, R.S. (2007). A systematic review of evaluation in formal continuing medical education. *Journal of Continuing Education in the Health Professions*, 27(1): 16–27.

Westhead, P. and Storey, D. (1996). Management training and small firm performance: Why is the link so weak? *International Small Business Journal*, 14(4): 13–24.

Wright, P.M. and McMahan, G.C. (1992). Theoretical perspectives for strategic human resource management. *Journal of Management*, 18(2): 295–320.

Chapter 5

Selection and Training for Teamwork

Implications for Diverse, Virtual, and Human-Machine Teams

Samantha Dubrow,[1,*] *Sarah Resnick*[2] and *Anwesha Choudhury*[2]

Sixty percent of new venture failures can be, at least partially, explained by challenges with the entrepreneurial team (de Mol et al. 2015, Eisenhardt 2013). Teamwork plays a critical role in entrepreneurship because of the volatility and dynamism faced when beginning new ventures. Entrepreneurs typically work together in fluid environments with high uncertainty that requires multiple individuals from diverse backgrounds to combine their efforts. New venture performance is therefore reliant on the ability of those individuals to effectively work as a team.

Several definitions of the term "team" have been offered over the years. One comprehensive definition that encompasses several important components states that a team has, "(a) two or more individuals who (b) socially interact (face-to-face or, increasingly, virtually); (c) possess one or more common goals; (d) are brought together to perform organizationally relevant tasks; (e) exhibit interdependencies with respect to workflow, goals, and outcomes; (f) have different roles and responsibilities; and (g) are together embedded in an encompassing organizational system, with boundaries and linkages to the broader system context and task environment" (Kozlowski and Ilgen 2006, p. 76). Organizational teams can take different forms based on their interdependence, similarity, task consistency, and membership stability (Tannenbaum and Salas 2020).

[1] The MITRE Corporation, 400 8th NW Apt 903, Washington, DC, 20009, 610-291-6476.
[2] Binghamton University School of Management, Academic Building A, 4400 Vestal Pkwy East, Binghamton, NY 13902.
Emails: Sresnic3@binghamton.edu; achoudh6@binghamton.edu
*`Corresponding author: sdubrow@gmail.com

Teams are typically evaluated based on their performance, or effectiveness, which can encompass productivity, efficiency, and quality (Mathieu et al. 2019, Salas et al. 2008). Team performance and effectiveness are impacted by team inputs such as team composition, processes such as communication and coordination, and emergent states such as mutual trust and cohesion. Thus, team performance, both in new ventures and large established businesses, is dependent on individual teamwork competencies ranging from interpersonal flexibility to social intelligence (Brinckmann and Hoegl 2011). Large established businesses, on the other hand, have typically been able to rely on selecting employees who have strong technical expertise and rely on leaders to integrate work from individual contributors.

The purpose of this chapter is to highlight the specific teamwork skills and processes that will be necessary for teams to perform effectively across a variety of entrepreneurial domains, ranging from new ventures to intrapreneurial teams in large businesses, as the nature of work continues to change. Over the years, teams have become increasingly diverse both in terms of their surface-level diversity (e.g., gender, race, national origin) and deep-level diversity (e.g., functional expertise, disciplinary background). Additionally, teams worldwide have become more virtual, especially since the beginning of the COVID-19 pandemic. Finally, human-machine teams are creeping into the workplace, fundamentally changing the nature of how teams interact. Each of these changes to work is resulting in a developing need for teamwork skills, in addition to technical capabilities. At the end of the chapter, tools are provided for both selecting and training for teamwork skills. These selection and training methods can be implemented by all companies, even with limited resources.

The Growing Importance of Teamwork

Team processes are at the core of teamwork and include all activities that team members engage in, either together or independently (Kozlowski and Ilgen 2006). In the academic literature, teams are commonly described using the IPMO model of inputs, processes, mediators (emergent states) and outputs (Marks et al. 2001). Inputs include team composition, context, and task. Team composition and context include the diversity of team members' personalities, cultures, disciplinary backgrounds, and species (e.g., human-only vs. human-machine teams) and team "virtualness". Team task inputs include the technical requirements for completing the work and the task type (e.g., physical vs. intellectual tasks; Devine 2002). Team processes include communication, collaboration, adaptation, and conflict. Inputs and processes affect team emergent states (i.e., "cognitive, motivational, and affective states of teams" that develop over time as the team continues to work together;" Marks et al. 2001, p. 358). Important emergent states include shared cognition and mutual trust, which can help to optimize future subsequent teamwork processes that contribute to team outcomes. Finally, outcomes include team performance and individual team members' satisfaction with the team (Kozlowski and Ilgen 2006).

Skills and processes required to facilitate team performance and team effectiveness are often split into two categories: taskwork and teamwork. Taskwork includes individual functions that team members complete to accomplish their work, including interacting with tools and systems (Cannon-Bowers et al. 1995, de Mol et al. 2015). Teamwork includes interpersonal processes through which team members coordinate with one another interdependently to complete team tasks. Teamwork includes the interactions between team members that are needed to combine their taskwork to create a shared outcome.

The terms teamwork and taskwork can refer to both processes (what a team does) and KSAOs (knowledge, skills, abilities, and other attributes of individual team members). Taskwork KSAOs are the generic and specific competencies individuals need to do their jobs and are an important part of most organizational selection processes. Interdisciplinary teams typically have a diverse composition of taskwork KSAOs. Teamwork KSAOs are also needed to integrate the taskwork of team members (Thayer et al. 2014). Generic teamwork KSAOs are the capabilities required on most tasks, such as communication skills and collective orientation. Specific teamwork KSAOs are capabilities that are important for members of a particular team, such as cognitive flexibility on human-machine teams and cultural intelligence on international teams.

Taskwork KSAOs are usually the focus of selection and training, especially in small businesses with limited resources. However, teamwork KSAOs are a critical predictor of team performance, especially in dynamic and complex environments (DeChurch and Mesmer-Magnus 2010). When individuals with strong teamwork skills come together and spend time getting to know each other they can successfully develop expectations of one another and establish norms for working together that will promote successful team performance (Dyer 1984, Salas et al. 2005). Teamwork processes are increasingly important in large teams with high interdependence where teamwork breakdowns are both common and likely to have a strong impact on team performance (LePine et al. 2008).

Core Teamwork Processes and Emergent States

Team processes of communication, coordination, mutual performance monitoring, adaptability, and conflict management are critical for team performance, especially as the nature of work changes. As a team process, communication is defined as "the exchange of information between a sender and a receiver irrespective of the medium (Salas et al. 2005, p. 561). Communication is a key behavioral process in all teams (Mak et al. 2017). Teams are most likely to have strong communication practices when team members have communication and information sharing skills, team orientation, and cross-cultural communication competence. Individuals with communication and information sharing skills are able to share ideas with their teammates openly and at appropriate times (Brinckmann and Hoegl 2011, Cannon-Bowers et al. 1995, Mesmer-Magnus and DeChurch 2009). Individuals with team orientation (i.e., a drive toward behaviors that benefit the team over themselves individually) go out of their way to incorporate all team members in open communication and information sharing (Salas et al. 1995).

Team *coordination* is "the act of managing interdependencies among team members to achieve a goal" (Mak et al. 2017, p. 9). When teams coordinate effectively, they can synchronize their taskwork, balance their workloads, and integrate their work into a final product. Coordination is itself a teamwork skill. Teams will also be able to facilitate coordination if they are made up of members with perspective taking, information sharing, planning and task coordination, and collaborative problem-solving skills.

Teams working on complex tasks in dynamic environments need to engage in teamwork processes of mutual performance monitoring followed by backup behaviors. *Mutual performance monitoring* occurs when team members identify mistakes of other team members, and backup behaviors are subsequently used to correct those mistakes (Salas et al. 2005). Mutual performance monitoring is a skill that can be selected for and trained and includes an individual's ability to develop an active mental model of the team and identify when and how to engage in backup behaviors to keep team performance on track (Mathieu et al. 2014). Social perceptiveness is another skill that can help team members maintain awareness of each other's actions and reactions and forecast their future behaviors to prepare to engage in any necessary backup behaviors (Morgeson et al. 2005).

Team *adaptability* is the process of adjusting strategies as the team environment and task requirements change over time (Salas et al. 2005). Team members start by identifying a cue that there is a change that needs to be reacted to, then work together to develop a course of action for altering their strategy accordingly, and finally implement that new strategy. Adaptability is an individual skill that can be selected for and trained (Mathieu et al. 2014). Two other skills important for team adaptability are tolerance for ambiguity and openness to experience. Individuals with a high *tolerance for ambiguity* are able to manage and adapt to unexpected circumstances (Carter et al. 2015). *Openness to experience* is a personality trait of individuals who are curious, risk takers, and comfortable with change (Caliguri and Tarique 2012).

Some forms of conflict can be functional for team performance. Specifically, conflict around how a task should be performed, and which tasks should be prioritized can have a positive impact on team performance, especially for tasks that require team creativity (Kozlowski and Ilgen 2006). Relationship conflict, on the other hand, is almost always negatively related to team performance (Kozlowski and Ilgen 2006). Thus, the process of *conflict management* is required for teams to navigate and resolve disagreements effectively. Conflict resolution skills can be selected for and trained (O'Neill et al. 2012). Additionally, teams will benefit from individuals with team orientation and cross-cultural competencies to engage in productive conflict management.

In addition to team processes, team *emergent states* are important teamwork factors that develop over time, result in both team inputs (e.g., team member KSAOs), and affect team outcomes. Diverse, virtual, and human-machine teams will benefit from various emergent states, including shared cognition and mutual trust. *Shared cognition* is a critical emergent state in all teams, and the need for strong shared cognition is increased in teams working in complex environments with diverse and interdependent team members. Meta-analyses have demonstrated that team cognition explains significant variance in team performance, even after controlling for team processes and team member motivation (DeChurch and Mesmer-Magnus 2010). Team cognition includes the shared knowledge of a team, including team mental models and teamwork knowledge, both of which are positively related to team performance (Mathieu et al. 2014, Normand 2019).

Finally, *mutual trust* is "the shared belief that team members will perform their roles and protect the interests of their teammates" (Salas et al. 2005, p. 561). Teammates can develop trust in each other's benevolence, competence, and integrity (Chen and Dhillon 2003). Benevolence includes a trust that a teammate has the team's best interest in mind. Competence is the trust in teammates' capabilities to do their taskwork. Integrity includes trust that team members will act as expected. Individuals with perspective taking skills, cultural intelligence, and a propensity to trust are most likely to develop mutual trust with their teammates (Hertel et al. 2005, Magpili and Pazos 2018).

Throughout this chapter, three types of teams are discussed to highlight how the need for teamwork skills and processes is growing as the nature of work changes and new challenges emerge. These team types include: (1) *diverse teams* with members from different functional and/or cultural backgrounds), (2) *virtual teams* with members who are generally not collocated and do not usually meet face-to-face, and (3) *human-machine teams,* which include interdependent, but non-human, team members. The following sections demonstrate how the team processes and states discussed above are uniquely important in diverse, virtual, and human-machine teams.

Teamwork Skills for Working in Diverse Teams

Team diversity can take many different forms. First, teams have *surface-level diversity* when team members have different overt characteristics, including gender and race (and in human-machine teams, species). *Deep-level diversity* comes when team members have different KSAOs, including their technical backgrounds, personalities, and attitudes (Mathieu et al. 2019). There are several benefits of team diversity, particularly for entrepreneurial teams working on innovative solutions that require combinations of various types of unique expertise from component team members. For teams focused on innovation as their performance outcome, both functional and educational diversity of team members can have positive impacts (Bell et al. 2011, Mathieu et al. 2014).

Team diversity can also create challenges for team performance, particularly when members lack teamwork competencies such as interpersonal skills and perspective taking. There are often coordination challenges when working across diverse teams. For example, individuals are more likely to share information they already have in common than to share unique information (Mesmer-Magnus and DeChurch 2009). However, interdisciplinary teams have to share their unique information to effectively integrate their work, particularly on projects where their work is highly interdependent.

Diverse teams require clear and open communication with an emphasis on unique information sharing. Often, members of diverse teams fail to share unique information with one another, either because they do not believe it is relevant for others to know or because they do not believe it is worth the effort to explain all the disciplinary context required to understand the unique information (DeChurch and Mesmer-Magnus 2010). However, diverse teams are often diverse in the first place because they have a team task that is complex enough to require team members from different functional backgrounds to work together and integrate their knowledge. Thus, communication skills are required for diverse teams to be successful. In startups and small businesses especially, individuals from diverse backgrounds need to share information about their unique past experiences to optimize their collaborative decision making.

Diverse teams also require team members with collaborative problem-solving skills to achieve effective coordination. Individuals who can recognize the potential obstacles to coordination in diverse teams will be able to address those obstacles and go out of their way to integrate the full range of skills and perspectives that need to be considered on the team.

Members of diverse teams need to ensure that all contributions to the team are considered equally, making mutual performance monitoring a critical process. Additionally, for backup behavior to be successful, individual contributors must be accepted by all teammates, meaning that diverse team members should be open to listening to criticism and accepting help from others who may be different from them. Startup teams often work in high-risk situations where single decisions can make or break the entire business. Thus, team members from different backgrounds should monitor each other's thought processes and decisions to help the team adjust its course of action quickly when needed. Given the need for quick adjustments to keep up with changes in dynamic environments, team adaptability is also a critical process for startups. Tolerance for ambiguity and openness to experience are seen as important characteristics for members of diverse teams, including cross-cultural teams for them to be adaptable (Carter et al. 2015).

Diverse teams are prone to relationship conflict, but relationship conflict can be avoided when team members have cross-cultural competencies and are able to communicate and coordinate with others who are different from them. For diverse teams, team orientation has been found to reduce the negative correlation between diversity and conflict resolution (Mathieu et al. 2014).

Diverse teams may face challenges in developing a shared cognition if they do not understand what information is and is not overlapping across team members with diverse functional backgrounds. Additionally, even if team members communicate openly, commonly shared information may be interpreted differently by team members with different functional and cultural backgrounds, limiting the sharedness of team mental models.

Diverse teams can struggle to develop mutual trust, as humans are less likely to trust people who are different from themselves. However, individuals with cultural intelligence and global mindsets are more willing to actively learn about their teammates and are patient to develop trust over time. Members of diverse teams who can engage in perspective taking by putting themselves in the shoes of their teammates can also increase mutual trust for both them and their teammates (Burke et al. 2010, Hertel et al. 2005, Thayer et al. 2014).

Teamwork for Virtual Teams

Before the 2020 COVID-19 pandemic, nearly 30 years of research explored the differences between face-to-face and virtual teams. Now, virtual teams are typically just thought of as "teams." Much of the research published during the pandemic does not even acknowledge that the teams used in their research studies were, in fact, virtual. However, virtual teams do have some key differences from face-to-face teams, and those differences have negative impacts on team performance if members lack critical teamwork skills.

Virtual teams are physically defined as groups of "people or stakeholders working together from different locations and possibly different time zones, who are collaborating on a common project and use information and communication technologies (ICTs) intensively to co-create" (Garro-Abarca et al. 2021 p. 2). Virtual teams exist across a spectrum of "virtualness," such that teams with higher virtualness receive fewer social cues during interpersonal interactions. A team's level of virtualness is dependent on the richness and synchronicity of team member communication (Kirkman and Mathieu 2005). Teams with low synchronicity and richness have the highest virtualness.

Synchronicity is the extent to which individuals are communicating in real time. Emails have low synchronicity, instant messages have moderate synchronicity, and verbal conversations have high synchronicity. Richness is the amount of valuable information that is carried across a communication platform. Emails have low richness, telephone calls have moderate richness, and video calls are thought to have high richness. However, "researchers agree that videoconferencing cannot fully substitute for, and is less rich than face-to-face communication" (Kirkman and Mathieu et al. 2005, p. 702). Research has demonstrated that videoconferencing is somewhat limited in its richness because of the inability to make eye contact with others. Cameras are typically placed above or below computer screens, meaning that video conference participants look like they are making eye contact only when they are looking somewhere slightly off screen (McNelley 2005).

Many startups are made up of virtual teams. Whether individual contributors at a startup are located in different countries or the startup has not yet invested in office space, it is unlikely that early-stage startup teams are consistently collocated. Thus, the distinction between taskwork and teamwork skills is particularly important in virtual teams. Many people argue that individual taskwork productivity has increased since the start of the pandemic (Jarrahi and Sawyer 2013). However, the co-creation of knowledge and products by teams requires additional teamwork processes, such as conflict management, collective decision making, and communication, which can be inhibited when teams are unable to meet face-to-face.

Virtual teams, especially those with low richness and synchronicity, can have communication process challenges (Kirkman and Mathieu 2005). Virtual teams often miss critical communication practices such as dropping by someone's office to ask a question and having watercooler conversations. Therefore, virtual teams require more explicit effort toward communication processes. Communication needs to be managed in virtual teams in a way that allows all team members to maintain collective situational awareness without crowding too much of the team's time with meetings that often result in redundant and irrelevant information sharing when too frequent or too large (Endsley 1995).

Virtual teams have been found to have stronger coordination when team members have telecooperation skills (Hertel et al. 2005). Telecooperation skills include self-management and interpersonal and intercultural sensitivity and are particularly important in virtual startup teams with members from different countries. Individuals with telecooperation skills also have a higher propensity to trust and are therefore more willing to coordinate with others in different locations or from different cultural backgrounds. Perspective taking is also important for coordination in virtual environments when low richness and synchronicity exacerbate coordination challenges (Magpili and Pazos 2018). When team members engage in perspective taking they are able to increase their awareness of how others should be contributing to the team and will understand how to coordinate the taskwork skills of members from across the teams.

Virtual teams should actively engage in mutual performance monitoring because they have fewer cues for when there is slippage in the performance of teammates and less knowledge of when backup behaviors are needed. Members of virtual teams can also use social perceptiveness skills to identify when teammates need backup behaviors, even when they are not collocated.

In virtual teams, there is an additional need for intentional collaboration to maintain adaptability when there are environmental or task changes. Individuals who are adaptable, have high tolerance for ambiguity, and are open to new experiences are likely to come together to adapt effectively, regardless of whether they are collocated.

Virtual teams may struggle to engage in conflict management when team members do not receive social cues indicating that a conflict is arising and needs to be resolved. Team members with high team orientation are more likely to try to overcome this challenge by going out of their way to monitor the team's behaviors, integrate diverse team member contributions, and arrange virtual meetings to resolve arguments.

Virtual teams also need to work to actively develop shared cognition because their communication is often asynchronous and informal conversations are less common than on face-to-face teams. One method that can be used to develop shared cognition is to establish team member roles (Hertel et al. 2005). When specific roles are established, team members know who to look to for different areas of expertise and when the team needs to adapt to dynamic situations. Roles are important to establish on virtual teams where they are less likely to emerge naturally through continuous social interaction.

Mutual trust is developed over time as teams gain experience working together and can perceive each other's actions. By working together and engaging in processes such as communication, coordination, and backup behaviors, teammates can start to develop expectations of one another, followed by mutual trust. Virtual teams tend to spend less time together than face-to-face teams and receive fewer social cues (e.g., body language) that are often used to develop trust. Therefore, virtual teams need to share information with one another early and often so that mutual trust can begin to develop.

Teamwork of the Future: Working in Human-Machine Teams

According to a 2018 report from the World Economic Forum, humans perform approximately 70% of workplace tasks. That same report predicts that over half of tasks currently performed by humans will be performed by machines by 2025. Contemporary academic work in job automation has found that "automation, information technology, and technological progress, in general, are encroaching upward in the task domain and beginning to substitute strongly for the work done by professional, technical, and managerial occupations" (Autor 2015 p. 21). Due to its speed, cost, and reliability as compared to human labor, organizations will continue to transform workplace tasks with technology (Chigbu and Nekhwevha 2020). However, despite some predictions of extensive job loss to this technology, single job tasks likely become automated rather than whole occupations (Arntz et al. 2017). As autonomous machines continue to automate human workplace tasks but not others, those machines are likely to become interdependent collaborators with humans, meaning that human-machine teams will become increasingly more common over the coming years. The combination of needing to work interdependently with machine teammates and the growing reliance on machines to perform the taskwork parts of currently human jobs implies that the importance of teamwork skills will increase over time.

Scholars have yet to agree on a definition for human-machine teams, or on what qualifications a machine needs to be considered a "teammate" rather than a "tool." For this chapter, human-machine teams fall within the same definition as a traditional team, noting that the term "individuals" can be about either human or non-human contributors. Like human teammates, machine teammates have some level of autonomy and work interdependently with other members of the team. While human-machine teams are not yet common in every workplace, they do exist across a few domains. For example, artificial intelligence is used to support analysts through tasks that have complex cognitive requirements by filtering through available data and providing the most relevant sources for the analysts to work with (Dubrow and Orvis 2020). Robots are also becoming common teammates. Search-and-rescue teams use remote control robots to complete physical tasks such as room clearing and breaking through walls (You and Robert 2017). Some human-robot teams, such as those tasked with building cars, have shown up to 85% more productivity than human-only and robot-only teams (Tsai et al. 2022).

New machines are being designed with features that help them to be better teammates to human team members by being observable, predictable, and directable (Major and Shah 2020). Observability allows human teammates to physically see what a machine is doing, so humans can learn how it is acting and infer how decisions are made. Observability helps to increase the predictability of machines, which is an important factor for allowing human teammates to forecast what actions the machine may perform next. Directability exists when machines can be, at least partially, controlled by humans. While these are all important design features for new machines to be good teammates to humans, humans also need to have certain skills to be effective teammates to their machine counterparts (Dubrow and Orvis 2020). Human-machine teams need co-awareness between human and machine teammates to work together interdependently (Drury et al. 2022). Because cognition does not exist in the same way for humans and machines, co-awareness is dependent on some overt translation of information that the other can understand and incorporate into shared work.

Startup teams tend to use more innovative practices than many large corporate and government organizations because their smaller size allows for quicker change. Startups are likely to take advantage of early-stage AI automation if it can be used to reduce expensive human capital costs. For example, startups already use various AI technologies to automate human resources, sales, advertising, customer support, and marketing. These technologies can reduce the number of staff startups need to hire.

Communication will look quite different in human-machine teams compared to human-only teams. For example, communication will not necessarily occur through spoken or written language. Some artificial intelligence is built using natural language processing software (see Wagner, this volume) and can seemingly communicate in the same way as humans, but because machine cognition is different from human cognition, the way machine teammates interpret communication may not be the same as how human teammates expect the machines to interpret that communication. Humans and machines will need to work together over time to learn how to communicate effectively together.

Perspective taking is critical for coordination in human-machine teams (Dubrow and Orvis 2020). Humans engaging in perspective-taking on human-machine teams can use the observability, predictability, and direct ability of machines to see the situation from the machine's point of view and make coordination decisions accordingly (Galinsky et al. 2005, Dubrow and Orvis 2020).

For the foreseeable future, human members of human-machine teams need to pay relatively close attention to machine teammates through mutual performance monitoring processes. For example, human teammates working for startups need to ensure AI technologies are not making risky financial decisions, such as exceeding an advertising budget to maximize prospective customer reach. Machines with higher levels of autonomy are less likely to inform human teammates about decisions or ask for permission before performing actions (Parasuraman et al. 2000). Thus, human teammates need mutual performance monitoring skills to confidently allow machines to operate autonomously when appropriate while also engaging in backup behaviors to change or redirect the behavior of machine teammates when needed. As artificial intelligence capabilities develop, humans will also need to be willing to accept criticism and backup behaviors from machine teammates.

When it comes to working with machines, researchers have found that propensity to trust and openness to new experiences were important characteristics for early adopters of automatic transmission (Gill et al. 2005). The same is likely to be true for accepting machines as teammates in human-machine teams. When humans have the propensity to trust machines, and individuals adaptable enough to think critically about making decisions regarding whether technology is acting as intended, they are better at collaborating with machine teammates (Basu and Singhal 2016). Humans with tolerance for ambiguity and flexibility will be more effective when it comes to adapting in teams that include machines with high autonomy, because they will be able to adapt to unexpected shifts in their machine teammates in addition to shifts in the environment (Dubrow and Orvis 2020).

When machines fail to perform as expected, humans are likely to get frustrated and lose both trust and the motivation to cooperate with their machine teammates. Without effective conflict management processes, humans with negative early impressions of their machine teammates may fail to improve their attitudes toward the machines. Effective conflict management on human-machine teams would include both human and machine teammates and help team members to understand how machine decisions are made and how humans may be able to direct their machine teammates more effectively when needed.

Research has only recently attempted to approach the challenge of understanding shared cognition in human-machine teams. You and Robert (2017) argue that shared mental models between humans and machines are cognitive mediators in a Human-machine team input-process-mediator-output framework. Drury and colleagues (2022) go a step further to reimagine cognition in a way that encompasses awareness of both humans and machines, even though machines do not have cognition in the same way that humans do. They explain that team situational awareness develops when large amounts of information about both the task and the team are identified, filtered, and analyzed. For shared cognition to develop in human-machine teams, both human and machine teammates need to be transparent in a way that reveals the information that needs to be processed collectively. While machines can be designed to have transparency (Chen 2020, Lyons and Havig 2014), humans also need to go out of their way to share information in a way that machine teammates can interpret before shared cognition between human and machine teammates can be developed.

Trust is one of the most common topics in human-machine teaming research to date. Recently, scholars have begun to emphasize the need to have accurate, or calibrated, trust in machines, rather than high trust in machines (Schaefer et al. 2017). If human team members misinterpret machine abilities and develop trust that is undeserved, humans will fail to intervene when needed. Trust denigration is also a large challenge in human-machine teams, such that when humans have one negative experience with a machine teammate they may start to intervene and control machine actions unnecessarily, or they may stop using the machine altogether. Finally, mutual trust in human-machine teams implies the need for not only humans to trust machines but for machines to also trust humans. Only then will human-machine teams be able to truly work together interdependently through mutual performance monitoring and effective backup behaviors.

Until this point, the importance of teamwork processes and emergent states for diverse, virtual, and human-machine teams have been explained, and KSAOs that can be used to promote the development of these processes and states have been offered.

The following sections discuss practical selection and training methods that can be used to create teams that have the most necessary teamwork KSAOs to be successful in their team, whether it is diverse, virtual, and/or includes machine teammates.

Selection for Teamwork Skills

Some KSAOs needed for success working in diverse, virtual, and human-machine teams are traits foundational to individuals that can be selected for, while others can be developed through both training and teamwork experience. The following section describes several selection methods that can be used to identify individuals who will have teamwork skills described in the sections above that are critical for effective teamwork in diverse, virtual, and/or human-machine teams. These data-driven selection methods include the Stevens and Campion (1993) Teamwork Test, the Big Five personality test, and scales for tolerance for ambiguity, collective orientation, and cultural intelligence. Finally, structured interview protocols for social skills are described.

Teamwork Test

The most widely cited method for assessing teamwork skills is the Teamwork Test for KSAO requirements for teamwork (Stevens and Campion 1993, 1999). Teams composed of members with strong teamwork KSAOs are expected to perform more effectively than those with weak teamwork KSAOs (Kozlowski and Ilgen 2006). The Teamwork Test explains additional variance in teamwork behaviors compared to other traditional predictors of work performance (e.g., general mental ability; personality; Mathieu et al. 2014). It includes both interpersonal KSAOs and self-management KSAOs. Interpersonal KSAOs include conflict resolution, collaborative problem solving, and communication; self-management KSAOs include goal setting and performance management and planning and task coordination. Table 1 provides example questions.

Conflict resolution KSAOs include the ability to encourage productive conflict (e.g., task conflict, process conflict) while discouraging unproductive conflict (e.g., relationship conflict). Individuals with strong conflict resolution skills can help their teams engage in productive negotiation and come to compromised solutions quickly. Collaborative problem solving KSAOs include the ability to identify which situations require collective decision making and which situations can be solved by a single individual. Team members with strong communication KSAOs can utilize their networks to gather and share information openly with the appropriate team members. Goal setting and performance management KSAOs include the ability to set and monitor team goals that are challenging, but achievable. Finally, planning and task coordination KSAOs include the ability to coordinate team members and synchronize and integrate their work.

The Teamwork Test includes a series of multiple-choice behavioral questions (see Table 1 for example items). While the subcategories of the Teamwork Test may be useful for development and training purposes they should not be used in isolation

Table 1. Stevens and campion teamwork test sample items.

Question	Multiple Choice Options
1. Suppose that you find yourself in an argument with several co-workers about who should do a very disagreeable, but routine task. Which of the following would likely be the most effective way to resolve this situation?	A. Have your supervisor decide, because this would avoid any personal bias. *B. Arrange for a rotating schedule so everyone shares the chore.* C. Let the workers who show up earliest choose on a first-come, first-served basis. D. Randomly assign a person to do the task and don't change it.
2. Your team wants to improve the quality and flow of the conversations among its members. Your team should:	*A. Use comments that build upon and connect to what others have already said.* B. Set up a specific order for everyone to speak and then follow it. C. Let team members with more to say determine the direction and topic of conversation. D. Do all of the above.
3. Suppose you are presented with the following types of goals. You are asked to pick one for your team to work on. Which would you choose?	A. An easy goal to ensure the team reaches it, thus creating a feeling of success. B. A goal of average difficulty so the team will be somewhat challenged, but successful without too much effort. *C. A difficult and challenging goal that will stretch the team to perform at a very high level, but attainable so that effort will not be seen as futile.* D. A very difficult, or even impossible goal so that even if the team falls short, it will at least have a very high target to aim for.

Note. Adopted from Stevens and Campion 1993, 1999. Answers in *italics* are the "correct" answers.

for selection; only the overall Teamwork Test score should be used for selection purposes (Stevens and Campion 1999, O'Neill et al. 2012). The subscales of the test have been shown to have low reliability and therefore should not be used for selection or formal employee assessment purposes (McClough and Rogelberg 2003, O'Neill et al. 2012). Overall, the Teamwork Test can be used during selection to identify potential team members that will have better communication, collaborative problem solving, conflict management, and coordination skills.

Personality Testing

The most common and reliable personality traits used when selecting individuals for jobs and consistently correlate with work performance are referred to as the Big Five. The Big Five personality (also referred to as the Five Factor Model in other chapters) traits include extraversion, agreeableness, conscientiousness, emotional stability (also referred to as neuroticism), and openness to experience. Extraverted individuals are typically sociable, talkative, and active collaborators when working on teams (Morgeson et al. 2005). Agreeable individuals are likely to go along with others on the team and are unlikely to share strong adverse opinions that go against the rest of the group. Conscientious individuals are detail oriented and focused on getting the work done and can help teams stay on track. Conscientious individuals tend to have strong performance monitoring skills. Individuals high in emotional stability handle stress and dynamic work situations well. Finally, individuals high in openness are willing to explore new experiences and tend to have strong cognitive flexibility and adaptability, both of which are especially important traits of humans working on human-machine teams (Dubrow and Orvis 2020).

Tolerance for Ambiguity

Herman and colleagues' (2010) Tolerance for Ambiguity can be used to identify individuals with tolerance for ambiguity, which is an important trait for people working on human-machine teams where they may be interacting with artificially intelligent systems that update continuously, are autonomous, and are not always clear in how their decisions are made. There are four dimensions for tolerance for ambiguity, including valuing diverse others, change, challenging perspectives, and unfamiliarity (Herman et al. 2010).

Collective Orientation

Team orientation can be measured using the Collective Orientation Scale (Driskell et al. 2010). Individuals with high collective orientation tend to prefer working with teams and actively seek input from their teammates. Team orientation is important for all work that requires teamwork and is especially important in situations where teamwork can be challenging, such as in diverse, virtual, and human-machine teams.

Cultural Intelligence

Cultural intelligence is important for members of cross-cultural teams and can also be important on human-machine teams where humans are working with autonomous systems that do not share cultural norms and have different social and behavioral indicators that need to be accounted for. The Cultural Intelligence Scale (Bücker et al. 2016) can be used to help select team members who can pick up on cues from others from different backgrounds and can develop knowledge and skills while working with others who are different from themselves.

Structured Interview

In addition to survey-based assessments, structured interviews can be utilized for a selection of individuals with strong teamwork skills. Active listening skills, communication skills, social perceptiveness, self-monitoring, altruism, warmth, patience and tolerance are all skills that can be assessed using structured interviews (Morgeson et al. 2005).

Active listening skills include waiting for others to finish speaking, avoiding interrupting others, and asking questions. Communication skills include active consideration for who someone is communicating with and how they are communicating (e.g., virtually or face-to-face). Social perceptiveness is someone's awareness of others and the ability to predict how others will react to certain situations. Self-monitoring is the ability to understand how one's behavior affects others, and to adjust behaviors accordingly. Individuals high in altruism are likely to engage in team backup behaviors and are likely to help others without being asked. Warmth is similar to agreeableness and is correlated with cooperation and coordination with others. Finally, individuals with strong patience and tolerance are willing to accept criticism and adapt to feedback from others (Tannenbaum and Salas 2020).

Structured interviews include both situational and behavioral questions. Situational questions include an example of a scenario the interviewee may face, and the interviewee is meant to explain how they would handle that situation. Behavioral interview questions assess what someone has done in a specific scenario, rather than what they might do. Interviewers can ask about a specific instance when the interviewee encountered a scenario, and how they responded to that scenario. Behavioral interview questions usually start with, "Tell me about a time when…" Table 2 provides a summary of the selection tools and associated KSAOs.

Training for Teamwork Skills and Processes

Team training can be used to develop both individual- and team-level taskwork and teamwork skills and processes. Unlike foundational human traits such as personality and cognitive abilities, team members can be trained to develop certain skills, including communication and collaborative problem solving. Training can also be used to develop team states such as mutual trust and team mental models.

Training has been defined as "a set of tools and methods that, in combination with required competencies and training objectives, form an instructional strategy" (Salas and Cannon-Bowers 1997). Team training methods have been found to positively affect team learning, processes, emergent states, and performance outcomes (Hughes et al. 2016, Kozlowski and Ilgen 2006). Research has shown that up to 19% of the variance in team outcomes can be explained by training, and scholars have argued that this is likely an underestimate of training impact (Hunter and Schmidt 1990, Salas et al. 2008).

Training can be focused on taskwork, teamwork, or both. Taskwork training is focused on the development of technical skills, which may need to be completed within a team setting (Mak et al. 2017). Taskwork training might include learning to use collaborative tools such as Gitlab, Slack, and Google Drive. Taskwork trainings are usually targeted at individuals but can sometimes occur in team settings (Mathieu et al. 2008). Teamwork training, on the other hand, is used to help members build skills required to collaborate with one another effectively (Ellis et al. 2005). Ideally, teamwork training occurs in group settings where all team members can learn and practice new skills together in real time (Mathieu et al. 2008).

There are many forms of team training, all of which can be adapted to a team's specific needs. Four training methods ideal for developing the teamwork skills

Table 2. Selection tools and associated KSAOs.

Selection Tool	Description	Associated KSAOs	References
Teamwork Test for Knowledge, Skills, and Abilities (Overall Score)	Series of multiple choice behavioral questions to identify individuals with stronger teamwork skills	Conflict resolution	Stevens and Campion 1993, 1999, O'Neill et al. 2012
		Collaborative problem solving	
		Communication skills	
		Goal setting and performance management	
		Planning and task coordination	
Big-5 Personality Scale	A survey measure to assess the most common and reliable personality traits found in individuals	Extraversion	John et al. 1991
		Agreeableness	
		Conscientiousness	
		Emotional stability	
		Openness	
Tolerance for ambiguity scale	A survey measure to assess an individual's ability to function and collaborate with others in situations with limited or unclear information	Tolerance for ambiguity	Herman et al. 2010
Collective Orientation Scale	A survey measure to identify individuals who tend to prefer working with teams and actively seeking input from their teammates	Team orientation	Driskell et al. 2010
Cultural Intelligence Scale	A survey measure to identify individuals who can effectively work with others who are different from themselves	Cultural intelligence	Bücker et al. 2016
Structured interview	A meeting where individuals are asked both situational and behavioral interview questions to assess a wide range of KSAOs	Active listening	Morgeson et al. 2005
		Communication skills	
		Social perceptiveness	
		Self-monitoring	
		Altruism	
		Warmth	
		Patience and tolerance	

include (1) coordination and adaptation training, (2) team self-correction training, (3) cross-training, and (4) perspective taking training. Finally, team building, which is slightly different from team training, can be used to improve teamwork processes in in-tact teams.

The following sections describe each training type, how they are conducted, their relationships with positive team outcomes, and what skills and processes they should be used to help enhance (see Table 3 for a summary).

Table 3. Training methods and associated skills and processes.

Training Method	Description	Skills/Processes	References
Coordination and Adaptation Training	Intensively addresses team processes through activities such as role playing to improve team performance	Communication Collaborative decision making Team adaptation Team mental models Mutual trust Conflict management	Hamilton 2009, Mak et al. 2017, Salas et al. 2001, Thayer et al. 2014
Team Self Correction Training	Trains teams how to identify and address communication and collaboration breakdowns as they occur, often through classroom instruction or computer-based training	Communication Collaborative problem solving Mutual performance monitoring Team mental models	Smith-Jentsch et al. 2008, Thayer et al. 2014
Cross-Training	Trains team members on each other's taskwork responsibilities, often with hands-on methodology	Teamwork mental models Taskwork mental models Team backup behaviors Information sharing Mutual performance monitoring	Kozlowski and Ilgen 2006, Thayer et al. 2014, Volpe et al. 1996
Perspective Taking Training	Trains team members to develop their individual teamwork skills, often in lecture format	Coordination Mutual trust	Magpili and Pazos 2018
Team Building	Enhances team member social relationships and role clarity, often through facilitated collaborative activities	Mutual trust Conflict resolution Goal setting Collective problem-solving	Klein et al. 2009, Salas et al. 2008, Shuffler et al. 2018, Thayer et al. 2014

Coordination and Adaptation Training

Coordination and adaptation training, also known as Crew Resource Management (CRM) training, is one of the most popular techniques for teamwork training and is used across a variety of domains ranging from military teams to healthcare teams and has been shown to improve team performance as well as team processes such as communication, collaborative decision making, and team adaptation (Salas et al. 2001, Thayer et al. 2014). Teams that go through coordination and adaptation training also tend to have strong emergent states including mutual trust, positive attitudes toward the team, and accurate team mental models (Hamilton 2009, Mak et al. 2017). Accurate team mental models that arise from coordination and adaptation training have positive impacts on team performance (Hamilton 2009).

Coordination and adaptation training uses techniques such as role playing to teach team members how to enhance their coordination skills (Kozlowski and Ilgen 2006). This training method focuses on the teamwork competencies that directly impact coordination, including mutual performance monitoring and team backup behaviors (Kozlowski and Ilgen 2006). Implementation of team coordination and adaptation training should be used to develop communication, cooperation, and conflict management skills and processes.

Team Self-Correction Training

The purpose of team self-correction training is to help teams identify when their collaboration and communication are breaking down and implement strategies to adapt and rebuild teamwork accordingly (Smith-Jentsch et al. 2008, Thayer et al. 2014). Team self-correction training can be conducted in classroom settings guided by an instructor or through computer-based training. Team self-correction training has been shown to improve both similarity and accuracy of team mental models (Smith-Jentsch et al. 2008), and should be used to help improve team communication, collaborative problem solving, strategy formulation and planning, and mutual performance monitoring.

Cross-Training

Cross-training is used to train team members on each other's roles and responsibilities and is used to help build transactive memory systems and team interaction models (Kozlowski and Ilgen 2006). During cross-training, team members are directly exposed to the responsibilities of each team member. Oftentimes, team members practice fulfilling each other's roles during the training. Cross-training is especially important in interdisciplinary teams where team members have different technical expertise and understandings of the team task. It can be used in combination with team self-correction training to facilitate communication and collaboration within the team during the training session (Kozlowski and Ilgen 2006). This combination allows for the enhancement of important team processes such as open and unique information sharing, mutual performance monitoring, and team backup behaviors (Thayer et al. 2014, Volpe et al. 1996).

Perspective-Taking Training

While perspective-taking is a skill that can be selected for, team training can also help improve the skill, resulting in increased team coordination and mutual trust. Perspective-taking training is a targeted training method that helps to improve individual teamwork skills. This training method typically consists of a recorded or live lecture related to perspective-taking, including what it is, what its elements are, and what the benefits of engaging in perspective-taking are (Magpili and Pazos 2018). Perspective-taking training can be paired with other team training methodologies and should be done before team-building activities so members can practice perspective taking during those activities (Magpili and Pazos 2018).

Team Building

Team building is different from team training in the sense that team training targets specific tasks and competencies, while team building allows team members to get to know each other better socially and provides active process interventions to help improve intact team behaviors (Thayer et al. 2014). Team building can either be formal or informal, and acts as a team process intervention by helping to improve role clarity and social relationships (Klein et al. 2009, Thayer et al. 2014). Typically, team building focuses on four categories: (1) interpersonal relations, (2) role clarification, (3) collective problem-solving, and (4) goal setting (Shuffler et al. 2018). Team building directly impacts team processes and emergent states such as team member affect and role clarity (Shuffler et al. 2018). However, team training typically has a more direct impact on team performance compared to team building (Salas et al. 2008). Team building can be used to help solve interpersonal challenges and improve team conflict resolution capabilities, which is especially important in diverse and human-machine teams. Thus, team building should be used to help develop mutual trust in teams, as well as to improve the cognitive, affective, process, and performance outcomes (Shuffler et al. 2018).

Conclusion

Startups are increasingly relying on teams to perform effectively, drive innovation, and gain competitive advantage. Teamwork processes are important to leverage the synergistic benefits of the individual taskwork processes, such that the whole becomes greater than the sum of its parts. In light of workplace changes driven by societal and technological shifts, the importance of teamwork is likely to be even greater in future. Startup leaders will need to understand how to build, manage, and motivate teams to successfully grow and scale their businesses. This chapter highlights this need and provides an overview of important teamwork skills and processes that organizations can select and train for using empirically validated methods. This chapter also encourages future scientific studies examining how the context of entrepreneurship and startups influence team performance. Through this, academic insights as well as practical guidelines are provided to enable organizations to capitalize on their teams for organizational effectiveness.

References

Arntz, M., Gregory, T. and Zierahn, U. (2017). Revisiting the risk of automation. *Economics Letters*, 159: 157–160.

Autor, D.H. (2015). Why are there still so many jobs? The history and future of workplace automation. *Journal of Economic Perspectives*, 29(3): 3–30.

Basu, C. and Singhal, M. (2016, March). Trust dynamics in human autonomous vehicle interaction: A review of trust models. In *2016 AAAI Spring Symposium Series*.

Bell, S.T., Villado, A.J., Lukasik, M.A., Belau, L. and Briggs, A.L. (2011). Getting specific about demographic diversity variable and team performance relationships: A meta-analysis. *Journal of Management*, 37(3): 709–743.

Brinckmann, J. and Hoegl, M. (2011). Effects of initial teamwork capability and initial relational capability on the development of new technology-based firms. *Strategic Entrepreneurship Journal*, 5(1): 37–57.

Bücker, J., Furrer, O. and Weem, T.P. (2016). Robustness and cross-cultural equivalence of the Cultural Intelligence Scale (CQS). *Journal of Global Mobility: The Home of Expatriate Management Research*, 300–325.

Burke, C.S., Shuffler, M.L., Salas, E. and Gelfand, M. (2010). Multicultural teams: Critical team processes and guidelines. *Going Global: Practical Applications and Recommendations for HR and OD Professionals in the Global Workplace*, 27: 46–71.

Caligiuri, P. and Tarique, I. (2012). Dynamic cross-cultural competencies and global leadership effectiveness. *Journal of World Business*, 47(4): 612–622.

Cannon-Bowers, J.A. and Salas, E. (1997). A framework for developing team performance measures in training. In *Team Performance Assessment and Measurement*, 57–74. Psychology Press.

Cannon-Bowers, J.A., Salas, E., Tannenbaum, S.I. and Mathieu, J.E. (1995). Toward theoretically based principles of training effectiveness: A model and initial empirical investigation. *Military Psychology*, 7(3): 141–164.

Carter, D.R., Seely, P.W., Dagosta, J., DeChurch, L.A. and Zaccaro, S.J. (2015). Leadership for global virtual teams: Facilitating teamwork processes. In *Leading Global Teams*, 225–252. Springer, New York, NY.

Chen, J.Y., Flemisch, F.O., Lyons, J.B. and Neerincx, M.A. (2020). Guest editorial: Agent and system transparency. *IEEE Transactions on Human-Machine Systems*, 50(3): 189–193.

Chen, S.C. and Dhillon, G.S. (2003). Interpreting dimensions of consumer trust in e-commerce. *Information Technology and Management*, 4(2): 303–318.

Chigbu, B.I. and Nekhwevha, F.H. (2020). The extent of job automation in the automobile sector in South Africa. *Economic and Industrial Democracy*, 29(3): 3–30.

de Mol, E., Khapova, S.N. and Elfring, T. (2015). Entrepreneurial team cognition: A review. *International Journal of Management Reviews*, 17(2): 232–255.

DeChurch, L.A. and Mesmer-Magnus, J.R. (2010). The cognitive underpinnings of effective teamwork: A meta-analysis. *Journal of Applied Psychology*, 95(1): 32.

Devine, D.J. (2002). A review and integration of classification systems relevant to teams in organizations. *Group Dynamics: Theory, Research, and Practice*, 6(4): 291.

Driskell, J.E., Salas, E. and Hughes, S. (2010). Collective orientation and team performance: Development of an individual differences measure. *Human Factors, 52(2), 316*–328.

Drury, J. L., Klein, G. L., Booker, L., Ryall, K., and Dubrow, S. (2022). Reimagining situation awareness and option awareness for human-machine teaming. *IEEE CogSIMA*, 1–7.

Dubrow, S. and Orvis, K.L. (2020). Human-machine teaming: What skills do the humans need? *The Interservice/Industry Training, Simulation and Education Conference (I/ITSEC) Published Proceedings*, 1–11.

Dyer, J.L. (1984). Team research and team training: A state-of-the-art review. *Human factors Review*, 26: 285–323.

Eisenhardt, K.M. (2013). Top management teams and the performance of entrepreneurial firms. *Small Business Economics*, 40(4): 805–816.

Ellis, A.P., Bell, B.S., Ployhart, R.E., Hollenbeck, J.R. and Ilgen, D.R. (2005). An evaluation of generic teamwork skills training with action teams: Effects on cognitive and skill-based outcomes. *Personnel Psychology*, 58(3): 641–672.

Endsley, M.R. (1995). Toward a theory of situation awareness in dynamic systems. *Human Factors*, 37(1): 32–64.

Galinsky, A.D., Ku, G. and Wang, C.S. (2005). Perspective-taking and self-other overlap: Fostering social bonds and facilitating social coordination. *Group Processes & Intergroup Relations*, 8(2): 109–124.

Garro-Abarca, V., Palos-Sanchez, P. and Aguayo-Camacho, M. (2021). Virtual teams in times of pandemic: Factors that influence performance. *Frontiers in Psychology*, 12: 232.

Gill, H., Boies, K., Finegan, J.E. and McNally, J. (2005). Antecedents of trust: Establishing a boundary condition for the relation between propensity to trust and intention to trust. *Journal of Business and Psychology*, 19(3): 287–302.

Hamilton, K.L. (2009). *The Effect of Team Training Strategies on Team Mental Model Formation and Team Performance under Routine and Non-routine Environmental Conditions.* The Pennsylvania State University.

Herman, J.L., Stevens, M.J., Bird, A., Mendenhall, M. and Oddou, G. (2010). The tolerance for ambiguity scale: Towards a more refined measure for international management research. *International Journal of Intercultural Relations*, 34(1): 58–65.

Hertel, G., Geister, S. and Konradt, U. (2005). Managing virtual teams: A review of current empirical research. *Human Resource Management Review*, 15(1): 69–95.

Hughes, A.M., Gregory, M.E., Joseph, D.L., Sonesh, S.C., Marlow, S.L., Lacerenza, C.N., Benishek, L. E., King, H.B. and Salas, E. (2016). Saving lives: A meta-analysis of team training in healthcare. *Journal of Applied Psychology*, 101(9): 1266–1304.

Hunter, J.E. and Schmidt, F.L. (1990). Dichotomization of continuous variables: The implications for meta-analysis. *Journal of Applied Psychology*, 75(3): 334–339.

Jarrahi, M.H. and Sawyer, S. (2013). Social technologies, informal knowledge practices, and the enterprise. *Journal of Organizational Computing and Electronic Commerce*, 23(1-2): 110–137.

John, O.P., Donahue, E.M. and Kentle, R.L. (1991). The big five inventory—versions 4a and 54.

Kirkman, B.L. and Mathieu, J.E. (2005). The dimensions and antecedents of team virtuality. *Journal of Management*, 31(5): 700–718.

Klein, C., DiazGranados, D., Salas, E., Le, H., Burke, C.S., Lyons, R. and Goodwin, G.F. (2009). Does team building work?. *Small Group Research*, 40(2): 181–222.

Kozlowski, S.W. and Ilgen, D.R. (2006). Enhancing the effectiveness of work groups and teams. *Psychological Science in the Public Interest*, 7(3): 77–124.

LePine, J.A., Piccolo, R.F., Jackson, C.L., Mathieu, J.E. and Saul, J.R. (2008). A meta-analysis of teamwork processes: Tests of a multidimensional model and relationships with team effectiveness criteria. *Personnel Psychology*, 61(2): 273–307.

Lyons, J.B. and Havig, P.R. (2014, June). Transparency in a human-machine context: Approaches for fostering shared awareness/intent. In *International Conference on Virtual, Augmented and Mixed Reality*, 181–190. Springer, Cham.

Magpili, N.C. and Pazos, P. (2018). Self-managing team performance: A systematic review of multilevel input factors. *Small Group Research*, 49(1): 3–33.

Major, L. and Shah, J. (2020). *What to Expect when You're Expecting Robots: The Future of Human-robot Collaboration*. Hachette UK.

Mak, S. (2017). *Adaptive Guidance with Teams: Shifting from Taskwork to Teamwork.* Michigan State University. Psychology.

Marks, M.A., Mathieu, J.E. and Zaccaro, S.J. (2001). A temporally based framework and taxonomy of team processes. *Academy of Management Review*, 26(3): 356–376.

Mathieu, J., Maynard, M.T., Rapp, T. and Gilson, L. (2008). Team effectiveness 1997–2007: A review of recent advancements and a glimpse into the future. *Journal of Management*, 34(3): 410–476.

Mathieu, J.E., Gallagher, P.T., Domingo, M.A. and Klock, E.A. (2019). Embracing complexity: Reviewing the past decade of team effectiveness research. *Annual Review of Organizational Psychology and Organizational Behavior*, 6: 17–46.

Mathieu, J.E., Tannenbaum, S.I., Donsbach, J.S. and Alliger, G.M. (2014). A review and integration of team composition models: Moving toward a dynamic and temporal framework. *Journal of Management*, 40(1): 130–160.

McClough, A.C. and Rogelberg, S.G. (2003). Selection in teams: An exploration of the teamwork knowledge, skills, and ability test. *International Journal of Selection and Assessment*, 11(1): 56–66.

McNelley, S. (2005). Immersive group telepresence and the perception of eye contact. *Irvine, CA: Digital Video Enterprises.*

Mesmer-Magnus, J.R. and DeChurch, L.A. (2009). Information sharing and team performance: A meta-analysis. *Journal of Applied Psychology*, 94(2): 535.

Morgeson, F.P., Reider, M.H. and Campion, M.A. (2005). Selecting individuals in team settings: The importance of social skills, personality characteristics, and teamwork knowledge. *Personnel Psychology*, 58(3): 583–611.

Normand, S.L. (2019). Exploration of taskwork and teamwork skills of selected military personnel (Order No. 13886705). Available from ProQuest Dissertations & Theses Global; Social Science

Premium Collection. (2282004714). Retrieved from http://proxy.binghamton.edu/login?url=https://www.proquest.com/dissertations-theses/exploration-taskwork-teamwork-skills-selected/docview/2282004714/se-2?accountid=14168.

O'Neill, T.A., Goffin, R.D. and Gellatly, I.R. (2012). The knowledge, skill, and ability requirements for teamwork: revisiting the teamwork-KSA test's validity. *International Journal of Selection and Assessment*, 20(1): 36–52.

Parasuraman, R., Masalonis, A.J. and Hancock, P.A. (2000). Fuzzy signal detection theory: Basic postulates and formulas for analyzing human and machine performance. *Human Factors*, 42(4): 636–659.

Salas, E. and Cannon-Bowers, J.A. (1997). Methods, tools, and strategies for team training. pp. 249–279. *In*: M.A. Quinones and A. Ehrenstein (eds.). *Training for a Rapidly Changing Workplace: Applications of Psychological Research* (Washington, DC: American Psychological Association).

Salas, E. and Cannon-Bowers, J.A. (2001). The science of training: A decade of progress. *Annual Review of Psychology*, 52(1): 471–499.

Salas, E., DiazGranados, D., Klein, C., Burke, C.S., Stagl, K.C., Goodwin, G.F. and Halpin, S.M. (2008). Does team training improve team performance? A meta-analysis. *Human Factors*, 50(6): 903–933.

Salas, E., Prince, C., Baker, D.P. and Shrestha, L. (1995). Situation awareness in team performance: Implications for measurement and training. *Human Factors*, 37(1): 123–136.

Salas, E., Sims, D.E. and Burke, C.S. (2005). Is there a "big five" in teamwork? *Small Group Research*, 36(5): 555–599.

Schaefer, K.E., Straub, E.R., Chen, J.Y., Putney, J. and Evans III, A.W. (2017). Communicating intent to develop shared situation awareness and engender trust in human-agent teams. *Cognitive Systems Research*, 46: 26–39.

Shuffler, M.L., Diazgranados, D., Maynard, M.T. and Salas, E. (2018). Developing, sustaining, and maximizing team effectiveness: An integrative, dynamic perspective of team development interventions. *Academy of Management Annals*, 12(2): 688–724.

Smith-Jentsch, K.A., Cannon-Bowers, J.A., Tannenbaum, S.I. and Salas, E. (2008). Guided team self-correction: Impacts on team mental models, processes, and effectiveness. *Small Group Research*, 39(3): 303–327.

Stevens, M.J. and Campion, M.A. (1993). The Teamwork-KSA Test. *IL: NCS–London House. Chicago.*

Stevens, M.J. and Campion, M.A. (1999). Staffing work teams: Development and validation of a selection test for teamwork settings. *Journal of Management*, 25(2): 207–228.

Tannenbaum, S. and Salas, E. (2020). *Teams that Work: The Seven Drivers of Team Effectiveness*. Oxford University Press.

Thayer, A.L., Rico, R., Salas, E. and Marlow, S.L. (2014). Teams at work. In *An Introduction to Contemporary Work Psychology*, 434–457. John Willey & Sons.

Tsai, C.Y., Marshall, J.D., Choudhury, A., Serban, A., Hou, Y.T.Y., Jung, M.F., Dionne, S.D. and Yammarino, F.J. (2022). Human-robot collaboration: A multilevel and integrated leadership framework. *The Leadership Quarterly*, 33(1): 101594.

Volpe, C.E., Cannon-Bowers, J.A., Salas, E. and Spector, P.E. (1996). The impact of cross-training on team functioning: An empirical investigation. *Human Factors*, 38(1): 87–100.

You, S. and Robert, L. (2017). Emotional attachment, performance, and viability in teams collaborating with embodied physical action (EPA) robots. *Journal of the Association for Information Systems*, 19(5): 377–407.

Chapter 6

Human Capital Due Diligence

Leveraging Psychometric Testing for Wiser Investment Decisions

Nikki Blacksmith,[1,*] *Kelly Diouf*[2] and *Maureen E McCusker*[3]

In 2021, the value of venture capital (VC) dealmaking in the US reached an all-time high of over 600 billion U.S. dollars (KPMG 2022). This astronomically high dealmaking is matched by a startup failure rate that is just as striking: nearly 90% of all startups, regardless of funds received, fail in the first 5 years (Patel 2015). The failure rate for startups receiving VC funds is only slightly less bleak: about 75% (or three out of every four) of venture-backed startups do not succeed (Gage 2012). Investors bet large sums of money on the startup's success; when a startup fails, VCs lose 100% of their invested funds (Othman 2022). Approximately 30–40% return the original capital and 10–20% produce outsized returns (Gage 2012).

The statistics are depressing, but the reasons underlying the failure yield potentially promising solutions. Preliminary research suggests that the evaluative process investors use to make new venture funding decisions may be contributing to low financial returns and biased outcomes (Tinkler et al. 2015). While it is well understood amongst investors that a startup is nothing without its founders, few turn to objective data to measure the qualifications of the startup team during due diligence. Why? Research has uncovered several reasons, but this chapter argues that the primary reason is the lack of tools and knowledge (Smart 1999). Human capital due diligence

[1] American University, Blackhawke Behavior Science, 1709 21st St NW #52, Washington, DC 20009, 630-888-8150.

[2] Blackhawke Behavior Science, 5825 Marbury Road, Bethesda, MD 20817, 310-948-5268.

[3] American University, Blackhawke Behavior Science, 3325 Stuart Ave, Richmond, VA 23221, 703-609-0257.

Emails: kelly@blackhawke.io; Maureenemccusker@gmail.com

* Corresponding author: nikkiblacksmith@gmail.com

is likely novel, non-traditional, and challenging for those without a background in psychological science. The good news is that researchers and practitioners do not have to reinvent the wheel; various tools within the organizational sciences (specifically personnel selection research) can be easily adapted selection processes in the entrepreneurship ecosystem to mitigate the risk involved in investing in startups.

The purpose of this chapter is to illustrate how leveraging tools, findings, and research from the science of hiring (i.e., personnel selection science to research) can provide investors with the information needed to make wiser, data driven decisions (Smart 1999). As Tinkler et al. (2015 p. 1) argue, “like hiring, venture capitalists’ decisions operate in an evaluative context that necessitates a decision where limited information about candidates exists.” However, to the authors’ knowledge, the science of personnel selection research has yet to be systematically applied to investment decision making in the VC industry. This is surprising, given the similarities between the two. Organizations seek to select people that will drive performance. VC firms seek to select founders or founding teams that will drive startup performance. As such, this chapter explains how scientists and practitioners alike can leverage a century’s worth of research on selection (Terman 1918, Thurstone 1925, Sackett et al. 2017) in the workplace to improve success rates of venture capital decisions.

This chapter focuses on decisions made during the due diligence phase, with a laser focus on human capital due diligence (HCDD, Harding and Rouse 2007, Smart 1999). The phrase “human capital due diligence” is used to represent the process by which investors evaluate the people of startups, namely the founders, founding team, or existing employees, as a criterion in determining whether or not (i.e., level of risk) to invest in a startup. The HCDD process is parallel to the hiring process in organizations. In both contexts, there is limited information available from which the decision-makers can draw judgments about the applicants. In this chapter, the term “investor” is used to subsume angel investors, venture capitalists, venture studios, or other funding firms that select startups in which to invest. The term “venture capital” refers to the funds invested in startups, regardless of what specific type of investor provides the financial capital.

Shortcomings of Current Assessment Methods for Investment Decisions

While new venture investors use several types of data (e.g., market size, financial projections, user growth) to make investment decisions (Kirsch et al. 2009), rarely are there interventions in place to collect human capital data to evaluate the degree to which the founder or founding team are likely to succeed. This is not to say that inventors do not value human capital; in fact, most investors identify the founding team as the most important criterion for their decision (Block et al. 2021, Franke et al. 2008, Gompers et al. 2020). However, the current methods used to make evaluative judgments about human capital are susceptible to bias and are poor predictors of future performance (Kuncel et al. 2013).

Current HCDD Methods are Susceptible to Stereotypes and Bias

Many investors (similar to hiring managers) believe they know a successful entrepreneur when they see one. Furthermore, they believe they can get a better sense of a founder's performance potential by speaking with them than they could get from objective human capital data (Dana et al. 2013, Highhouse 2008). Employee selection decisions (i.e., hiring for jobs) are notorious for containing bias because they are often made based on gut instinct — intuition (Epstein 2010 Highhouse 2008, Landy 2008). Judgments without objective data are strongly influenced by commonly held stereotypes and are susceptible to cognitive bias (Kahneman 2011, Kahneman and Tversky 1979, Landy 2008). People use mental shortcuts when making decisions, often resulting in biased outcomes (Hastie and Dawes 2009, Kahneman and Tversky 1979). Venture capital decision making is no different; investors are susceptible to and influenced by cognitive bias (Balachandra 2019, Forrester 2014, Shane 2008). Numerous types of bias exist and likely influence investors' decisions. However, due to space constraints, we only touch on two types to provide examples: availability and similarity bias.

Availability Bias. Due to how founders are typically portrayed in the media and historical stories, there are deeply ingrained stereotypes about the profile of successful entrepreneurs, leaving investment decisions open to availability bias (Appel and Weber 2021, Shane 2008). Availability bias occurs when people rely on information that can be recalled most easily (Folkes 1988). Popular media portrays founders as young, white males working from a garage, likely because Hewlett and Packard, high profile entrepreneurs, developed their first product in a shed (Erlanger and Ortega 2018). They are often portrayed as college dropouts turned tech entrepreneurs because famous entrepreneurs like Bill Gates, Mark Zuckerberg, and Steve jobs, dropped out of college to pursue their ideas and founded huge, successful startups. Due to the nature of human instinct, investors are more likely to invest in founders that look and seem like the stereotypical founder that stem from the profiles depicted in media and mainstream society.

Similarity Bias. Many investors prefer to meet founders that are already within their network and can find a way to get a "warm introduction." Working with someone from a similar background or network likely feels less risky to investors. However, this approach opens the door to similarity bias — people naturally have an affinity for others that are similar to them (Shane and Cable 2002, Murnieks et al. 2011). An estimated 30% of deals come from leads from VCs' former colleagues or work acquaintances, 20% of deals come from referrals by other investors, and 8% from referrals by existing portfolio companies (Gompers et al. 2020).

Current Methods are the Least Predictive

The most used by investors for evaluating human capital include unstructured interviews and reference calls. These methods of HCDD are time-consuming, costly, and offer little predictive value (Gompers et al. 2020, Schmidt and Hunter 1998). Moreover, when founding teams are evaluated with unstructured interviews and reference calls, it becomes difficult to make fair, nondiscriminatory decisions. Instead of relying on unstructured interviews, reference calls, and gut instinct, investors ought to identify the factors that truly impact performance through the collection of objective data. Typically, very little objective, empirical data is collected on the human capital of the startup in the due diligence process. With objective and standardized processes like assessments, every startup team gets evaluated using the same methods, and bias is mitigated. This offers a fairer, just, equitable, and informed way of allocating investor funds based on objectivity that reduces biases, increases diversity, and puts entrepreneurs who are most likely to succeed in positions to succeed. The good news is we don't have to start from scratch in developing better decision aids for HCDD; we can leverage insights from personnel selection in the organizational sciences.

Applications of Personnel Selection Research for Venture Capital Decision-making

Over 100 years of personnel selection research exists (Ployhart et al. 2017, Ryan and Ployhart 2014), offering scientists and practitioners opportunities to understand and improve venture capital decisions (Smart 1999). Personnel selection is the process of selecting the best candidate using quantitative data and the scientific method (Ployhart et al. 2017). The fundamental question for companies selecting people for a job is, "*will this job candidate perform highly in this role in the future so that they're worth investing company money to pay them for this work*?" This question is equivalent to the question venture capitalists ask when selecting companies to invest in: "*Will this entrepreneurial team (or solopreneur) perform highly in the future so that it's worth investing my money to achieve a return on my investment*?" There are two primary differences between hiring decisions and VC investment decisions: (1) VC investments are often (but not always) higher financial stakes, and (2) the VC typically needs to evaluate a team instead of a single individual. Both decisions involve evaluating the degree to which people can perform in a way that will give a return on the investment. Therefore, the personnel selection methods used to make hiring decisions can be used to make venture capital decisions.

Assessment Methods

Some of the most predictive methods for conducting human capital due diligence include structured interviews, work samples, assessment centers, and psychometric assessments. Interviews are one of the most commonly used methods in the hiring and HCDD contexts. However, most interviews are unstructured, meaning every candidate receives a different set of questions. *Structured interviews* are similar to the traditional interview but consist of a standardized set of questions and are defined by explicit content (Campion et al. 1997, Hough and Oswald 2000). Standardized questions ensure that every applicant is evaluated in the same manner, making it more conducive to comparing across candidates. *Work samples* refer to activities or exercises that simulate an actual work task (Roth et al. 2005). For example, if a job requires that the employee give presentations often, a candidate would give a presentation to the hiring staff as part of the application process. *Assessment centers* are methods that present the candidate with a series of activities and exercises (e.g., role play, leaderless group projects) designed to measure multiple knowledge, skills, abilities, and other characteristics, (KSAOs; Lievens 1998, Woehr and Arthur 2003). Assessment centers are typically reserved for the evaluation of senior-level leaders due to the high price per candidate (Hough and Oswald 2000). *Psychometric assessments* are tests or surveys that measure psychological traits such as cognitive ability, personality, and cognitive styles (Guion 2011, Ployhart et al. 2017, Ryan and Ployhart 2014). These assessments typically have self-report questions about behaviors and tendencies or questions with right or wrong answers. Psychometric assessments are useful for screening when there are a large number of applicants because they can be administered more efficiently than structured interviews (Independent 2018). They are also less costly than assessment centers. Research has demonstrated that using multiple assessment methods in the selection process can increase the accuracy of the decision (Ryan and Ployhart 2014).

Designing A Selection Tool: A Tutorial and an HCDD Case Study

Countless "off-the-shelf" psychometric assessments exist, some of the most prevalent include the NEO-PRI, which measures the Big Five personality traits (Costa and McCrae 2008); the Strengthsfinder (Rath 2007); and the Hogan Personality Inventory (Boyle 1992). These assessments, while valuable, have been designed for general purposes — to appeal to the masses — rather than for any specific job or context. Research demonstrates that when assessments are contextualized or focused on a narrow purpose their predictive capabilities increase (Shaffer and Postlethwaite 2012). That is, the more general the intended audience of a psychometric assessment, the lower its potency to predict; the inverse holds for the more specifically designed psychometric assessments. Several assessments that have been designed specifically for predicting entrepreneurship performance including Entrepreneurial Aptitude Test (Favretto et al. 2003), META (Ahmetoglu 2015, Ahmetoglu et al. 2011), the Entrepreneur DNA (Founder Institute, n.d.), and the BEPE battery (Cuesta et al. 2018). There is a standard process used to develop customized assessments. Below we share a step-by-step approach (albeit dangerously oversimplified for purposes of clarity) to developing psychometric assessments by discussing the development and validation of the Entrepreneurship Value Profile (EVP; Blacksmith et al. 2023b). The EVP is a psychometric assessment developed specifically to measure entrepreneurial human capital. Among its myriad uses, investors administer the tool when conducting venture due diligence to assess the likelihood that an entrepreneur or startup will succeed, based on a set of validated performance criteria.

Step 1: Define Performance

The first step to designing a psychometric assessment is to define the performance domain; it is critical for investors to understand what they are trying to predict (i.e., define performance). The performance domain is typically investigated and defined via a work analysis. Chapter 2 provides an overview and tutorial for conducting a work analysis. Alternatively, investors can look for and leverage scientific research that has already defined the performance domain (Blacksmith et al. 2023a, Chandler and Jansen 1992, Mitchelmore and Rowley 2009, Robles and Zárraga-Rodriguez 2015).

For the purposes of developing the EVP, Blacksmith et al. (2023a) developed a model of entrepreneurial performance that conceptualizes the criterion space using performance behaviors. Performance behaviors are the actions people enact to complete job demands and tasks (i.e., the assignments needed to accomplish organizational goals) (Bartram 2005, Binning and Barrett 1989). First, Blacksmith et al. (2023a, 2023b) identified the critical tasks and responsibilities of entrepreneurship (common across all startups) in order to understand the important performance behaviors. The focus was on startups because the entrepreneurship process is central to their purposes. To identify performance behaviors several steps were taken including reviewing decades of scientific research and literature, studying popular press literature and media, and interviewing subject matter experts (Blacksmith et al. 2023a, 2023b). All the information gathered was synthesized and organized into broad categories or themes. This thematic analysis yielded 12 dimensions of entrepreneurship performance that covered cognitive, behavioral, relational, and motivational aspects. The 12 dimensions (referred to as the 12 Pillars of Entrepreneurship Performance) were Vision, Strategy, Resourcefulness, Execution, Innovation, Decision Making, Collaboration, Direction, Influence, Autonomy, Intensity, and Tenacity.

Step 2: Identify and Map KSAOs

The next step in designing selection assessments is to identify KSAOs that are predictive of the performance dimensions identified in the first step (e.g., Begley and Boyd 1987, Gao et al. 2020, Lumpkin and Dess 1996). In other words, test developers need to determine what KSAOs should be assessed by investigating which KSAOs will lead to high performance. A more detailed discussion of KSAOs is outlined in Chapter 5 of this volume. For example, if a job requires a great deal of writing, some important KSAOs might be verbal ability, reading comprehension, and conscientiousness. Chapter 2 offers more details about how test developers identify and map KSAOs to the performance behaviors that need to be predicted.

With regard to the EVP, following the identification of the 12 pillars, Blacksmith et al. (2023b) scoured the scientific literature to identify traits that have been shown to predict each dimension. For example, the Execution dimension comprised behaviors related to devising and implementing business plans, core business functions, and operational systems. KSAOs predictive of Execution included conscientiousness (i.e., being diligent, and organized) because developing operational systems requires carefully constructing detailed processes and procedures (Zhao and Siebert 2006) and action-orientation (i.e., the tendency to take abstract ideas and take action to make them a reality) was identified as another important KSAO (Locke and Baum 2007).

Step 3: Write Items

After the performance and the associated KSAOs (i.e., predictors) are identified, the content domain of each KSAO must be defined. In other words, there needs to be clear conceptualization that provides boundaries for writing the items (i.e., test questions). For example, if one is attempting to conceptualize openness to experience, one should describe example behaviors that represent the construct (e.g., welcoming novel ideas). Because KSAOs are intangible and unobservable, items must be focused on that which is observable: related behaviors. The behaviors are indirect indicators of the KSAO. That is, if someone is high on openness to experience, they would endorse the behavior of "welcoming novel ideas." Item writing is not something any person can do; industrial-organizational psychologists are trained to develop measurements of intangible concepts (e.g., personality) that are accurate, fair, and unbiased. Thus, it is highly recommended that a trained professional be involved when creating the items for the assessment. While it is beyond the scope of this chapter to provide in-depth details on item writing, there exist ample resources devoted to best practices in writing psychometric items and tests (e.g., DeVellis 2017, Guion 2011, Ghiselli et al. 1981).

When developing the EVP, Blacksmith et al. (2023b) followed industry recommendations and standards (e.g., AERA, APA, and NCME 2014, Crocker and Algina 1986, Deshon 2001, 2002, DeVillis 2017, Kline 2005, Thurstone 1925). Items were written to maximize clarity, precision, and parsimony and mapped conceptually to the KSAO content domain. To define the content domain the item writers consulted previous research and empirical literature (Kline 2005). More items than needed for the final version were written with the intent of removing multiple items by keeping the items with the strongest psychometric properties. Items were then reviewed, revised, and approved by subject matter experts who were uninvolved in the initial item writing process.

Step 4: Validity Analysis

Once the final version of an assessment has been developed, the next step in developing psychometric assessments is to gather validity evidence (Binning and Barret 1989, Campbell and Fiske 1999, Cronbach and Meehl 1955, Messick 1995). Validity refers to the degree to which the information gathered from the results of an assessment reflects what they are supposed to measure. To establish validity, evidence from several sources and statistical analyses are necessary. Validity is on a spectrum from low to high; it is not an all-or-nothing concept. Therefore, many forms of validity evidence need to be gathered to demonstrate that the test measures what you want it to measure.

To continue the example from the previous steps, items for the first draft of the EVP assessment were administered to hundreds of people to gather data. Using that data, the psychometric properties of the items were evaluated. Those items with the best psychometric properties were retained for the final version. The final test underwent further validity tests using advanced statistical techniques validity tests, as well as pilot field studies. In the pilot field tests, the EVP was administered to existing entrepreneurs with an eye toward understanding themselves, as well as investors with an eye toward understanding existing or potential portfolio companies to use in the HCDD process. Piloting the assessment on groups who would actually use and take the survey is another form of validity evidence (i.e., to gauge the degree to which the assessment is realistic and usable). A best practice for assessing validity is to continue to gather and periodically test the validity of the test, as the more data, the stronger the evidence for test validity, especially in industries as dynamic as the entrepreneurial ecosystem. Accordingly, Blacksmith et al. (2023b) will continue to collect validity evidence over time.

Step 5: Implementation

Once enough evidence has been collected to establish adequate validity, the test is used in practice. The implementation process varies depending on the purpose and type of assessment. Generally, it includes digitizing the assessment, incorporating the assessment as one step of a larger selection process, providing education to those who will use the test, and clearly communicating the test to all stakeholders. It may also include integrating the assessment system or data with the human resources information system. Below we provide a specific example that offers more details about how a psychometric assessment can be used for HCDD.

Supplementary Step: "So What?"

When used by investors, the EVP is administered to founders and entrepreneurs of prospective portfolio companies while the investor is conducting due diligence. As mentioned, the purpose of the assessment for investors is to inform wiser investment decisions. How to translate data from the assessment into coherent results and insights is one of the greatest challenges for test-developers. Doing so differentiates a good test from an excellent one. Once investors have the results from the assessment, they receive a report about the founding team. The report synthesizes, visualizes, and presents the results of the assets and liabilities of each of the members of the founding team, as well as the collective founding team. The results enable investors to determine the level of risk associated with investing in the individuals and team. In the spirit of awareness, the founders also receive a report based on their EVP results that help them gain self-awareness and self-manage their performance. Placing the report in the hands of both the investors and the founder enables transparent conversations about mitigating human capital liabilities and leveraging/maximizing human capital assets. For example, if the report shows that a founding team has several members who are fit to carry out visioning tasks but has no members who are fit to carry out execution tasks, it can be concluded that the gap in the execution pillar in the team is substantial. Such results may suggest any or all of the following: (1) the team is a riskier investment since execution is a critical pillar of entrepreneurial performance and without it, the team is likely to struggle; (2) development of execution abilities via coaching and training is paramount for the startup to succeed; (3) hiring or bringing on additional talent with execution skills to fill the gap can help offset the risk and drive human capital performance.

Important Considerations for Psychometric Assessments

Over the last century, psychologists have conducted thousands of studies to inform the development of valid assessments and determine how to use assessments for decision-making ethically. Authoritative institutions such as the American Psychological Association (APA), National Council on Measurement in Education (NCME), and the American Educational Research Association (AERA) introduced the guidelines, *Standards for Educational and Psychological Testing* (APA, NCME, and AERA 2014), to set forth best practices for the development of valid assessments and offer insights. The Society for Industrial-Organizational Psychology (SIOP) expanded those standards to focus on the selection process and developed *Principles for the Validation and Use of Personnel Selection Procedures* (SIOP 2018). Below are important considerations when using selection methods such as those described in the previous section.

Investigate Validity Evidence

Without evidence of validity, conclusions drawn from assessment scores will likely be misleading. The usefulness of an assessment is only as good as the validity of the assessment itself. One-way test users can check the validity of an assessment is to review its technical report to determine if this method is valid for making their decisions. The technical report must detail how the development and validation of the assessment were based on rigorous scientific evidence.

Gain User Training and Education

For most assessments on the market, an individual needs to be trained in how to use the assessment and interpret the data (SIOP 2018). If the investor does not want to or have the means to take their time to attend a training or certification program, another option would be to hire a coach or a psychologist trained in that particular assessment.

Focus on Systems Not Tools

Like any selection decision, investing should not be solely based on a single aspect or a single assessment. For better decision making, inventors ought to integrate psychometric assessments in the overall due diligence process. That is, human capital is only one aspect of the due diligence process and should not be used as a stand-alone criterion. Investors must integrate human capital data with the rest of the data they gather during their due diligence process.

Ensure Effective Communication

Stakeholders need to understand why and how the psychometric assessment is being used. Therefore, it is critical for investors to share with their partners and other VC staff to ensure everyone is aligned and understands the new process. Communicating transparently to prospective founders is also necessary.

Conclusion

Human capital due diligence should be treated as the most important part of the due diligence process, just as founders and teams are the most important decision criteria. Investors often lament the difficulty of assessing people and resort to their gut instinct to make decisions (Smart 1999). However, a century of research dedicated to understanding how to measure traits predictive of future performance exists from which the VC industry and entrepreneurship scientists can adapt. From this large body of personnel selection research, various tools emerged to examine psychological aspects of people that predict their performance at work. Of these tools, psychometric assessments provide some of the most objective, quantitative data. This chapter explained the utility of psychometric tests measuring human capital and provided information on how investors can gather and leverage objective data when conducting human capital due diligence to make wiser, more data-driven investment decisions.

References

Ahmetoglu, G. (2015). *The Entrepreneurial Personality: A New Framework and Construct for Entrepreneurship Research and Practice*. Goldsmiths, University of London.

Ahmetoglu, G., Leutner, F. and Chamorro-Premuzic, T. (2011). EQ-nomics: Understanding the relationship between individual differences in trait emotional intelligence and entrepreneurship. *Personality and Individual Differences*, 51(8): 1028–1033.

American Educational Research Association, American Psychological Association, and National Council on Measurement in Education. (2014). *Standards for Educational and Psychological Testing*. Washington, DC: American Educational Research Association.

Appel, M. and Weber, S. (2021). Do mass mediated stereotypes harm members of negatively stereotyped groups? A meta-analytical review on media-generated stereotype threat and stereotype lift. *Communication Research*, 48(2): 151–179.

Balachandra, L. (2019). How gender biases drive venture capital decision-making: Exploring the gender funding gap. *Gender in Management: An International Journal*, 35: 261–273. https://doi.org/10.1108/GM-11-2019-0222.

Bartram, D. (2005). The great eight competences: A criterion-centric approach to validation. *Journal of Applied Psychology*, 90: 1185–1203.

Begley, T.M. and Boyd, D.P. (1987). Psychological characteristics associated with performance in entrepreneurial firms and smaller businesses. *Journal of Business Venturing*, 2(1): 79–93.

Binning, J.F. and Barrett, G.V. (1989). Validity of personnel decisions: A conceptual analysis of the inferential and evidential bases. *Journal of Applied Psychology*, 74: 478–494. https://doi.org/10.1037/0021-9010.74.3.478.

Blacksmith, N., McCusker, M.E., Diouf, K. and Wagner, D. (2023a). *Entrepreneurship Performance in Startups: Development of a Behavioral Taxonomy.* Poster presented at the 38th Annual Conference for the Society for Industrial and Organizational Psychology, Boston, MA.

Blacksmith, N., McCusker, M.E., Wagner, D. and Diouf, K. (2023b). The Entrepreneurship Value Profile (EVP). [Unpublished Technical Report].

Block, J.H., Hirschmann, M. and Fisch, C. (2021). Which criteria matter when impact investors screen social enterprises? *Journal of Corporate Finance*, 66: 101813. https://doi.org/10.1016/j.jcorpfin.2020.101813.

Boyle, G.J. (1992). Hogan personality inventory. *Psychological Test Bulletin*, 5(2): 130–138.

Campbell, D.T. and Fiske, D.W. (1959). Convergent and discriminant validation by the multitrait-multimethod matrix. *Psychological Bulletin*, 56: 81–105. https://doi.org/10.1037/h0046016.

Chandler, G.N. and Jansen, E. (1992). The founder's self-assessed competence and venture performance. *Journal of Business Venturing*, 7: 223–236.

Campion, M.A., Palmer, D.K. and Campion, J.E. (1997). A review of structure in the selection interview. *Personnel Psychology*, 50(3): 655–702.

Conlan, B. (2021). Harvard Business School Professor Says 65% of Startups Fail for One Reason. Here's How to Avoid It. *Entrepreneur.* Retrieved from: https://www.entrepreneur.com/leadership/harvard-business-school-professor-says-65-of-startups-fail/370367#:~:text=In%20his%20book%2C%20The%20Founder%E2%80%99s%20Dilemma%2C%20best-selling%20author,who%20has%20ever%20tried%20to%20start%20a%20business.

Costa Jr, P.T. and McCrae, R.R. (2008). *The Revised Neo Personality Inventory* (NEO-PI-R). Sage Publications, Inc.

Crocker, L. and Algina, J. (1986). *Introduction to Classical and Modern Test Theory*. Orlando, FL: Holt, Rinehard, and Winston.

Cronbach, L.J. and Meehl, P.E. (1955). Construct validity in psychological tests. *Psychological Bulletin*, 52(4): 281–302. https://doi.org/10.1037/h0040957.

Cuesta, M., Suárez-Álvarez, J., Lozano, L.M., García-Cueto, E. and Muñiz, J. (2018). Assessment of eight entrepreneurial personality dimensions: Validity evidence of the BEPE battery. *Frontiers in Psychology*, 9: 2352.

DeShon, R.P. (2001). Generalizability theory. *In*: F. Drasgow and N. Schmitt (eds.). *Advances in Measurement and Data Analysis*. San Francisco, CA: Jossey-Bass.

DeShon, R.P. (2002). A generalizability theory perspective on measurement error corrections in validity generalization. *In*: K.R. Murphy (ed.). *Validity Generalization: A Critical Review*. Mahwah, NJ: Erlbaum.

DeVellis, R.F. (2017). *Scale Development: Theory and Applications*. Thousand Oaks, CA: California.

Epstein, S. (2010). Demystifying intuition: What it is, what it does, and how it does it. *Psychological Inquiry*, 21(4): 295–312.

Erlanger and Ortega. (2018). The origins of Silicon Valley's garage myth. *Fast Company.* Retrieved from: https://www.fastcompany.com/90270226/the-origins-of-silicon-valleys-garage-myth.

Evans, J. and St. B.T. (2011). Dual-process theories of reasoning: Contemporary issues and developmental applications. *Developmental Review*, 31: 86–102. https://doi.org/10.1016/j.dr.2011.07.007.

Evans, J., St. B.T. and Stanovich, K.E. (2013). Dual-process theories of higher cognition: Advancing the debate. *Perspectives on Psychological Science*, 8, 223–241. https://doi.org/10.1177/1745691612460685.

Favretto, G., Pasini, M. and Sartori, R. (1994). Attitudine Imprenditoriale e misura psicometrica: Il TAI (Test di Attitudine Imprenditoriale). Risorsa Uomo, (2003/3–4).

Folkes, V.S. (1988). The availability heuristic and perceived risk. *Journal of Consumer Research*, 15(1): 13–23.

Founder Institute. (n.d.). Entrepreneur DNA Assessment. The Founder Institute. Retrieved from: https://fi.co/dna.

Franke, N., Gruber, M., Harhoff, D. and Henkel, J. (2008). Venture capitalists' evaluations of start–up teams: Trade–offs, knock–out criteria, and the impact of VC experience. *Entrepreneurship Theory and Practice*, 32(3): 459–483. https://doi.org/10.1111/j.1540-6520.2008.00236.x.

Gage, D. (2012). The venture capital secret: 3 out of 4 start-ups fail. *Wall Street Journal*, 20.

Gao, Q., Wu, C., Wang, L. and Zhao, X. (2020). The entrepreneur's psychological capital, creative innovation behavior, and enterprise performance. *Frontiers in Psychology*, 11: 1651.

Ghiselli, E.E., Campbell, J.P. and Zedeck, S. (1981). *Measurement Theory for the Behavioral Sciences* (Vol. 8). San Francisco: WH Freeman.

Gompers, P.A., Gornall, W., Kaplan, S.N. and Strebulaev, I.A. (2020). How do venture capitalists make decisions?. *Journal of Financial Economics*, 135(1): 169–190. https://doi.org/10.1016/j.jfineco.2019.06.011.

Guion, R.M. (2011). *Assessment, Measurement, and Prediction for Personnel Decisions.* New York, NY: Routledge.

Harding, D. and Rouse, T. (2007). Human due diligence. *Harvard Business Review*, 85(4): 124.

Hastie, R. and Dawes, R.M. (2009). *Rational Choice in an Uncertain World: The Psychology of Judgment and Decision Making*. Sage Publications.

Hough, L.M. and Oswald, F.L. (2000). Personnel selection: Looking toward the future—Remembering the past. *Annual Review of Psychology*, 51(1): 631–664.

Highhouse, S. (2008). Stubborn reliance on intuition and subjectivity in employee selection. *Industrial and Organizational Psychology*, 1(3): 333–342.

Independent. (2018). 75% of Fortune 500 companies use psychometric testing in recruitment. Independent. com Retrieved from: https://www.independent.com.mt/articles/2019-06-05/business-news/75-of-Fortune-500-companies-use-psychometric-testing-in-recruitment-6736209146.

Kahneman, D. (2011). *Thinking Fast and Slow*. New York, NY: Farrar, Straus, and Giroux.

Kahneman, D. and Tversky, A. (1979). Prospect theory: An analysis of decision under risk. *Econometrica*, 47: 263–291. https://doi.org/10.2307/1914185.

Kirsch, D., Goldfarb, B. and Gera, A. (2009). Form or substance: the role of business plans in venture capital decision making. *Strategic Management Journal*, 30(5): 487–515. https://doi.org/10.1002/smj.751.

Kline, T.J. (2005). *Psychological Testing: A Practical Approach to Design and Evaluation*. Sage Publications.

KPMG. (2022). Global venture capital annual investment shatters records following another healthy quarter. Retreived from: https://kpmg.com/xx/en/home/media/press-releases/2022/01/global-venture-capital-annual-investment-shatters-records-following-another-healthy-quarter.html.

Kuncel, N.R., Klieger, D.M., Connelly, B.S. and Ones, D.S. (2013). Mechanical versus clinical data combination in selection and admissions decisions: A meta-analysis. *Journal of Applied Psychology*, 98(6): 1060.

Landy, F.J. (2008). Stereotypes, bias, and personnel decisions: Strange and stranger. *Industrial and Organizational Psychology*, 1(4): 379–392.

Lievens, F. (1998). Factors which improve the construct validity of assessment centers: A review. *International Journal of Selection and Assessment*, 6(3): 141–152.

Locke, E.A. and Baum, J.R. (2007). Entrepreneurial motivation. pp. 93–112. *In*: J.R. Baum, M.R.A. Frese, R.A. Baron (eds.). *The Psychology of Entrepreneurship.* Erlbaum: Mahwah, NJ, USA.

Lumpkin, G.T. and Dess, G.G. (1996). Clarifying the entrepreneurial orientation construct and linking it to performance. *Academy of Management Review*, 21(1): 135–172.

Mitchelmore, S. and Rowley, J. (2009). Entrepreneurial competencies: A literature review and development agenda. *Entrepreneurial Behavior and Research*, 16: 92–111.

Murnieks, C.Y., Haynie, J.M., Wiltbank, R.E. and Harting, T. (2011). 'I like how you think': Similarity as an interaction bias in the investor–entrepreneur. *Journal of Management Studies*, 48(7): 1533–1561.

Othman, A. (2022). How much can you lose on a failed startup investment? *AngelList.* Retrieved from: https://www.angellist.com/blog/failed-startup-investment-return.

Patel, N. (2015). 90% Of Startups Fail: Here's What You Need To Know About The 10%. Retrieved from: https://www.forbes.com/sites/neilpatel/2015/01/16/90-of-startups-will-fail-heres-what-you-need-to-know-about-the-10/?sh=59352d296679.

Ployhart, R.E., Schmidtt, N. and Tippins, N.T. (2017). Solving the supreme problem: 100 years of selection and recruitment at the Journal of Applied Psychology. *Journal of Applied Psychology. XXXXX* https://doi.org/.

Rath, T. (2007). *StrengthsFinder 2.0.* Simon and Schuster.

Robles, L. and Zárraga-Rodríguez, M. (2015). Key competencies for entrepreneurship. *Procedia Economics and Finance*, 23: 828–832.

Roth, P.L., Bobko, P. and McFarland, L.A. (2005). A meta-analysis of work sample test validity: Updating and integrating some classic literature. *Personnel Psychology*, 58: 1009–1037.

Ryan, A.M. and Ployhart, R.E. (2014). A century of selection. *Annual Review of Psychology*, 65: 693–717.

Sackett, P.R., Lievens, F., Van Iddekinge, C.H. and Kuncel, N.R. (2017). Individual differences and their measurement: A review of 100 years of research. *Journal of Applied Psychology*, 102: 254–273. https://doi.org/10.1037/apl0000151.

Schmidt, F.L. and Hunter, J.E. (1998). The validity and utility of selection methods in personnel psychology: Practical and theoretical implications of 85 years of research findings. *Psychological Bulletin*, 124(2): 262. https://doi.org/10.1037/0033-2909.124.2.262.

Shaffer, J.A. and Postlethwaite, B.E. (2012). A matter of context: A meta-analytic investigation of the relative validity of contextualized and noncontextualized personality measures. *Personnel Psychology*, 65(3): 445–494.

Shane, S. (2008). *The Illusions of Entrepreneurship: The Costly Myths That Entrepreneurs, Investors, and Policy Makers Live By*. New Haven: Yale University Press.

Shane, S. and Cable, D. (2002). Network ties, reputation, and the financing of new ventures. *Management Science*, 48(3): 364–381.

Smart, G.H. (1999). Management assessment methods in venture capital: An empirical analysis of human capital valuation. *The Journal of Private Equity*, 2(3): 29–45.

Society for Industrial and Organizational Psychology (Division 14 of the American Psychological Association) (2018). Principles for the Validation and Use of Personnel Selection Procedures. Bowling Green, OH: Society for Industrial and Organizational Psychology.

Terman, L.M. (1918). The use of intelligence tests in the army. *Psychological Bulletin*, 15(6): 177.

Tinkler, J.E., Whittington, K.B., Ku, M.C. and Davies, A.R. (2015). Gender and venture capital decision-making: The effects of technical background and social capital on entrepreneurial evaluations. *Social Science Research*, 51: 1–16. https://doi.org/10.1016/j.ssresearch.2014.12.008.

Thurstone, L.L. (1925). A method of scaling psychological and educational tests. *Journal of Educational Psychology*, 16: 433–451. https://doi.org/10.1037/h0073357.

Watson, P.C. (1960). On the failure to eliminate hypotheses in a conceptual task. *Quarterly Journal of Experimental Psychology*, 12(3): 129–140.

Woehr, D.J. and Arthur Jr, W. (2003). The construct-related validity of assessment center ratings: A review and meta-analysis of the role of methodological factors. *Journal of Management*, 29(2): 231–258.

Zhao, H. and Seibert, S.E. (2006). The Big Five personality dimensions and entrepreneurial status: A meta-analytical review. *Journal of Applied Psychology*, 91: 259–271.

Part II

Startup Situations, Environments, and Support Systems

Chapter 7

Opportunity or Threat? Entrepreneurs' Well-Being and Performance in the Data-Driven Era

Yik Kiu Leung[1,*] and *Christine Yin Man Fong*[2]

Data science undoubtedly opened opportunities for human beings to create better lives. Information and communication technologies (ICTs) have become an inseparable part of work and personal lives in the 21st century. With smart devices, people are instantly connected to the rest of the world and can work anytime and anywhere they want. The advancement of technology has also created a new wave of digitization in everyday life. On an organizational level, companies are using digital tools, technology, and ecosystem to improve customers' experience, share data with business partners, and innovate their products. On a personal level, people are opting for e-payment, tracking their daily activities with wearable devices, and live-stream music and movies. However, while the positive side of big data is well-discussed and well-studied, the negative side of big data, especially how data science may be harmful to entrepreneurs' well-being is often overlooked.

While ICTs have brought much convenience to people's lives and become a core part of the economy, like all other aspects of life, it also comes with a dark side. Importantly, the dark side of ICT is often not as apparent as other problems. For example, it is hard to say whether entrepreneurs should set a hard boundary between work and home and avoid interaction with work-related contact with others during non-work hours. On the one hand, ICTs may improve the productivity of work but on the other hand, they may induce anxiety and tension in users who need to adjust to new ICTs.

[1] Jheronimus Academy of Data Science, Tilburg University, Mariënburg Building Room 2.06, Sint Janssingel 92, s-Hertogenbosch, 521.

[2] Warandelaan 2, 5037 AB, Tilberg, The Netherlands.

Email: y.m.fong@tilburguniversity.edu

* Corresponding author: y.k.leung@tilburguniversity.edu

This chapter discusses the impact of the advancement of ICTs on the well-being of entrepreneurs. Particularly, the chapter first focuses on the technostress resulting from recent development in the data-driven era. This chapter then discuss personality and demographic factors that play a role in technostress. Finally, a self-help tool for entrepreneurs to manage technostress by proactively crafting their work is provided.

Technostress and Entrepreneurship in the Data-Driven Era

Stress associated with ICTs (i.e., technostress) was first proposed by clinical psychologist Craig Bord in 1984. He defined technostress as a "modern disease of adaptation caused by an inability to cope with new computer technologies in a healthy manner" (p. 16). There are many ways in which ICTs may negatively affect the well-being of entrepreneurs. For example, entrepreneurs may find it difficult to focus on their work with the constant interruptions of messages, reminders, and notifications from their smart devices. To make it worse, work norms nowadays commonly expect working individuals to stay connected via different communication tools, such as emails and phone messages (Best et al. 2006). For instance, working individuals who multitask with ICTs for task-relevant and urgent activities are seen as more competent, socially attractive, and dedicated (Bell et al. 2005). While some studies have shown that distraction can be helpful in monotonic (Atchley and Chan 2011) and repetitive activities (Becic 2009), it often hampers the performance of tasks that require a great deal of attention. Importantly, cognitively demanding tasks, such as long-term strategic planning and new product development, are often vital to the success of a business. Thus, continuous partial attention to different distractions is likely to hinder performance and induce stress on entrepreneurs (McFarlane and Latorella 2002).

In addition to constant interruption, ICTs are also blurring the boundaries between personal life and work. This is especially true during the COVID-19 pandemic when people are forced to work from home and thus further blurring the boundary between personal and work life. While a weak boundary between work and personal life may improve the efficiency in managing demands from both sides, it may also give rise to work-family conflict – a form of inter-role conflict that occurs when resources (e.g., time, energy) and demands (e.g., work tasks, house chores) from the two domains conflict with each other (Greenhaus and Beutell 1985). This is particularly relevant for growth-oriented entrepreneurs who tend to have a strong passion for their work and thus are more willing to sacrifice their personal life in exchange for their work time.

Furthermore, since the nature of the work of entrepreneurs is highly flexible, entrepreneurs often have the autonomy to decide what, when, and where to work. Unlike a typical 9 to 5 work schedule, entrepreneurs typically work in a more fragmented schedule. For instance, an entrepreneur may start the day with some emailing and texting, work on a project for a couple of hours, pick up the kids from the childcare center, have a video call with a business partner while driving, and ends the day by catching up with the remaining task when the kids have slept, and the entrepreneur can finally focus on the project. This example demonstrates that while a flexible work schedule allows an entrepreneur to manage the family and work roles flexibly, it may encourage entrepreneurs to work during non-work hours and thus increase the chance of having work-family conflict and burnout.

The digitization of peoples' lives generates a huge volume of data with great variety and velocity – commonly referred to as "big data". In this new era of digitalization and data-rich business environment, entrepreneurs are often forced to adapt and react to the demand and fast-changing business world. While the unprecedented proliferation of data and advancement of ICTs have enabled an entirely new business landscape, extracting value from raw data (e.g., assessing a new market opportunity, and optimizing business operations) can be a daunting process. At the very least, the data needs to be collected, stored, prepared, analyzed, and interpreted before it can be used to generate useful insights for decision-making and value creation.

However, entrepreneurs may not have the data literacy required to transform data into value. The lack of data literacy may in turn lead to information overload and thus technostress. A report published by Quik and Accenture (2020) indicated that about 79% believe that they do not have the data literacy necessary to work with data comfortably and safely. Moreover, 74% of the respondents found themselves overloaded and 61% felt stressed because of the overload of data and information. Importantly, professionals who feel overwhelmed by the demand for working data tend to procrastinate and avoid working with data in the future. While entrepreneurs are not the only occupation group affected by these emerging demands and challenges, they often lack the kind of resources and support those salaried workers get from their companies.

To sum up, the transformation triggered by technological advancement with digitization and big data is constantly changing the business landscape, ways of communication, consumer preference, lifestyle, etc. Such a dynamic environment not only requires constant adaptation but also creates an overload of information and distraction, which encourage multi-tasking and divide attention and thus further burden entrepreneurs.

Measurement of Technostress

Technostress is commonly measured using self-report measures. One of the most used measures of technostress is the measure developed by Ragu-Nathan et al. (2008). This measure includes two second-order constructs: technostress creators (i.e., factors that create stress from ICTs) and technostress inhibitors (i.e., factors that reduce stress from ICTs). Technostress creators and technostress inhibitors can be further divided into five and six sub-constructs, respectively. Specifically, technostress creators include (1) techno-overload, (2) techno-invasion, (3) techno-complexity, (4) techno-insecurity, and (5) techno-uncertainty whereas technostress inhibitors include (1) literacy facilitation, (2) technical support provision, (3) involvement facilitation, (4) job satisfaction, (5) organizational commitment, and (6) continuance commitment.

However, since this measure was not developed specifically for entrepreneurs, some of the constructs may not be suitable for entrepreneurs or research on entrepreneurship. For example, the construct of technical support provision is less applicable to entrepreneurs as many of them work in small and medium-sized businesses in which IT support staffs are not available. Therefore, entrepreneurs who are interested in assessing their level of technostress should focus on the constructs that are relevant to them. For instance, the construct of techno-invasion can be very relevant for entrepreneurs because of the flexible and dynamic work nature of entrepreneurs. Sample item of this construct includes "I have to be in touch with my work even during my vacation and weekend time to keep current on new technologies." This is a good example of work-family conflict/ interference discussed earlier. Another good example is techno uncertainty. A sample item of this construct is "I have to constantly update my skills to avoid being replaced". To sum up, entrepreneurs who are interested in their level of technostress may assess it using the suggested measure.

The Person-Environment Perspective of Technostress

Like all other types of stress, technostress includes two parts: an external stimulus also known as a stressor, and the psychological (e.g., fear, anxiety), as well as biological responses (e.g., surging heart rate, muscle tension), associated with that stimulus. The appraisal of the stressor plays an important role in the effect of the psychological and biological consequences induced by the stressor. Because of this reason, the outcome of the same stressor can lead to very different outcomes depending on the characteristics of the person.

The Person-Environment fit model of stress is one of the most widely used models in explaining the dynamic between people and their environment (Holland 1997). This model suggested that stress is a result of a lack of correspondence between the characteristics of the person (e.g., values, desires, personalities) and the environment (e.g., demand, support). For example, an entrepreneur who grew up with ICTs may find the fast-changing demand of technology easy to manage versus an entrepreneur who has always been communicating with his customers with audio calls might find the need to interact with customers via social media platforms or create a website for his business difficult.

Previous research on technostress suggests that personality traits play a central role in the appraisal of a stressor. One of the most researched personality models is the Big Five personality also known as the Five-Factor Model. The Big Five personality includes five different personality traits: Openness, Conscientiousness, Extraversion, Agreeableness, and Neuroticism. For example, in a survey study with 152 senior managers, Srivastava et al. (2015) found that managers with a relatively high level of agreeableness are more likely to experience job burnout when experiencing technostress. In contrast, managers with a relatively high level of extraversion are less likely to suffer from technostress.

Recognizing the importance of considering the interaction between personal and situational attributes, Khedhaouria and Cucchi (2019) conducted a fuzzy-set configurational analysis. Unlike regression analysis which sees personality traits within the same model as competing predictors, fuzzy-set configurational analysis allows researchers to identify a specific combination of different factors (e.g., personality traits and stressors) at the same time without compromising its predictive power. As a result, researchers can identify different profiles (i.e., a combination of different factors) that is most relevant to certain outcome variables (e.g., burnout). Here, Khedhaouria and Cucchi (2019) found that extraversion is one of the most profound personality traits in predicting burnout due to technostress. This is interesting as extraversion is commonly known as a protective factor of psychological well-being. One explanation is that senior managers who are extroverted are more sociable and thus willing to stay connected with others. However, this may create work-family conflict and leave little time for other work tasks. Furthermore, consistent with previous research, they found that neuroticism also played a prominent role across different configurations. However, when compared to the general public, entrepreneurs tend to score higher on conscientiousness and openness but lower on neuroticism and agreeableness with no significant difference in extraversion (Zhao and Seibert 2006). As such, there is no evidence that entrepreneurs are particularly vulnerable to technostress on a personal level with Big Five personality traits.

In another study, Marchiori et al. (2018) examine the role of demographic characteristics, such as age, gender, and educational level by surveying 927 employees who rely heavily on ICTs for their main business processes. They found that older workers with long professional experiences have greater difficulties in the face of technological complexity. Furthermore, while women reported a higher level of techno-complexity and techno-uncertainty, men reported a higher level of techno-overload and techno-invasion. Finally, there is no evidence that the education level of workers is related to the level of technostress in general.

Overall, entrepreneurs are not particularly vulnerable to technostress on a personal level. To a certain extent, the characteristics of typical entrepreneurs (i.e., high openness, conscientiousness, and low agreeableness and neuroticism) may even protect them from technostress. This is not surprising as the work environment of entrepreneurship is inherently risky, dynamic, and fast-changing. Therefore, individuals who are not comfortable or able to adapt to such a dynamic, risky, and fast-changing work environment are likely to exit entrepreneurship.

The previous section discussed different risk and protective factors of entrepreneurs. In the next section, this chapter provides guidance to tackle challenges in the data-driven era as an entrepreneur.

Entrepreneurs' Well-being: A Job Crafting Approach

Given the challenge and stress that entrepreneurs face in the data-driven era is enormous, effectively designing their work and career is vital to their well-being and success. As a type of work (re)design, job crafting refers to proactive strategies to change individuals' work demands and resources to better fit their job with personal interests, goals, and abilities. The central idea of job crafting is that individuals proactively design and redesign their work roles and boundaries, by (1) increasing job tasks that are challenging while decreasing job tasks that are hindering, (2) increasing job resources, (3) as well as perceiving/viewing their works in a more meaningful way (Tims et al. 2012, Wrzesniewski and Dutton 2001).

The aim of job crafting is that an individual can craft "a way of working" that he/she truly enjoys. A way of working that not only allow entrepreneurs to excel and perform in their companies but also align the job with their interests, strengths, and work goals. Previous research on job crafting consistently shows that individuals can create meaningful work via job crafting (Hulshof et al. 2020) which in turn leads to better well-being (e.g., a lower level of stress and a higher level of engagement) and work performance (Rudolph et al. 2017).

Job Crafting as a Way for Entrepreneurs to Cope with Technostress

Given the benefits of job crafting, organizational psychologists and management scholars have paid much attention to job crafting research and its application. While job crafting has been well-studied in the organizational behavior field with employee samples, and various job crafting interventions have been developed to help employees to increase their crafting behaviors, job crafting is seldom explored in the field of entrepreneurship. It may be fruitful to bring insights and knowledge of job crafting theory to the field of entrepreneurship to enhance entrepreneurs' well-being.

Job crafting is not about making dramatic changes in one's work, but consciously making small changes in tasks, work relationships, and how one perceives their work. It can help individuals to create better work that they find meaningful and enriching, even though that work may remain busy, difficult, and unstable. Job crafting also pushes the entrepreneur to reflect on what is important in their work and to redirect their energy from those work aspects that are hindering, to those work aspects that are truly enjoyable and meaningful to them.

However, this may sometimes mean that job crafting requires entrepreneurs to reduce some hindering work tasks (i.e., task crafting) and relationships (i.e., relationship crafting) that they would like to avoid, but these tasks and relationships may be important to their companies and thus cannot be avoided. In this case, entrepreneurs can still craft their jobs without avoiding these tasks, such as by attaching more meaning to these tasks (i.e., cognitive crafting) to form a more positive perception of these tasks. For example, an entrepreneur is suffering from technostress because they feels the pressure of keeping up with the latest data science technique to adapt to the ever-change business development. This entrepreneur could then try to create meaning for this task. Specifically, in case they is a green entrepreneur who cares a lot about sustainability. They can see working on these data science techniques not only for business survival but also for creating a path to help the next generation to enjoy a greener and more sustainable planet. By attaching meaning to this unpleasant task, this entrepreneur can reduce their technostress.

Furthermore, job crafting requires minimal money, time, and energy for the entrepreneur to perform, and it does not involve big risks as it does not require the entrepreneur to change their work dramatically. Yet, research consistently shows that it can help individuals to cultivate meaningful experiences in their work, and to improve work-related well-being, such as work-related enjoyment, vigor, dedication, and absorption (Tims et al. 2012, Tims et al. 2016). Below, three forms of job crafting that can be used by entrepreneurs to deal with technostress are highlighted.

Task Crafting

With the aim to craft a work that better fits an individual's needs and interests, task crafting is about actively and consciously optimizing one's work role by adding and/or dropping work tasks within their work boundary. These changes in work tasks should be small and not dramatically impact the function of the entrepreneur's work role and the daily operation of their company. This can be altering the amount of time and energy that they spend on their work tasks. For example, when an entrepreneur sees an interesting funding opportunity that can help his company to be more sustainable, they may give up persuading a difficult client to close the deal but use the time and energy they saved to compete for this funding opportunity that they finds meaningful.

Relational Crafting

Relational crafting refers to shaping one's social relationships at work within their work boundaries. This can involve changing who they are interacting with, as well as the quality and quantity of these social interactions. An example of relational crafting can be building up close relationships with a group of resourceful peers that are working in a similar profession as the entrepreneur does, and this entrepreneur feels comfortable sharing their stress and concern with them, as well as seeking advice and support from them. Another example of relational crafting is to identify and reduce the social relationship that they finds hindering, such as giving up difficult clients that drain their energy and at the same time do not benefit their long-term success.

Cognitive Crafting

Cognitive crafting is a specific form of job crafting. Unlike task and relational crafting, cognitive crafting does not require entrepreneurs to make actual changes to their jobs but to cognitively redefine their work roles (i.e., how they perceive their work). For example, an entrepreneur may redefine their work role as creating a startup that helps the next generation to enjoy a cleaner and green planet, instead of creating a startup that solely focuses on developing technology and making money. By changing their perspective about what the entrepreneurs are working on, they can find more meaning in their work, even though this work is still hectic and difficult for them.

Theory and Research

As found by a few meta-analyses (e.g., Lichtenthaler and Fischbach 2019, Rudolph et al. 2017) and qualitative meta-synthesis (e.g., Lazazzara et al. 2019), job crafting is positively associated with a wide range of important individual (e.g., meaningful work, work engagement, person-job fit and organizational outcomes (e.g., performance). More importantly, job crafting can be increased by job crafting intervention as shown in a meta-analyses study (Oprea et al. 2019). While job crafting is examined extensively with employee samples, it is seldom examined with entrepreneur samples. Unlike employees, entrepreneurs do not have their managers and HR departments to help them with their job designs. Thus, compared with employees, job crafting may be more important to entrepreneurs, as they may have even higher responsibility and autonomy to design their own works. It may be interesting for future research to investigate how entrepreneurs can increase their job crafting behaviors, and how job crafting behaviors may enhance their well-being and company success. Relevant questions to be asked may include (1) will online intervention or self-help manuals specifically designed for entrepreneurship help them enhance their job crafting behaviors? (2) which type of job crafting behavior (e.g., task, relational or cognitive crafting, or a new type of crafting such as time-spatial job crafting and strengths – interests-based job crafting) is most effective in reducing technostress for entrepreneurs? (3) do characteristics of an entrepreneur (e.g., age, personality, industry) influence the effectiveness of job crafting in reducing technostress?

Steps to Engage in Job Crafting

The previous session discussed research in job crafting. Now, this section will look into how entrepreneurs can practically apply job crafting to redesign their work to become more meaningful in a stepwise manner. Following the steps, some tips for the reader are provided.

Step 1: Job Analysis

As outlined in Chapter 2, job (work) analysis aims to help understand how entrepreneurs are spending their energy and time on their everyday work tasks. List out all the work tasks that they are currently engaging in and place them according to the energy and time that they require from the entrepreneurs, from a lot to very little. Starting from the top of the list, reflect on each of the tasks and classify if they are challenging or hindering.

A *challenging task* refers to work demands that may cost one's effort but provide them with reward, enjoyment, and personal growth. An example of a challenging task is to identify the latest and most exciting product development in one's field. Another example can be pitching a brilliant business idea to a venture capital firm (VC) for funding.

A *hindering task* refers to work demands that not only cost one's effort but also are not enjoyed by them and hinder one's ability to achieve their work goal and personal growth. An example of a hindering task is following the ever-changing AI technology development day and night because the entrepreneur is afraid of missing out on some important developments. Another example of hindering tasks is constantly replying to phone messages from clients and suppliers which distracts the entrepreneur from the core work tasks.

Step 2: Person Analysis

List out the entrepreneurs' personal strengths, work goals, as well as obstacles that they are facing in their work. Strength can be a great business idea that they own, their data science skills, their connections with supportive partners, clients, and VCs, or funding that they have already acquired. To gain awareness of their strengths, using psychometric tools, (e.g., META; Ahmetoglu et al. 2011, 2018) is an effective way. Work goals can be multidimensional, such as company growth, revenue, reputation, personal skill development, sustainability, or even work-life balance.

Here is a case example of obstacles that can prevent an entrepreneur from utilizing their strength to achieve their work goals. They has a great business idea and a supportive team that they loves to work with. Their work goal is to work with their team to put this idea into practice by acquiring funding to launch this exciting artificial intelligence (AI) project within a year. However, this entrepreneur identifies two obstacles, the first one is that their time and energy are occupied by searching for various funding opportunities, but they never commits to finish and submit one single proposal to any of these opportunities because they does not know how to write the technical part of the proposal. Yet almost all funding applications require these technical elements. The second obstacle is that she is constantly distracted by the non-stop pop-up messages from her phone thus preventing her from having some quality time to focus on the funding proposal.

Step 3: Job-Person Analysis

Couple the tasks that the entrepreneurs identify in the job analysis (Step 1) with the strength, work goals, and obstacles that they identify in the personal analysis (Step 2). In this process, entrepreneurs can try to decide to increase their time and energy spent on challenging tasks that fit their strengths and work goals. Besides, entrepreneurs are encouraged to reduce their hindering tasks, especially if these tasks do not fit their strengths, are not that important to their work goals, and require a lot of time and energy from them. For example, if an entrepreneur's hindering *task* is to take care of the technical part of a project, and their *strength* is that they knows a trustworthy person who is very knowledgeable about data science techniques and has a good relationship with him/her. Then, the entrepreneur can pair up this hindering demand with this personal strength, i.e., inviting this person to join this project and take over the technical part of the project, or consult this person in terms of which way they should go to deal with the technical issues. By doing so, the entrepreneur can utilize their strengths to reduce the negative impact of the hindering demand, while allowing more time and energy to spend on challenging tasks that they enjoy and can contribute to their work goals.

Another example of *hindering tasks* is that an entrepreneur is constantly disturbed by emails and phone messages, and their *strength* is their skills in programming. They can then utilize their programming skill to write a small program that can temporarily block their phone notification whenever they opens the folder of the funding proposal. This can prevent email and phone messages from disturbing him/her from focusing on the proposal, which is their core work task and is meaningful to their long-term career.

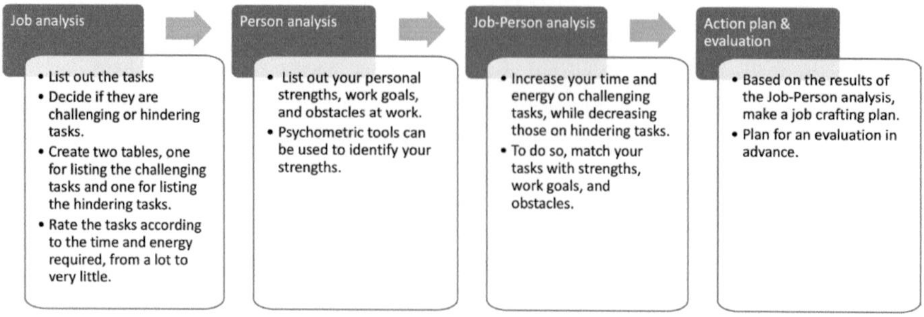

Figure 1. Four-step approach to engaging in job crafting.

Step 4: Action Plan and Evaluation

According to the results of the Job-Person analysis, make a specific and actionable plan that includes concrete tasks that he/she can achieve within a week. This goal can be as simple as "I will meet this person to discuss if he/she is interested in joining my project and if he/she can help with the technical part on Monday." Another example can be "I will create the app to block the phone message and use it whenever I work on the funding proposal by Wednesday."

Plan time in advance to reflect on one's job crafting strategies at the end of the week. Reserve time for this reflection in the calendar. These reflections should focus on the tasks that list in one's action plan. Go to the list and check if they have completed each of the tasks that they list on the action plan.

If yes, the entrepreneurs can take a moment to appreciate their achievement, and then reflect on how these changes in their work benefit their personal and career goals. Recognize the achievement and decide if they would like to continue in the coming week.

If not, the entrepreneurs did not complete a particular task on the action plan. It is alright. They should not feel frustrated, but they can reflect on what prevented them from following the action plan. Try to clear the obstacles and do it again next week. They may repeat Steps 1 to Step 4 a few times, or until they have achieved the benefit of the job crafting strategies that they are targeting.

Tip 1: Account for Personal Life Too

When crafting their jobs, entrepreneurs are probably wise to focus not only on work-related needs and interests but also take their personal and family-related needs and interests into account. For example, they might be very interested in following the latest news and project development via social media day and night, but it may hinder them from detaching themselves from work while they are at home spending time with their family.

Tip 2: Crafting a Meaningful Work Life

When formulating work goals for job crafting, it is a good time to reflect on whether wider types of work goals are desirable to enhance well-being. Attracting funding, generating revenue, and building a reputation may be important goals for entrepreneurs, but once their startups are up and running, they may have the luxury to ask for more. It may be the time for them to work on a new field that does not focus solely on growth, profit, and reputation but a new field that they find meaningful, fulfilling, or simply a joy. This can be, for example, a mentor-mentee program where they can share their successful experience and help the next generation to excel in the industry.

Tip 3: Job Crafting as a Habit

Job crafting as a job design and redesign strategy is not a one-time thing. To benefit from a job crafting strategy, an entrepreneur needs to engage in job crafting over time to deal with the ever-changing business environments and their personal needs and interests.

In conclusion, job crafting provides a relatively accessible yet effective strategy for designing their work as an entrepreneur. By reflecting on their strengths, work goals, and obstacles, as well as the five steps proposed above, entrepreneurs can achieve meaningful work in the future.

References

Ahmetoglu, G., Akhtar, R., Tsivrikos, D. and Chamorro-Premuzic, T. (2018). The entrepreneurial organization: The effects of organizational culture on innovation output. *Consulting Psychology Journal*, 70(4): 318–338. https://doi.org/10.1037/CPB0000121.

Ahmetoglu, G., Leutner, F. and Chamorro-Premuzic, T. (2011). EQ-nomics: Understanding the relationship between individual differences in trait emotional intelligence and entrepreneurship. *Personality and individual differences*, 51(8): 1028–1033.

Atchley, P. and Chan, M. (2011). Potential benefits and costs of concurrent task engagement to maintain vigilance: A driving simulator investigation. *Human Factors*, 53: 3–12.

Becic, E. (2009). Aging and the effects of conversation with a passenger of a caller on simulated driving performance. *ProQuest.*

Bell, C.S., Compeau, D.R. and Olivera, F. (2005). Understanding the social implications of technological multitasking: A conceptual model. *SIGHCI 2005 Proceedings*, 2.

Best, S.J., Krueger, B.S. and Ladewig, J. (2006). Privacy in the information age. *International Journal of Public Opinion Quarterly*, 70: 375–401.

Brod, C. (1984). *Technostress: The Human Cost of the Computer Revolution.* Reading, MA: Addison-Wesley.

Greenhaus, J.H. and Beutell, N.J. (1985). Sources of conflict between work and family roles. *Academy of Management Review*, 10: 76–88.

Holland, J.L. (1997). *Making Vocational Choices: A Theory of Vocational Personalities and Work Environments* (3rd ed.). Lutz, FL: Psychological Assessment Resources.

Hulshof, I.L., Demerouti, E. and Le Blanc, P.M. (2020). Day-level job crafting and service-oriented task performance: The mediating role of meaningful work and work engagement. *Career Development International*, 25(4): 355–371.

Khedhaouria, A. and Cucchi, A. (2019). Technostress creators, personality traits, and job burnout: A fuzzy-set configurational analysis. *Journal of Business Research*, 101: 349–361

Lazazzara, A., Tims, M. and De Gennaro, D. (2020). The process of reinventing a job: A meta–synthesis of qualitative job crafting research. *Journal of Vocational Behavior*, 116: 103267.

Lichtenthaler, P.W. and Fischbach, A. (2019). A meta-analysis on promotion-and prevention-focused job crafting. *European Journal of Work and Organizational Psychology*, 28(1): 30–50.

Marchiori, D.M., Mainardes, E.W. and Rodrigues, R.G. (2018). Validation of the ISS-QUAL and the role of gender, age and education on it service quality in the public sector. *Information Technology and Management*, 19: 217–230.

McFarlane, D.C. and Latorella, K.A. (2002). The scope and importance of human interruption in human–computer interaction design. *Human-Computer Interaction*, 17: 1–61.

Oprea, B.T., Barzin, L., Vîrgă, D., Iliescu, D. and Rusu, A. (2019). Effectiveness of job crafting interventions: A meta-analysis and utility analysis. *European Journal of Work and Organizational Psychology*, 28: 723–741.

Qlik and Accenture. (2020). The human impact of data literacy. https://www.accenture.com/_acnmedia/PDF-115/Accenture-Human-Impact-Data-Literacy-Latest.pdf.

Ragu-Nathan, T.S., Tarafdar, M., Ragu-Nathan, B.S. and Tu, Q. (2008). The consequences of technostress for end users in organizations: Conceptual development and empirical validation. *Information Systems Research*, 19(4): 417–433.

Rudolph, C.W., Katz, I.M., Lavigne, K.N. and Zacher, H. (2017). Job crafting: A meta-analysis of relationships with individual differences, job characteristics, and work outcomes. *Journal of Vocational Behavior*, 102: 112–138.

Srivastava, S.C., Chandra, S. and Shirish, A. (2015). Technostress creators and job outcomes: Theorising the moderating influence of personality traits. *Information Systems Journal*, 25: 355–401.

Tims, M., Bakker, A.B. and Derks, D. (2012). Development and validation of the job crafting scale. *Journal of Vocational Behavior*, 80: 173–186.

Tims, M., Derks, D. and Bakker, A.B. (2016). Job crafting and its relationships with person-job fit and meaningfulness: A three-wave study. *Journal of Vocational Behavior*, 92: 44–53.

Wrzesniewski, A. and Dutton, J.E. (2001). Crafting a job: Revisioning employees as active crafters of their work. *Academy of Management Review*, 26: 179–201.

Zhao, H. and Seibert, S.E. (2006). The big five personality dimensions and entrepreneurial status: A meta-analytical review. *Journal of Applied Psychology*, 91(2): 259.

Chapter 8

Using Data to Build More Diverse, Equitable, and Inclusive Startups

Victoria Mattingly, PhD,[1,*] *Sertrice Grice, MS*,[2]
Anthony Roberson, MS[3] and *Kelsie Colley, ABD*[4]

Following multiple sexual harassment allegations, fostering a toxic "bro culture," harassing his own drivers, and taking measures that seemingly supported the "Muslim-ban" enacted by the Trump administration in early 2017, Uber's co-founder and first CEO, Travis Kalanick, was ousted. Despite Kalanick's departure, the #DeleteUber campaign nonetheless resulted in short-term (i.e., more than 200,000 lost customers) and long-term damage (i.e., poor reputation and lack of trust) the company is still recovering from to this day (Siddiqui 2019, Wong 2017).

After dis-banning its diversity, equity, and inclusion team and function, the cryptocurrency startup, Basecamp, lost a third of its staff due to the perception that the company does not value its people, especially those from underrepresented groups (Uhereczky 2021).

Women only receive about 3% of venture capital funding, with Women of Color receiving less than 1%. Relatedly, white men control 93% of venture capital (VC) funds, whereas only 0.2% of VC partners are Black or Latino women (Houser and Kisska-Schulze In Press).

The above examples are just the tip of the iceberg when it comes to the current state of the entrepreneurship ecosystem as related to diversity, equity, and inclusion (DEI), especially with the treatment of vulnerable and underrepresented groups. The

[1] 100 S. Commons, #102, Pittsburgh, PA 15201.
[2] 4000 Old Sturbrudge Drive, Apex, NC27539, 785-844-0903.
[3] 5805 South 98th Plaza #1B, 201-463-5746.
[4] 2740 W Elizabeth, 616-443-1166.
* Corresponding author: victoria@mattinglysolutions.com

startup world was built for a select few; for example, startups funded by the top VCs are about 90% male, 72% white, 35% based in Silicon Valley, and 14% Ivy League-educated (Billington 2021). It is likely that familiarity bias is at play—VCs are most likely not discriminating against more diverse startups explicitly, rather unconsciously preferring startup teams that remind them of themselves (e.g., a variant of similarity bias discussed in Chapter 6). Breaking such biases, which is a key focus of DEI efforts, could be a powerful lever for diversifying the startup world by encouraging more investments in startups led by people with different identities, tapping into currently overlooked potential.

Investing in DEI efforts could not only solve the demographic homogeneity of the startup world, but also prevent future DEI-related crises from occurring, and the subsequent fallout that follows (e.g., losing investors, top talent, and customers who chose to take their business to competitors with values that align more closely to their own). Committing to DEI from the very early stages of building and scaling a startup all the way through to exit could result in securing otherwise untapped long-term investments, talent, and customers. Startups could also save millions of dollars of legal and PR fees after some otherwise preventable human-centered crisis, preserving a company's resources and reputation as an organization that takes care of *all* of its people, not just the chosen few who won the identity lottery (i.e., white, educated, affluent, and male).

This chapter will focus on how startup leaders can effectively use measurement and evidence-based best practices to do DEI work well by aligning it to the unique needs of a startup's most important asset—its people. The goal is to educate and inspire founders, funders, people leaders, and any other key decision-makers within the entrepreneurship ecosystem to use data to embed DEI into how startups are founded, built, and scaled.

This chapter will begin by sharing evidence and rationale as to why startups are in a prime position to benefit from weaving DEI into the fabric of their organization. Next, it will introduce the science of DEI, establishing key terminology with an emphasis on why inclusion should be operationalized as a *behavioral* construct. It will also share how data can and should be used to drive DEI work in startups. This chapter will provide insights from DEI consulting work and the scientific literature to share common methodological techniques used to turn DEI data into a science, resulting in measurable and meaningful outcomes. Finally, the chapter will conclude with two case studies about startups taking a more data-driven approach to DEI, as well as provide additional resources[1] that can be used right away.

The primary goal for this chapter is to leave readers feeling more confident in their ability to use data to build more diverse, equitable, and inclusive startups from the onset, rather than waiting until the next DEI-related crisis occurs and/or before becoming too big to pivot and change for the better. A secondary goal is to

[1] For those out there who want to take a deeper dive into the topics mentioned in this chapter and apply best practices for collecting and analyzing DEI data, download this resource list at www.mattinglysolutions.com/dei-data-startups.

inspire scholars and researchers to better refine key DEI constructs and continue developing methodological best practices that will enable DEI practitioners to bring more scientific rigor to this space, especially in startups.

Why Should Startups Care about DEI?

Entrepreneurs and venture capitalists should be tuned in to DEI for reasons such as litigation prevention, the ethical case of caring for others and reducing harm, and the impact DEI has on strategy, finances, talent, and culture (e.g., Blackwell et al. 2017, Cassells and Duncan 2020, Rohwerder 2017, Turner 2018). Getting DEI right is imperative to the long-term success of a startup. Fortunately, startups have a leg up compared to their larger, more well-established enterprise counterparts. Startups are in an advantageous position for implementing DEI best practices due to their agility, speed to scale, and fluidity of a culture that is yet to solidify an otherwise difficult-to-change, non-inclusive status quo (Ely et al. 2011, Paternoster et al. 2014).

When it comes to advancing DEI efforts, startups are like jetboats. Compared to the "Titanic" bigger, more established organizations, startups are small and agile—they can swiftly change directions (like deciding to embed DEI into their people operations function as they build it) and easily pivot when they miss the mark (like realizing that their first few board members all share the same identity and committing to diversifying its board moving forward). Speed is another way startups are like DEI jetboats. Although DEI is a long-term journey, a high-growth startup is in a unique position to quickly diversify its talent compared to larger companies with a relatively stable workforce that has little to no turnover. When a startup cannot hire fast enough, there is no excuse not to do all in its power to fill those empty seats with people who represent the population it serves.

For example, imagine a startup recognizing its lack of gender diversity and deciding to actively recruit more women. After investing in more gender-inclusive recruiting practices, they track how the demographics of their applicants have changed over time. While more women applicants were indeed applying—and getting hired!—their data also showed that the number of BIPOC (Black, Indigenous, and People of Color) applicants were now *decreasing*, across *all* genders. As it turns out, only focusing on gender diversity hurt this startup's image among BIPOC candidates (e.g., only seeing white women in recruitment materials or at hiring events). Fortunately, the startup is still hiring rapidly and can quickly integrate more intersectional approaches to its recruitment methods and begin balancing out the numbers. Because of its speed and agility, the startup jetboat can immediately pivot its recruitment strategy to appeal to both women *and* BIPOC applicants and more effectively diversify its homogeneous workforce. And gender and race/ethnicity are only the start (more on identities and demographics startups should measure later in this chapter).

Now, apply the same recruitment example to a corporate Titanic. Deciding to make any change to the recruitment process would require far more influencing, multi-level approvals, planning, coordination, and overall effort to pull off. The process of adapting a well-established selection system would require more than just resources, it would take much longer time to implement and see any tangible results. And when change does not happen soon enough, the ship is left vulnerable to hitting DEI icebergs, as the examples illuminated at the beginning of this chapter.

On the other hand, committing to DEI early on can help startups in the long run, rather than trying to clean up a toxic culture once it is too late. Startups have much to gain by getting DEI right from the onset, including the competitive advantage of a more committed and innovative workforce, bigger financial gains, attracting and retaining top talent, and fostering healthier workplace cultures compared to startups who do not choose to embed DEI into their overall business strategy.

Competitive Advantage

Strategically speaking, doing DEI right would set a startup apart from its competitors who choose not to commit to building more diverse, equitable, and inclusive organizations. In fact, many organizations that do have some DEI efforts in place are often viewed as not doing enough, as the market is demanding more from companies in terms of DEI (Blanche 2022). This means that startups that do not take DEI seriously may lose out on investors, customers, and top talent, therefore falling behind compared to their competitors.

Startups that get DEI right also yield benefits related to innovation and overall employee performance. Experts suggest that "Corporate America is missing out on one of the biggest opportunities of our time for driving innovation and growth: creating business value by advancing racial equity" (Blackwell et al. 2017, p. 2). Diverse management teams (e.g., differences in background, personality, and values), tend to be more creative than homogenous management teams (Torchia et al. 2015). Research has found a positive relationship between team diversity and out-of-box thinking (van Dijk et al. 2012), which is crucial when considering how paramount a startup's ability to innovate is to its success.

Financial Outcomes

Companies in the top quartile for diversity financially outperform competitors who lack gender and racial diversity, and the impact of diversity on finance metrics is only increasing in strength (Dixon-Fyle et al. 2020). Research has shown that as racial diversity increases at the executive level of companies, there is also an increase in earnings (Hunt et al. 2015). Also, one cannot overlook the financial benefit of reducing turnover and increasing productivity—two outcomes organizations can expect from investing in long-term DEI efforts (McFeely and Wigert 2019).

Also, investors are taking DEI more seriously. Organizations such as the Principles for Responsible Investment (2022) declared, "It is clear that DEI and financial performance are related." Deloitte (2017) found that organizations that are successful in their inclusive practices are twice as likely to surpass their financial goals. Inclusion is necessary to drive company performance and retain talent (Gaudiano 2020), the latter of which is addressed below.

Winning the War on Talent

In a post-COVID workplace, it is an employee-driven market. Top talent knows they have options of where they can go, so they are being selective with where to commit their time and are looking for workplaces that align with their values and will provide the best employee experience. A startup's commitment to DEI signals that the company cares about the well-being and success of ALL its employees, which gets people to stay around longer as a result.

Attracting Top Talent

Before getting into the effect DEI has on retaining top talent, it is useful to first address how DEI helps to get top talent into the door in the first place. Job seekers are intentionally seeking out organizations that prioritize DEI (Dauth et al. 2021, Madera et al. 2018, Williams and Bauer 1994). When companies make their DEI efforts known externally, they are likely to attract more prospective employees. This is partially why there has been increased transparency about organizations' DEI efforts, including publicly sharing DEI reports and developing public-facing DEI websites. Job seekers are also looking for evidence of DEI during the selection process, including the diversity of their selection panel and addressing DEI-related issues during the interview and negotiation process.

Retaining Top Talent

Successfully recruiting top talent would be futile if those people do not stay around. Belonging—the feeling that one is valued, respected, seen, and heard (Mattingly et al. 2022)—is a crucial component for retaining top talent, especially those from historically excluded identity groups. Experts suggest that "The superior outcomes you seek [with DEI] cannot be achieved without a sense of belonging" (Society for Human Resource Management 2021).

When it comes to retaining talent, belonging should be the primary goal of DEI efforts (Mattingly et al. 2022). Employees are more likely to feel like they belong when they see others who look like them (e.g., diversity), are treated fairly (e.g., equity), and are the recipients of behaviors that make them feel valued, respected, seen, and heard (e.g., inclusion). Employees who feel like they belong invest in their work and are more likely to stay at their organization (Carr et al. 2019). And research has found that when workers feel increased belonging, there is a 56% increase in job performance, a 50% decrease in turnover intention, and a 75% reduction in sick days compared to those who do not feel like they belong at their organization (BetterUp 2019).

Building Better Cultures

Getting DEI right can also result in healthier workplace cultures that set startups up for long-term success. Oftentimes, startups will focus strictly on growth and profit at the expense of the well-being of their people. This growth-at-all-costs mindset can lead to toxic workplace cultures well known—and criticized—in the startup world (see Chang 2019). When startups forget about the human side of their business, it can lead to inequitable practices, outcomes, and even lawsuits. On the contrary, when organizations take a human-centric approach by incorporating DEI into how they operate, they can expect an increase in employee engagement, job satisfaction, and overall employee experience (Deloitte 2017, Fernandes 2021).

The research is clear as to why startup founders, executives, and culture makers (e.g., Chief People Officer, Head of Culture, Fractional CHRO) should not only care about diversity, equity, and inclusion, but elevate DEI as a critical component of their long-term strategy. It is important not to wait to invest in DEI, though, as startups have a very small window to capitalize on their size, agility, and speed before they grow too much to retroactively embed DEI practices into a more solidified culture and organizational system. Before jumping into the *how* of integrating DEI into startup ecosystems, though, one must first establish *what* exactly DEI is and how it should be defined, measured, developed, and enforced—according to science.

What Exactly is Diversity, Equity, and Inclusion?

Diversity, equity, and inclusion (DEI), collectively, is how an organization fosters and maintains a workforce that has demographics representative of the broader population and maximizes the employee experience by enabling everyone to bring their best selves to work and thrive. The specific terms, though, are often conflated or [incorrectly] used interchangeably, when they should actually be treated as three distinct concepts. Diversity is the amount of difference in a group, especially when it comes to demographics and identity. Equity is "the intentional rebalancing of power dynamics to result in the fair treatment of all employees regarding the accessibility of information, opportunities, and resources" (Mattingly et al. 2022, p. 14). Finally, inclusion is the dynamic process when intentional actions make others (especially those from underrepresented groups) feel valued, respected, seen, and heard.

When it comes to the science of DEI as applied to startups, the scientific literature is sparse. For example, an EBSCO search of peer-reviewed articles with the terms "startups" or "startup" and "diversity" yielded five total results. Some scholarly work is beginning to address how startups can improve DEI outcomes by diversifying their workforce, (e.g., building more DEI-focused referral, screening, and interview processes across the startup ecosystem; Kaul 2022), yet nearly all evidence-based research on the impact of DEI on startup success comes in the form of white papers and consumer reports from consulting firms and other non-academic research institutions. There is a great opportunity for more DEI and startup scholars to grow this body of research.

There are numerous acronyms organizations use to describe DEI (see Table 1). Historically, the initial focus was on diversity, largely driven by affirmative action policies of the 1970s and increased pressure from the Equal Employment Opportunity Commission (EEOC) to diversify the workforce (Kelly and Dobbin 1998). Over time, as the concept of "managing diversity" emerged, it became apparent that diversity alone was not enough—an inclusive workplace was also necessary to reap the

Table 1. Common acronyms for diversity, equity, and inclusion.

Acronym	Full Title	Nuances
D&I	Diversity & inclusion	Emphasis on diversity, while also acknowledging the importance of inclusion
I&D	Inclusion & diversity	Emphasizing inclusion first, which will support a more diverse workforce
DEI	Diversity, equity, inclusion	Adding equity demonstrates a commitment to closing systemic gaps among historically excluded groups
DEIB	Diversity, equity, inclusion, belonging	Belonging emphasizes that the outcome of DEI efforts is for everyone to feel respected, valued, seen, and heard
IDEA	Inclusion, diversity, equity, access/anti-racism	Access typically refers to making spaces and resources more easily available to vulnerable groups; Anti-racism refers to the active effort of dismantling internalized and systemic racism, rather than simply not discriminating against another person because of their race
JEDI	Justice, equity, diversity, inclusion	Justice highlights fixing systems that produce privilege, oppression, and injustices
DEISJ	Diversity, equity, inclusion, social justice	Social justice is another variation of justice, with an emphasis on social issues
REDI	Race, ethnicity, diversity, inclusion	Race and ethnicity highlight the social construction of the terms and how societal structures implicate injustices against certain races/ethnicities

Note. This is not an exhaustive list of all DEI variations. These acronyms have evolved over time and will likely continue to do so.

benefits of a diverse workforce. And in the last 5–10 years, more organizations have begun to adopt the term "equity," recognizing the importance of updating policies and rebuilding systems that allow inequities between groups to persist.

In recent years, additional terms like "A" for accessibility, or "SJ" for social justice have been added to the DEI mix. In practice, however, it does not matter which acronym is used, but rather whether the specific terms provide a shared meaning, purpose, and direction for DEI as understood by everyone in the organization. Acronyms have also changed over time as organizations re-focus their DEI efforts. Language should evolve as strategy evolves, like how American Eagle Outfitters made the change from I&D to IDEA after deciding to double-down on creating more equitable practices and policies, as well as making accessibility a bigger priority for their ongoing DEI efforts.

The overarching goal of DEI is to build a more diverse workforce that provides fair treatment of all employees while holding its leaders and employees accountable for behaving inclusively, especially towards those from underrepresented and historically disadvantaged groups. The Society for Human Resource Management—the leading group of experts and thought leaders on issues impacting the workplace—asserts that data is a key player in successfully achieving DEI outcomes (Gurchiek 2021). DEI data include what can be counted, tracked, and evaluated against other key variables (e.g., retention, innovation) to determine who the people are, how they are being treated, and what systemic barriers are preventing certain identity groups from succeeding compared to others. Before jumping right into DEI data and measurement, though, it is crucial to first establish construct clarity—which means defining precise distinctions between different concepts (Suddaby 2010)—to ensure that there is a shared understanding of what these terms are, and therefore, how they should be quantified and qualified. Below are core definitions of diversity, equity, and inclusion, as well as how these constructs should be measured and enforced over time.

Diversity is Who We Are

Diversity is the presence and amount of difference among a group within a given setting. Diversity is how people see (or perceive) each other based on assumed or known aspects of a person's identity (Mattingly et al. 2022). It is important to note that diversity can either be surface-level, which is what a person can see on the outside, or deep-level, which includes less visible forms of identity such as an individual's beliefs, personality, or functional or educational background (Harrison et al. 1998).

While some of the most common demographic variables measured in the U.S. are surface-level identities (i.e., race/ethnicity, gender, age; Hayes et al. 2021), many organizations have also begun measuring deep-level identities with more novel and nuanced demographic variables (e.g., sexual orientation, caregiver status, and neurodivergence), accounting for a wider range of identity. Organizations have also started to focus on intersectionality, or "the interconnected nature of social categorizations such as race, class, and gender, regarded as creating overlapping and interdependent systems of discrimination or disadvantage" (Mattingly et al. 2022, p. 12). Table 2 displays a comprehensive list of common demographic variables used in DEI work, as recommended by the Society of Industrial Organizational Psychology (SIOP 2021).

When it comes to measuring diversity, the data organizations are allowed to collect is dependent on the countries in which their workforce resides. In the U.S., the EEOC requires that companies over the size of 100 employees collect data about their workforce's race/ethnicity and sex (EEOC n.d.). While not a legal requirement, startup investors are beginning to understand the financial benefits and hold founders responsible for diversifying their teams, even very small ones (Gompers and Kovvali 2018). Other countries like France and Germany legally prohibit collecting some of the demographic variables commonplace in the U.S., such as race/ethnicity. Policies are predicted to shift to allow more demographic data to be collected as the DEI space continues to expand globally (Oltermann and Henley 2020).

In addition to the legality of collecting data about employees' identities, there are also ethical considerations to keep in mind. For example, in some countries (e.g., Somalia and Saudi Arabia) it is illegal, and even punishable by death, to be gay or lesbian, which would make it legally dangerous to collect this data and

Table 2. Common DEI demographics and global considerations.

Identity Group + Definition	**Response Options**
Age – the length of time a person has been alive, typically measured in number of years. To avoid invertedly identifying people by their specific age, it is best to measure via age brackets (see response options)	1. 18–24 2. 25–34 3. 35–44 4. 45–54 5. 55–64 6. 65+
Caregiver status – The extent to which an individual provides care for another person who is not able to perform critical tasks necessary for everyday functioning	1. A child/children 2. Elderly individual(s) 3. Individual(s) with special needs 4. I am not a caregiver
Disability – a physical or mental impairment that substantially impacts how an individual performs one or more major life activities	1. Yes 2. No
Visible disability – an impairment that can be easily noticed by others. This can include certain body movements or facial features	1. A sensory disability 2. A mobility disability 3. A temporary disability due to illness/ injury 4. A disability not listed above
Non-apparent disability – an impairment that is not easily identified by others because they affect how an individual thinks or interacts with others	1. A learning disability 2. A long-term medical illness 3. A mental health disorder 4. A disability not listed above
Gender – Socially constructed expectations regarding the behaviors of men, women, and non-binary individuals. One's psychological sense regarding their gender may or may not align with a person's sex assigned at birth	1. Woman 2. Man 3. Genderqueer, nonbinary, or genderfluid 4. Prefer to self-describe
Military status – The extent to which an individual served, or is currently serving, in the active military	1. Veteran 2. Actively Serving 3. Neither actively serving nor a veteran
Neurodivergence – The extent to which an individual's neurological function differs from what is considered typical or normal *Note*: Neurodiversity should not be categorized as a disability.	1. I am neurodivergent 2. I am not neurodivergent
Race/Ethnicity – Ethnicity is the categorization of people who share a cultural background, such as language or location. Race is a socially constructed classification system based on shared phenotypical characteristics (i.e., skin color).	1. American Indian or Alaska Native 2. Asian 3. Black or African American 4. Hispanic, Latino/a/é, or Spanish 5. Middle Eastern or North African 6. Native Hawaiian or Other Pacific Islander 7. White 8. Prefer to self-describe

Table 2 contd. ...

...Table 2 contd.

Identity Group + Definition	Response Options
Religion – an individual's religious or spiritual beliefs and practices, or the religious group to which an individual belongs to	1. Agnostic 2. Atheist 3. Buddhist 4. Christian 5. Hindu 6. Jewish 7. Muslim 8. Spiritual, but not religious 9. Prefer to self-describe
Sexual orientation – An individual's disposition regarding sexual, affectionate, or romantic attractions and experiences with men, women, nonbinary people, etc.	1. Asexual 2. Bisexual or pansexual 3. Gay or lesbian 4. Heterosexual or straight 5. Prefer to self-describe
Transgender – An individual whose gender identity does not conform with their sex assigned at birth	1. Yes 2. No 3. Prefer not to say

Note: When asking employees to self-identify, always provide a "choose not to respond" option. This allows participants who are uncomfortable sharing their identity for any reason (e.g., fear of judgement, repercussions, confidentiality of data) an alternative option.

"out" members of this vulnerable identity group to authorities (Reality Check Team 2021). Because of the variability between countries around what diversity data can be collected, global DEI efforts often focus on gender as a starting point, as it is a goal that generalizes across most countries with a relatively easy-to-understand goal toward parity (i.e., about 50% across all organizations and leadership levels).

Measuring Diversity by Representation

Diversity should always be understood at the group level, not at the individual-level; there is no such thing as a "diverse person" or "diverse candidate," but rather an individual from some underrepresented or historically excluded group. Diversity should be thought of in terms of representation, ensuring that the company's workforce demographics represent the people in the market, industry, and the population of geographic areas in which the organization operates (Mattingly et al. 2022). This conceptualization of diversity in terms of representation is inclusive of all demographics. The key is, though, ensuring the proportion of those demographics is similar to the broader populations. For this reason, DEI interventions often focus on advancing underrepresented groups—not to give an upper hand, but rather to ensure that everyone has access to the same opportunities regardless of their identity or background. A more representative workforce raises the bar across *all* identity groups, improving the overall competence of the entire workforce (Chamorro-Premuzic 2019), as well as the additional benefits of DEI discussed in the previous section.

Fostering Diversity

When building startup teams, it is important to prioritize diversity from the very beginning. This is especially important for founders and early executive teams as research shows that representation in senior leadership trickles down to more diverse applicant pools and new hires, as well as increases engagement among employees who see themselves in senior leaders, and thus, as having a place in the future of the company (Johnston et al. 2022, Kazmi et al. 2022).

Startup leaders are in a prime position to quickly pivot their recruitment and hiring strategy once they realize they are at risk of having a homogenous workforce. Consider where jobs are being posted for recruiting new employees. What is the demographic make-up of the different talent pools? Is the organization's hiring process structured so that there is a representative pool of applicants before moving on to the next stage of the interview process? And as discussed earlier, what is being done to retain top talent once they are selected — especially those who belong to underrepresented and historically disadvantaged groups?

However, a word of caution to avoid tokenism –hiring a person solely to prevent criticism and give the appearance of diversity (Britannica n.d.). Tokenism has negative impacts on underrepresented employees. They may develop imposter syndrome and doubt leadership's faith in their abilities, which can also lead to anxiety and pressure to overperform (Cowie et al. 2018). Additionally, members from overrepresented groups may resent their peers for being selected. Instead of focusing on hiring more people from a certain group, take a step back and examine if the organization's hiring practices are inclusive (e.g., does the role need to have a degree requirement? Are the job descriptions inclusive? Have managers been trained on how to avoid bias?).

The above strategies to build and sustain a more diverse workforce are only as good as the method by which they are enforced. That is why it is so important to hold people accountable to DEI-related policies, practices, and procedures—which is what equity is all about.

Equity is Closing Systemic Gaps between Identity Groups

Equity refers to fair treatment, equitable opportunity, accessible information, and resources for all, achieved by the "intentional rebalancing of power" across identity groups, especially for those who have been historically left behind (Mattingly et al. 2022, p. 14). Ranking high in this construct means an organization has policies, practices, and procedures in place that provide all individuals (especially those from underrepresented groups) the opportunity to thrive. This is not the same as equality; DEI practitioners use the widely used image below (Fig. 1) to convey the difference between equity and equality (Craig 2016).

Though widely used, the "people on boxes" image has been criticized as not sufficiently representing the full picture of equity (Craig 2016). The simplicity of the picture ignores the systemic and historical roots of inequities, misplacing the problem (and corresponding solutions) at individual-level, identity-specific

Figure 1. A visual representation of the difference between equality and equity.

Note: The left side of the picture represents equality, where each person gets one box. The right side of the picture represents equity, where everyone gets the resources they need. This photo has been adapted and redrawn by many.

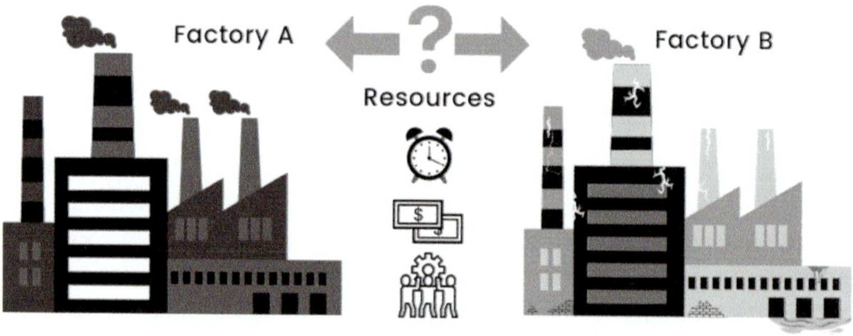

Figure 2. Visual representation of warehouses in need of different resources.

shortcomings (in this example, height). Instead of people on boxes, the concept could be better explained in terms of structural differences through the analogy of two manufacturing firms, depicted in the figure above (Mattingly et al. 2022).

Imagine a manufacturing company has two factories: Factory A and Factory B. Factory A is producing at a level of 90% effectiveness, while Factory B is at 50% effectiveness. There are $100,000 in resources that can be distributed amongst the factories to improve the company's margins specific to manufacturing productivity. To improve the company's overall productivity, it would be unwise to split these resources evenly between the two factories. It would be more advantageous to put more resources into Factory B where they are needed the most.

Applying this analogy to DEI in startups, consider the fact that women-founded startups only get a tiny sliver of VC investments (Houser and Kisska-Schulze In Press). The factories represent the two different worlds men and women founders live in, respectively—one (Factory A) that receives 97% of VC funding (male-led startup teams), and the other (Factory B) getting the remaining 3% (female-led startup teams). There is no research to suggest that men make inherently better founders. Quite the opposite: Women-founded startups generate up to 151% more revenue and sell their companies (or go public) much faster, and at higher values, than male-founded startups (Abouzahr et al. 2022). This means it is not women's fault they get less funding; the system (in this case, Factory B) was not built to enable women-led startups to get the funding they need to perform. The broken infrastructure represents the systemic barriers that impact the general success or failure of male and female employees within the distinct startup worlds, respectively. As a result of the factories being built in a way that benefits men, the men in Factory A have a higher production rate than the women in Factory B, not because of inherent ability tied to their gender, but because the system in which they are working systematically favors men over women. This disparity is why the women in Factory B need and deserve more resources: because they work in a system that was not built to meet their needs. This analogy is also backed by research. Studies have confirmed that the additional unpaid work women often complete, such as childcare, holds them back from moving forward in organizations (Cekala 2021)—an imbalance that has intensified throughout the COVID-19 pandemic (Del Boca et al. 2020). These types of systemic issues that impact groups differently are what must be examined and revised to reduce inequities in an organization.

Measuring Equity

When measuring equity, many organizations default to a pay equity analysis. This is because, for the most part, this is the easiest form of equity to measure and understand. At the most basic level, a pay equity analysis consists of reviewing the company's compensation data by identity, looking for disparities, and comparing it to other similar organizations (Nagele-Piazza 2020). This data can easily be found in the company's human resource information system (HRIS).

However, it is also important to go one step further and do a full organizational equity audit. Review policies, practices, and procedures by demographic and look for inequities. This may include recruitment practices, selection process, opportunities for development, promotion procedures, distribution of benefits, accommodations, and more. There is also power in perception. As the truism states, perception is reality. So, it is valuable to survey employees (as well as individuals that decide to leave the company) and gain insight into their perception of equity in the company. But be cautious. Once data is collected, it is critical to act on the findings. Collecting data without utilizing the data can lead to a decrease in employee trust (Shook et al. 2020).

Ensuring Equitable Practices. Fostering and enforcing equity is one in the same—policy change. Equity ensures that DEI is embedded and enforced throughout all organizational processes. As with any change in management efforts, when developing new, equitable policies, it is important to incorporate ways to hold people accountable for following the new guidelines. Set expectations and determine a process for checking in to see the impact of the new policies and procedures.

Inclusion Is What We Do

There are various conceptualizations of inclusion. While some researchers suggest inclusion is an emotional experience (Mor Barak 2015, Brimhall et al. 2014), others define inclusion as a behavior (e.g., Katz and Miller 2017). For this chapter, belongingness is defined as a *feeling* (i.e., affect) that one experiences as a consequence of others' inclusive *behaviors*. In other words, inclusion refers to the behaviors that make employees feel valued, respected, seen, and heard in their organization, on their team, and within their role (Mattingly et al. 2022).

Inclusive behaviors can be broken down into three levels based on the amount of effort required: everyday inclusion, inclusive leadership, and allyship (see Fig. 3). Everyday inclusion is the first level, and these are actions that are easy for leaders and employees to perform such as putting in the effort to pronounce someone's name correctly. The middle layer is inclusive leadership, which is all about the actions leaders take for those on their team. At the top, there is allyship, the gold standard of inclusive behaviors which includes working with a partner towards a shared goal of fairness, equity, and social justice. By defining inclusion as a behavior, inclusion becomes easier to measure objectively.

Figure 3. Three levels of inclusive behaviors.

Measuring Inclusion

When people hear inclusion, they tend to think it is a subjective concept, but by defining it as something a person does, it becomes an action that can be measured objectively. To go one step further, it should be measured through "other-report" data, not self-report. Self-report data would provide information about whether the *actor* of the reported behaviors perceives their behaviors are inclusive towards others. Other-report data illustrates the perception of the *recipient* of the behaviors, which is more powerful because it focuses more on the impact on the other person rather than on the intention of the actor. In other words, it should not matter how inclusive a founder or startup executive believes themself to be—what truly matters is how inclusive that founder's team, peers, or even customers find that founder to be.

There are several ways to capture objective inclusion data. One simple way is through a survey. Items (i.e., survey questions) would be directed to the recipient of the inclusive behaviors, and they should be asked in terms of frequency of the behavior, for example, "My manager uses inclusive language (e.g., gender-neutral words, pronouns)." Another valuable method is investigating observational data, such as reviewing meeting invites to see who is being included and excluded from various conversations. A formal organizational network analysis, technology that examines communication data to identify and understand how people interact in an organization, is a great way to dig deep into what silos exist in an organization and who is connected to who (Bento et al. 2020, Zielinski et al 2013).

Developing Inclusive Behaviors

Once it is clear who is being included and excluded, the focus can shift to training employees on how to be more inclusive. When developing training, it is important to consider the length of training, type of training, and time in between training, as all can affect on learning outcomes (Uslu et al. 2022). The key to a good DEI training program is that it is in fact a program, not a one-off training. Research shows that longer diversity training increases the probability of participants utilizing the skills and knowledge they learned on the job (Bezrukova et al. 2016). However, short training, such as "DEI moments" at the start of meetings or micro-learning, can be beneficial as long as they are recurring over time. What matters is determining what training structure best fits the company's culture, developing a clear plan forward, and of course, tying in accountability.

After developing a robust training program and its subsequent roll out plan, the next step is determining accountability metrics and impact measures. A training program is only useful if employees will be held accountable to use the skills they learned. An example of an accountability metric would be assigning a senior leader to ensure a certain completion rate (100% for mandatory training) of the employees within the function they lead. An example of an impact measure of inclusive leadership training would be asking leaders' subordinates to rate if their frequency of using the trained behaviors has increased compared to before the training was delivered. DEI training outcomes (e.g., behavior change and more positive sentiments, especially from those from underrepresented groups) can be also incorporated into the performance appraisal process and even tied into compensation.

It is also important to analyze the impact of any DEI training program. If the goal was knowledge-based, did participants learn the expected content? If it was centered around behavior change, was there an increase in desired behaviors post-training? Which behaviors have the biggest impact on making everyone—especially those from underrepresented and historically disadvantaged groups—feel like valued, respected, seen, and heard members of the startup community? If the expected impact is not occurring, how can the organization revisit the design of the training program to improve future results? Research has shown that all of these questions can be addressed by conducting a formal program evaluation, a process that should be used in any data-driven workplace training program (Kraiger et al. 1993, Alvarez et al. 2004).

How to Practice Data-Driven DEI

In today's digital-first workplace, data can—and should be—used to make all business-related decisions. Between the years of 2005 and 2010, the amount of data across the globe became 9 times larger (Gantz and Reinsel 2011), and has been ever growing (Blei 2012, Grimmer 2015). Not only has the creation of data increased, but the speed at which one can explore this data has, too (Shet et al. 2021). Successful startup leaders consistently use data to test hypotheses, validate strategic decisions, and pivot as needed to help grow their businesses (Ries 2011). If the best startups thrive on using data to guide their work, why should DEI be any different?

When it comes to making people and talent decisions, organizational leaders should use data that captures employees' voices, needs, and behaviors to understand the people within their startup, as well as the people their startup serves. Data sources could include meta-data from internal employee communication platforms (e.g., scraping data from Slack or Outlook to see who is being included—and excluded—in conversations and meetings), as well as performance, customer feedback, and employee review data (e.g., Glassdoor). In cases where it makes more sense to collect employee voice and behavior data rather than pull from archival data sources, it is crucial to building a data collection and management plan that protects the rights and confidentiality of the people submitting those data, which leads to ethical considerations.

Ethical Considerations

Before collecting DEI metrics and using the analyzed data to build more diverse, equitable, and inclusive startups, it is crucial to prioritize the safety and well-being of the employees who are being asked to share personal data. The sheer volume of data available to organizations leave many in the ethical gray area of privacy and confidentiality. To best navigate the grey area of DEI data collection, consider relying on the following best practices on professional ethical standards from a related field, such as psychology or human resources (American Psychological Association 2017, Society of Human Resource Management 2014).

In addition to the ethics of data collection, there are legal implications to consider. The number of employee lawsuits has increased over the years, with more cases being tied to how personal data is being managed and used (Tursunbayeva et al. 2021, Fernandez-Campbell 2018). Startups need to consider how they handle data and data protection to ensure employee trust, and privacy, and reduce the risk of litigation.

Accessibility and Privacy

Data related risks can occur in a variety of ways. One concern around data privacy is who has access to the data. Personal data can contain sensitive information that employees are uncomfortable having others know. One stark example in the DEI space is sexual orientation. If an employee willingly self-reports that they are gay, and the data are not protected in a way to maintain privacy and confidentiality of the employee's self-report, there is a risk of the employee being outed as gay without their consent.

Transparency and Consent

Another concern is the ethics behind using data from 'public' locations. It is not uncommon for recruiters to use social media to evaluate job candidates, which blurs the line between one's personal and professional identities (e.g., Wong 2021). Company emails, browser history, and other data sources could also be considered 'fair game' for employers to access. Some organizations may even consider scraping data from internal employee communication platforms, such as Slack, to see what topics were most commonly discussed among a top performing team, yet, it is very important to consider the possible legal and ethical controversy surrounding the use of that data without explicit consent from the team members or if the use of data goes against the platforms' terms of agreement (Krotov et al. 2020). If employees do not know their data was being collected and used, they may distrust and avoid using the technologies at work, and even potentially sue the company. This could be a crucial error financially and image-wise for any company as people have grown to be more critical of organizational misconduct (Rivera and Karlsson 2017).

Trained Experts

A well-designed DEI strategy can fail if data protections are not considered upfront. Since a startup may not have the resources for a full HR department, take a step back and consider who is responsible for the startup's DEI strategy and goals and whether they are knowledgeable on data management best practices. Ideally, this person would have advanced training in behavioral statistics, qualitative data, and data science communication and understand the nuances that come with small sample sizes. If there is not someone inhouse, consider bringing in support from outside the organization as needed. Industrial-organizational psychologists[2] are particularly well-suited to turn abstract concepts like inclusion into measurable constructs that can then be tied to other business metrics and outcomes. Whether the person in charge of data management is internal or external, transparency regarding data processes will help build trust in the processes and the conclusions drawn (Lawton 2021). Once the person overseeing the analyses is determined, the focus can shift to what data will be collected.

Data Collection

Data is typically broken down into two broad categories, quantitative (numbers-based) and qualitative (text and verbal data). It is not uncommon for DEI experts to start with quantitative data to identify inequities or significant gaps between how various identity groups are treated and feel. Once the experts know *what* is happening, then they can use qualitative data collection methods (e.g., focus groups) to assess *why* they found the quantitative findings they did and, more importantly, what to do next to close those previously identified gaps.

---

[2] Learn more about Industrial-Organizational psychology at www.siop.org or www.apa.org/ed/graduate/specialize/industrial.

Quantitative DEI Data Methods

Quantitative data is data that can be numerically quantified, such as averages, frequencies, percentages, and other statistical analyses. Quantitative data can come from a variety of sources such as representation data from a Human Resource Information System (HRIS), surveys, or the other-report data discussed previously. Specific to representation data, a good starting point for making diversity goals based on representation is by benchmarking demographics to the labor market and/or the geographic locations in which a startup's people reside. A simple comparison of the proportions of those in the organization and those reported to be in the workforce can illuminate the biggest gaps in terms of representation and set data-driven diversity goals for the future. Survey data and other behavioral feedback tools can be especially useful in assessing feelings of belongingness and the frequency with which inclusive behaviors are used, both key outcomes of DEI efforts.

A downside of using quantitative data methods in early-stage startups is that there is often too low of a sample size to run any meaningful statistical analyses. Overreliance on quantitative data may have side effects that risk confirming biases (e.g., asking leading questions or excluding certain viewpoints) and limit one's ability to understand employee experiences (Beckle et al. 2022). If scientists and practitioners only chase the "truth" found in traditional quantitative methodologies and measures, they may miss important aspects of what is happening within an organization that the numerical data does not capture. When studying startups with less than 50 employees, researchers are encouraged to lean into qualitative data collection strategies, which are detailed below.

Qualitative DEI Data Methods

Qualitative data is text or audio content that cannot be easily understood using numbers. Qualitative methods include focus groups, interviews, or even open-ended survey questions. Compared to quantitative data, qualitative methods often yield richer datasets that can be inductive and enable employees to share their experience beyond the limits of a survey item. For example, imagine a 15-person startup that has two working parents as employees. A quantitative survey was sent to all employees to explore how work schedules and flextime impact productivity. There was a question that asked if employees were satisfied with the organization's flextime policy. The average score for the item was a 92%, which at a surface level seems great. However, due to there are only being two working parents, looking at that overall score does not give insight into what parents may be experiencing. And in line with the ethics conversation, it is unethical to report the average score for working parents because there are only two and they could be easily identified as a result. Rather than simply ignoring the unique needs of working parents by not reporting their quantitative data, consider conducting a focus group with these individuals so they can provide confidential feedback about how the flextime policy could be updated to better support working parents. Without using this qualitative data collection technique, the organization would be doing a disservice to the sub-population of working parents who have much to benefit from a properly written and implemented flextime policy. A deeper look at the common forms of qualitative data collection is below.

Interviews. Interviews are one-on-one conversations used to gain insights into a person's beliefs, motivations, lived experiences, and perspectives about a specific topic. Speaking to people one-on-one allows them to speak transparently without fear of repercussions if they trust their interviewer. Having an external, third-party facilitator conduct the interviews may also help garner trust with participants (Wilkie 2018).

Interviews can also be useful if an organization only has a few employees. Interviews are the best data collection method for startups with less than 10 total employees. Example questions can include, "What are some behaviors that your direct supervisor engages in that helps make you feel like you belong here," and "Can you name some policy changes that you would like to see in the organization?" Making sure to probe further after each question helps provide even greater clarity into how employees feel about working at the startup.

Because of the labor-intensive nature of conducting interviews (and associated costs), though, researchers and consultants typically reserve this data collection method for the executive level. It is a chance to find out the priorities of senior leadership and what direction they want to go with the organization. Such executive interviews can also provide an opportunity for external consultants to build rapport with the executives, potentially leading to greater buy-in for DEI from the top leaders.

When it comes to the scientific integrity of collecting interview data, a structured interview process is recommended. Structured interviews standardize the questions asked and scoring protocols (e.g., rubrics) used to evaluate interviewees. Using a structured approach to interviews increases the validity and reliability of interviews while also reducing socio-cognitive biases and decreasing the differences between demographic groups (Mattingly et al. 2022, Woo et al. 2020). When analyzing interview data, general themes across all interviews can be extracted, as well as examples or stories that can be used to illustrate broader findings.

Focus Groups. Focus groups are a way to orally gather qualitative information from multiple employees at once, typically in groups of six to eight. Sometimes focus groups are developed based on identities such as gender identity, race, job level, or department. Often for startups and small organizations, it may be as simple as holding a few focus groups to ensure all employees can participate. During the focus group, the facilitator comes in with a set of questions – possibly driven by previously collected quantitative data – and asks for feedback from each participant. Participants can share their thoughts and build on each others' feedback as well. Similar to interviews, focus groups yield the best data when facilitated by a third party as it increases trust in confidentiality in employees.

Open-Ended Survey Items. While interviews and focus groups are ways to collect qualitative data through conversations, open-ended items are a way to collect qualitative data through writing. Open-ended items are useful for collecting a large amount of data quickly related to one or two questions. These items can be incorporated into longer surveys, pulse surveys, or sent out on their own.

Statistical Techniques

After collecting qualitative and quantitative data, the next step is analyzing that data. Statistics used to interpret DEI data can range from simple (e.g., means) to more complex (e.g., multi-level structural equation modeling, see Kozlowski and Klein 2000). Statistics can also be used to analyze text from sources like uploaded cover letters and job descriptions (e.g., Blei 2012). This chapter does not cover all areas of data, metrics, or statistical analyses related to DEI, but rather some high-level guidance and examples of how other startups put DEI data into action.[3]

After collecting data, the first step is to "clean" the data. This includes looking for missing data, duplicate entries, and other general preparations to prepare the data for analysis. After cleaning the data, it is time to run the statistical analyses on the quantitative data. Oftentimes startups focus on simple metrics such as averages or correlations due to the small sample size. Conducting more complex significance testing or multivariate testing can prove difficult when working with a small sample size. For instance, when trying to examine if a startup is engaging in adverse impact, which refers to different employment decision outcomes based on group differences, significance testing can be inaccurate or misleading because of such low sample sizes (Collins and Morris 2008).

It can be challenging to handle qualitative data like interview transcripts due to the lack of consistent best practices that are more common in quantitative data. Further, qualitative data can take a lot of time to read through and may be hard to make sense of if one is not familiar with thematic analysis, coding, and other qualitative fundamentals. Text mining approaches, like topic modeling, can be accomplished using free statistical software programs (e.g., R/R Studio) and assist in data-driven decision making. One of the most common qualitative data analysis methods used is thematic analysis (Kiger and Varpio 2020). This consists of reviewing the data for common themes. An illustration of how topic modelling can be used to make sense of employee voice data to advance DEI efforts can be found in the first case study presented below. The second case study provides an inspirational example of startups using data to advance DEI at the industry-level in the greater Baltimore area.

Case Study #1: Amplify Employee Voices with Topic Modelling

The Chief People Officer (CPO) for a 30-person tech startup called VSCJA was asked by their board of directors—mostly made up of key investors—to provide a progress report, including measurable outcomes, of their DEI efforts for the next board meeting. The CPO began with collecting and summarizing their diversity data. However, the CPO also realized there were several demographics that the company did not collect data for, so they noted the missing demographics in their summary

[3] For more on specific statistical techniques, variable types, and data visualization, see *The Practical Guide to HR Analytics*, Waters et al. 2018; for more on DEI specific data and metrics see *Inclusalytics*, Mattingly et al. 2022.

as an opportunity for them to add to their HRIS. VSCJA had not yet conducted an annual employee survey, so the CPO decided to collect qualitative data about inclusive behaviors and feelings of belonging. To do this, they sent the whole-company a one-item survey, asking "What behaviors could your fellow employees use to make you feel like a valued and respected person at this organization?" The survey closed with an 85% response rate—which is particularly impressive given that the typical workforce survey response rate is 77% (Workforce Science Associates n.d.).

Though getting such a high response rate from their employees was considered a success, collecting the data was only half the battle; the data had to be analyzed. The CPO decided to use a method called topic modeling to analyze the data (see Chapter 9 for additional methods to analyze qualitative data). Topic modeling can take unstructured data, like the typed responses to the open-ended survey question sent out by VSCJA, and propose several topics across all of the documents (i.e., open-ended responses).

Similar to model fit statistics in quantitative data, topic modeling goes through a process of exploring model fit metrics, such as semantic coherence (i.e., DEFINE), that provides decision cues about how many topics are in the data set. Topic modeling in R allows the use of 'stop words,' or words that should not be considered by the topic modeling process when searching for semantic connections in the dataset. Some common stop words are "um," "like," "so," "and," and so on; they are words that do not provide much meaning. Sometimes, it may be helpful to consider if some of the words from the item prompt (e.g., open-ended question) should be considered as stop words as well. For example, if most respondents started their written response with repetition or regurgitation of the item words, that does not provide much information on the core meaning of their responses. If that is the case, the words of the item text can be added to the list of stop words.

Once the CPO considered all stop words, they were ready to explore the model fit metrics. Although there are other metrics to consider when determining how many topics are in the dataset, the CPO focused on semantic coherence. In their R script, they requested a graph of what it would look like for a dataset that had 1 to 20 possible topics (see Fig. 4). To determine how many topics are present in the dataset, the CPO searched for the highest point in the graph, which estimated how many topics best fit the data. As the figure shows, there were about three to four possible topics that fit the data the best.

After examining the semantic coherence and other fit metrics, the CPO explored what those three and four topics entailed. The analysis provided the CPO with the top words for each topic, and the top documents (employee responses) that represented each topic. The CPO also examined a word cloud type visual to see how distinct words were between two topics. This supervised machine learning technique allowed for the development of theoretically grounded meaning for each topic. Through examining the topics, documents, and word cloud, the CPO developed a table to synthesize the findings from the topic modeling process (see Table 3). Overall, topic modeling can assist in wrangling large amounts of text data, help provide

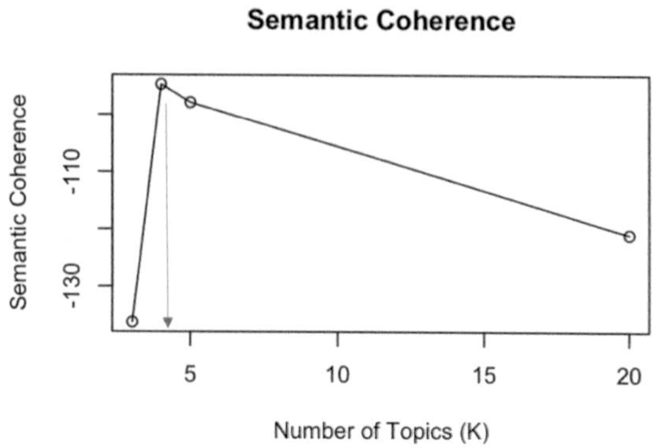

Figure 4. Graph of semantic coherence of open-ended items.

Table 3. Example of table synthesizing topic modeling data.

Topic & Top Words	Topic Example Text	Assigned Meaning by CPO
Topic 1: Measure, Data, Metrics, Account, Track	I think that this company wants to do well in DEI, but I am not sure we are tracking any diversity-related metrics to help hold us accountable. Metrics should be used to see where we are at now and where we are going. I …	*DEI data and measurement will demonstrate current DEI standing and help clarify goals*
Topic 2: Good, Happy, Fair, Unnecessary, Equal	I think my workplace has a good handle on DEI. I don't think we need to focus on making things more equal, I think they are already really fair. I am happy working here. It may be unnecessary to focus on DEI…	*Feels that workplace is equal and fair, happy with current DEI maturity*
Topic 3: Need, Include, Representation, Hiring, Recruitment	There is a lot of room for improvement here, especially in the space of recruiting and hiring BIPOC. We need to be better at including people outside of the majority group and representing them at higher levels in leadership…	*Specific call for DEI improvement when it comes to diversity and representation of underrepresented identity groups across all leadership levels*

clarity to the structure of the data, and provide a statistical approach to qualitative data, essentially providing DEI leaders with greater opportunities of amplifying employee voices. For example, to respond to the theme of enhancing "*diversity and representation of underrepresented identity groups across all leadership levels*," the CPO could advocate for funding to build and maintain a robust leadership development program focused on identifying high potential, diverse talent early on and providing them with the training, sponsorship, and opportunities to move up through the leadership ranks over time. As the workforce and leadership diversify over time, this organization can expect the financial benefits consistently experienced by more diverse organizations compared to more homogeneous competitors (e.g., Dixon-Fyle 2020).

Case Study #2: Using Data to Drive Industry-Level Startup Diversification

Despite Baltimore being 57.8% Black (U.S. Census Data 2020), this population only makes up 19.8% of the Baltimore startup ecosystem. Baltimore Tracks is a membership organization for tech startups who aspire to bring together leaders from local technology companies to commit to building a diverse workforce by facilitating dialogue, sharing resources and best practices, and taking action to ensure equitable and inclusive policies and practices at their respective companies. Mattingly Solutions, a DEI consulting firm, partnered with Baltimore Tracks to help them with this mission.

Mattingly Solutions consulted with Baltimore Tracks, guiding their coalition members toward a more strategic approach by developing a custom DEI survey to help them determine what their internal needs are in order to be more diverse, equitable, and inclusive. They then gave them support on how to interpret their survey findings and use them to develop a data-driven roadmap on how to move forward in an agile way. After conducting the survey, Mattingly Solutions helped a number of Baltimore Tracks' member companies develop reports summarizing their findings and identifying the next steps. They also assisted Baltimore Tracks in creating a holistic report, summarizing the results across all 28 of their member organizations that participated in the process. This report was designed to guide their mission forward.

Baltimore Tracks is using data to help their member organizations improve racial equity, which comes back to the core purpose of using data in DEI work: to drive meaningful change. As steering committee member Michael Castagnola said, "Our business isn't to force folks to do stuff, but to set the table in a way that companies will be able to take action…They'll get the data, have a way to aggregate and present it. That presentation should lead to opportunities to take next steps" (Baltimore Tracks 2022). Without following-through with action, there is no purpose in collecting data.

Conclusion

This chapter covered the why, what, and how startup leaders should use data to make decisions that result in more diverse, equitable, and inclusive organizations. Simply put, DEI is a set of variables that tells researchers and practitioners about the employee experience and, if leveraged correctly, builds their confidence in their decision-making. Meta-analytic evidence suggests that scientific evidence is more accurate than expert opinions (Antman et al. 1992). Therefore, take a scientist-practitioner approach to dig deeper into the relationships among these variables: explore the cumulative evidence demonstrating the positive impact of equitable decision-making, demographic representation, and inclusive practices.

Building DEI into the fabric of a startup from the very beginning will enable the organization to capitalize on the many benefits of DEI. The first step is to establish why DEI matters, connecting it to the startup's overall mission, vision, and values.

Next determine what information and actions are needed to accomplish those goals, measuring progress along the way. Remember, taking a data-driven approach to DEI can be simple, but it should be robust. Do not try to reinvent the wheel; research what is working for other startups and use resources and best practices recommended by experts. But most importantly, startups should push past the fear and anxiety of taking a data-driven approach to DEI and get started while they still have the agility and speed to build a more diverse, equitable, and inclusive organization.

Acknowledgements

This chapter wouldn't have been completed without the additional support provided to the authors along the way. First, thank you to Nikki Blacksmith for your insightful guidance and encouragement as our editor. We'd also like to thank Julie Chen and Camaria Lehman for your support with the chapter itself and for giving us the space to contribute to the diversity, equity, and inclusion (DEI) literature. Lastly, we'd each like to thank our family and friends for providing the emotional and motivational support sometimes necessary when wearing the many hats of DEI scientist-practitioners.

References

Abouzahr, K., Krentz, M., Harthorne, J. and Taplett, F.B. (2022, August 9). *Why Women-owned Startups are a Better Bet*. Retrieved from https://www.bcg.com/publications/2018/why-women-owned-startups-are-better-bet.

Alvarez, K., Salas, E. and Garofano, C.M. (2004). An integrated model of training evaluation and effectiveness. *Human Resource Development Review*, 3(4): 385–416. https://doi.org/10.1177/1534484304270820.

American Psychological Association. (2017). *Ethical Principles of Psychologists and Code of Conduct*. https://www.apa.org/ethics/code/index.

Antman, E.M., Lau, J., Kupelnick, B., Mosteller, F. and Chalmers, T.C. (1992). A comparison of results of meta-analyses of randomized control trials and recommendations of clinical experts: Treatments for myocardial infarction. *Jama*, 268(2): 240–248.

Baltimore Tracks. (2022). *Baltimore Tracks is Setting the Table for DEI Work in the Local Tech Community*. https://james414384.wixsite.com/baltimoretracks/post/baltimore-tracks-is-setting-the-table-for-dei-work-in-the-local-tech-community.

Beckel, J.L., Gardner, D.M. and Prasad, J.J. (2022). Decolonizing intervention assessment: Qualitative and interdisciplinary approaches to understanding "side effects". *Industrial and Organizational Psychology*, 15(1): 113–116.

Bento, F., Tagliabue, M. and Lorenzo, F. (2020). Organizational silos: A scoping review informed by a behavioral perspective on systems and networks. *Societies*, 10(3): 56.

BetterUp. (2019). *The Value of Belonging at Work: New Frontiers for Inclusion in 2021 and Beyond*. https://grow.betterup.com/resources/the-value-of-belonging-at-work-the-business-case-for-investing-in-workplace-inclusion.

Bezrukova, K., Spell, C.S., Perry, J.L. and Jehn, K.A. (2016). A meta-analytical integration of over 40 years of research on diversity training evaluation. *Psychological Bulletin*, 142(11): 1227–1274.

Billington, F. (2021). *The Vast Majority of Venture Dollars Go to White Male Founders, Report Finds*. Dot.La. https://dot.la/venture-capital-2650047652.html.

Blackwell, A.G., Kramer, M., Vaidyanathan, L., Iyer, L. and Kirschenbaum, J. (2017). *The Competitive Advantage of Racial Equity*. FSG. https://missioninvestors.org/sites/default/files/resources/The%20Competitive%20Advantage%20of%20Racial%20Equity.pdf.

Blanche, A. (2022). *DEI in 2022: Key Trends and Findings*. (2022). Culture Amp. https://www.cultureamp.com/blog/dei-2022-trends.

Blei, D.M. (2012). Probabilistic topic models. *Communications of the ACM*, 55: 77–84.

Brimhall, K.C., Lizano, E.L. and Barak, M.E.M. (2014). The mediating role of inclusion: A longitudinal study of the effects of leader–member exchange and diversity climate on job satisfaction and intention to leave among child welfare workers. *Children and Youth Services Review*, 40: 79–88.

Britannica. (n.d.). Tokenism. In *The Britannica Dictionary*. https://www.britannica.com/dictionary/tokenism.

Carr, E.W., Reece, A., Kellerman, G.R. and Robichaux, A. (2019). *The Value of Belonging at Work*. Harvard Business Review. https://hbr.org/2019/12/the-value-of-belonging-at-work.

Cassells, R. and Duncan, A.S. (2020). *Gender Equity Insights 2020: Delivering the Business Outcomes (No. GE05)*. Bankwest Curtin Economics Centre (BCEC), Curtin Business School.

Cekala, W. (2021). *W@W Survey: Child Care, Elder Care Top Demands for Working Women*. LiveCareer. https://www.livecareer.com/resources/special-reports/women-at-work/empowering-women-at-work.

Chamorro-Premuzic, T. (2019). *Why do so Many Incompetent Men become Leaders? (And How to Fix it)*. Harvard Business Press.

Chang, E. (2019). *Brotopia: Breaking up the Boys' Club of Silicon Valley*. Penguin.

Collins, M.W. and Morris, S.B. (2008). Testing for adverse impact when sample size is small. *Journal of Applied Psychology*, 93(2): 463–471.

Cowie, M.E., Nealis, L.J., Sherry, S.B., Hewitt, P.L. and Flett, G.L. (2018). Perfectionism and academic difficulties in graduate students: Testing incremental prediction and gender moderation. *Personality and Individual Differences*, 123: 223–228.

Craig, F. (2016). *The Evolution of an Accidental Meme*. Medium. https://medium.com/@CRA1G/the-evolution-of-an -accidental-meme-ddc4e139e0e4.

Dauth, T., Schmid, S., Baldermann, S. and Orban, F. (2021). Attracting talent through diversity at the top: The impact of TMT diversity and firms' efforts to promote diversity on employer attractiveness. *European Management Journal*.

Del Boca, D., Oggero, N., Profeta, P. and Rossi, M. (2020). Women's and men's work, housework and childcare, before and during COVID-19. *Review of Economics of the Household*, 18(4): 1001–1017.

Deloitte. (2017). *Inclusion Pulse Survey*. https://www2.deloitte.com/content/dam/Deloitte/us/Documents/about-deloitte/us-about-deloitte-inclusion-survey.pdf.

Dixon-Fyle, S., Dolan, K., Hunt, V. and Prince, S. (2020). *Diversity Wins: How Inclusion Matters*. McKinsey & Company. https://www.mckinsey.com/featured-insights/diversity-and-inclusion/diversity-wins-how-inclusion-matters.

EEOC. (n.d.). *Small Business Requirements*. https://www.eeoc.gov/employers/small-business/small-business-requirements#:~:text=If%20you%20have%20100%20or%20more%20employees%2C%20or,and%20the%20U.S.%20Department%20of%20Labor%20every%20year.

Ely, R.J., Ibarra, H. and Kolb, D.M. (2011). Taking gender into account: Theory and design for women's leadership development programs. *Academy of Management Learning & Education*, 10: 474–493.

Fernandes, N. (2021). *Belonging: The Intersection of DEI and Engagement*. Forbes. https://www.forbes.com/sites/forbeshumanresourcescouncil/2021/12/22/belonging-the-intersection-of-dei-and-engagement/?sh=5848059913e9.

Fernandez-Campbell, A. (2018). *Facebook, Amazon, and Hundreds of Companies Post Targeted Job ads that Screen Out Older Workers*. Vox. https://www.vox.com/policy-and-politics/2018/5/31/17408884/facebook-amazon-job-ads-age-discrimination-lawsuit.

Gantz, J. and Reinsel, D. (2011). Extracting value from chaos. *IDC Review*, 1142(2011): 1–12.

Gaudiano, P. (2020). *How Inclusion Improves Diversity and Company Performance*. Forbes. https://www.forbes.com/sites/paologaudiano/2020/07/13/how-inclusion-improves-diversity-and-company-performance/?sh=397ac37b56a6.

Gompers, P. and Kovvali, S. (2018, July 9). *Finally, Evidence that Diversity Improves Financial Performance*. Harvard Business Review. Retrieved from https://hbr.org/2018/07/the-other-diversity-dividend.

Grimmer, J. (2015). We are all social scientists now: How big data, machine learning, and causal inference work together. *PS: Political Science & Politics*, 48: 80–83.

Gurchiek, K. (2021). *Regularly Monitoring, using DE&I Data is Key to Closing Organizational Gaps*. Society for Human Resource Management. https://www.shrm.org/resourcesandtools/hr-topics/behavioral-competencies/global-and-cultural-effectiveness/pages/regularly-monitoring-using-dei-data-is-key-to-closing-organizational-gaps-.aspx.

Harrison, D.A., Price, K.H. and Bell, M.P. (1998). Beyond relational demography: Time and the effects of surface-and deep-level diversity on work group cohesion. *Academy of Management Journal*, 41(1): 96–107.

Hayes, A., Potters, C. and Beer, K. (2021). *Demographics*. Investopedia. https://www.investopedia.com/terms/d/demographics.asp#:~:text=The%20common%20variables%20gathered%20in,homeownership%2C%20and%20level%20of%20education.

Houser, K. and Kisska-Schulze, K. (in press). Disrupting venture capital: Carrots, sticks and artificial intelligence. *UC Irvine Law Review*.

Hunt, V., Layton, D. and Prince, S. (2015). *Why Diversity Matters*. McKinsey & Company. https://www.mckinsey.com/business-functions/people-and-organizational-performance/our-insights/why-diversity-matters.

Johnston, Z., Johnson, L.U., Bertram, D., Gafa, K., Maneethai, D., Obasare, R. and Khatib, Z. (2022). Representation does matter: indirect effects of representation on engagement [Poster]. *Society for Industrial and Organizational Psychology Annual Conference*, Seattle, WA, United States.

Katz, J.H. and Miller, F.A. (2017). Leveraging differences and inclusion pays off. *Organizational and Individual Change*, 49(1): 56.

Kaul, A. (2022). Of fruit flies, toads, and other hopeful monsters: thoughts on Levinthal's Evolutionary Processes and Organizational Adaptation. *Journal of Organization Design*, 1–4.

Kazmi, M.A., Samaniego, C., Jeff-Eke, E., Tellez, E., Barnes, D. and Spitzmueller, C. (2022). *Representation Matters. The Trickle-Down Effect of Underrepresented Minority Faculty* [Poster]. Society for Industrial and Organizational Psychology Annual Conference, Seattle, WA, United States.

Kelly, E. and Dobbin, F. (1998). How affirmative action became diversity management: Employer response to antidiscrimination law, 1961 to 1996. *American Behavioral Scientist*, 41(7): 960–984.

Kiger, M.E. and Varpio, L. (2020). Thematic analysis of qualitative data: AMEE Guide No. 131. *Medical Teacher*, 42(8): 846–854.

Kozlowski, S.W. and Klein, K.J. (2000). A multilevel approach to theory and research in organizations: Contextual, temporal, and emergent processes. pp. 3–90. *In*: K.J. Klein and S.W.J. Kozlowski (eds.). *Multilevel Theory, Research, and Methods in Organizations: Foundations, Extensions, and New Directions*. Jossey-Bass.

Kraiger, K., Ford, J.K. and Salas, E. (1993). Application of cognitive, skill-based, and affective theories of learning outcomes to new methods of training evaluation. *Journal of Applied Psychology*, 78(2): 311–328. https://doi.org/10.1037/0021-9010.78.2.311.

Krotov, V., Johnson, L. and Silva, L. (2020). Tutorial: Legality and ethics of web scraping. *Communications of the Association for Information Systems*, 47.

Lawton, G. (2021). *The Future of Trust will be Built on Data Transparency*. TechTarget. https://www.techtarget.com/searchcio/feature/The-future-of-trust-must-be-built-on-data-transparency.

Madera, J.M., Dawson, M. and Neal, J.A. (2018). Why investing in diversity management matters: Organizational attraction and person–organization fit. *Journal of Hospitality & Tourism Research*, 42(6): 931–959.

Mattingly, V., Grice, S. and Goldstein, A. (2022). *Inclusalytics: How Diversity, Equity, and Inclusion Leaders Use Data to Drive their Work*. Mattingly Solutions.

McFeely, S. and Wiger, B. (2019). *This Fixable Problem Costs U.S Businesses $1 Trillion Per Year*. Gallup. https://www.gallup.com/workplace/247391/fixable-problem-costs-businesses-trillion.aspx.

Mor Barak, M.E. (2015). Inclusion is the key to diversity management, but what is inclusion? *Human Service Organizations: Management, Leadership & Governance*, 39(2): 83–88.

Nagele-Piazza, L. (2020, February 28). *The Importance of Pay Equity*. SHRM. Retrieved from https://www.shrm.org/hr-today/news/hr-magazine/spring2020/pages/importance-of-pay-equity.aspx.

Oltermann, P. and Henley, J. (2020). *France and Germany Urged to Rethink Reluctance to Gather Ethnicity Data*. The Guardian. https://www.theguardian.com/world/2020/jun/16/france-and-germany-urged-to-rethink-reluctance-to-gather-ethnicity-data.

Paternoster, N., Giardino, C., Unterkalmsteiner, M., Gorschek, T. and Abrahamsson, P. (2014). Software development in startup companies: A systematic mapping study. *Information and Software Technology*, 56(10): 1200–1218.

Principles for Responsible Investment. (2022). *Diversity, Equity and Inclusion: Key Action Areas for Investors*. https://www.unpri.org/human-rights/diversity-equity-and-inclusion-key-action-areas-for-investors/9393.article.

Reality Check Team. (2021). *Homosexuality: The Countries Where it is Illegal to be Gay*. BBC News. https://www.bbc.com/news/world-43822234.

Ries, E. (2011). *The Lean Startup*. Crown Business.

Rivera, K. and Karlsson, P.O. (2017). *CEOs are Getting Fired for Ethical Lapses more than they used to*. Harvard Business Review. https://hbr.org/2017/06/ceos-are-getting-fired-for-ethical-lapses-more-than-they-used-to.

Rohwerder, B. (2017). *Impact of Diversity and Inclusion within Organisations*. K4D... https://assets.publishing.service.gov.uk/media/5ba51175e5274a54d5c39c19/109_Diversity_and_inclusion_within_organisations.pdf.

Shet, S.V., Poddar, T., Samuel, F.W. and Dwivedi, Y.K. (2021). Examining the determinants of successful adoption of data analytics in human resource management–A framework for implications. *Journal of Business Research*, 131: 311–326.

Shook, E., Sage-Gavin, E. and Cantrell, S. (2020, April 2). *How Companies can use Employee Data Responsibly*. Harvard Business Review. Retrieved from https://hbr.org/2019/02/how-companies-can-use-employee-data-responsibly.

Siddiqui, F. (2019). Internal data shows Uber's reputation hasn't changed much since #DeleteUber. *Washington Post*. https://www.washingtonpost.com/technology/2019/08/29/even-after-ubers-ipo-long-shadow-deleteuber-still-looms/.

SIOP. (2021). *Recommended Demographic Survey Items June 2021*. https://www.siop.org/Research-Publications/Items-of-Interest/ArtMID/19366/ArticleID/5699.

Society for Human Resource Management. (2021). *Code of Ethics*. https://www.shrm.org/about-shrm/pages/code-of-ethics.aspx.

Society for Human Resource Management. (2021). *Belonging is more than a seat at the table*. https://shrm-res.cloudinary.com/image/upload/v1634558334/Visionaries/Exclusive_New_Belonging_Research_Report.pdf.

Suddaby, R. (2010). Editor's comments: Construct clarity in theories of management and organization. *Academy of Management Review*, 35(3): 346–357.

Torchia, M., Calabrò, A. and Morner, M. (2015). Board of directors' diversity, creativity, and cognitive conflict: The role of board members' interaction. *International Studies of Management & Organization*, 45(1): 6–24.

Turner, A. (2018). *The Business Case for Racial Equity: A Strategy for Growth*. W.K. Kellogg Foundation. https://missioninvestors.org/sites/default/files/resources/Business%20Case%20for%20Racial%20Equity%202018.pdf.

Tursunbayeva, A., Pagliari, C., Di Lauro, S. and Antonelli, G. (2021). The ethics of people analytics: Risks, opportunities and recommendations. *Personnel Review*, 51(3): 900–921.

Uhereczky, A. (2021). *Why do Diversity and Inclusion Initiatives Fail? The Cautionary Tail of Basecamp*. Forbes. https://www.forbes.com/sites/agnesuhereczky/2021/05/09/why-do-diversity-and-inclusion-initiatives-fail-the-cautionary-tail-of-basecamp/.

U.S. Census Bureau. (2020). *Decennial Census: Race*. Retrieved from https://data.census.gov/cedsci/table?q=p1&g=0100000US_0400000US24_1600000US2404000&tid=DECENNIALPL2020.P1.

Uslu, D., Marcus, J. and Kisbu-Sakarya, Y. (2022). Toward optimized effectiveness of employee training programs: A meta-analysis. *Journal of Personnel Psychology*, 21(2): 49–65. https://doi.org/10.1027/1866-5888/a000290.

van Dijk, H., van Engen, M.L. and van Knippenberg, D. (2012). Defying conventional wisdom: A meta-analytical examination of the differences between demographic and job-related diversity relationships with performance. *Organizational Behavior and Human Decision Processes*, 119: 38–53.

Waters, S.D., Streets, V.N., McFarlane, L. and Johnson-Murray, R. (2018). The practical guide to HR analytics: Using data to inform, transform, and empower HR decisions. *Society For Human Resource Management.*

Wilkie, D. (2018). *Employee Engagement Surveys: Why do Workers Distrust Them.* Society for Human Resource Management. https://www.shrm.org/resourcesandtools/hr-topics/employee-relations/pages/employee-engagement-surveys.aspx.

Williams, M.L. and Bauer, T.N. (1994). The effect of a managing diversity policy on organizational attractiveness. *Group & Organization Management*, 19(3): 295–308.

Wong, J.C. (2017). *Uber CEO Travis Kalanick Resigns following Months of Chaos.* The Guardian. https://www.theguardian.com/technology/2017/jun/20/uber-ceo-travis-kalanick-resigns.

Wong, R. (2021). *Stop Screening Job Candidates' Social Media.* Harvard Business Review. https://hbr.org/2021/09/stop-screening-job-candidates-social-media#:~:text=Social%20media%20sites%20such%20as,because%20of%20what%20they%20found.

Woo, S.E., LeBreton, J., Keith, M. and Tay, L. (2020). Bias, fairness, and validity in graduate admissions: A psychometric perspective. *Perspectives on Psychological Science.*

Workforce Science Associates. (n.d.) *Less Means More: Employee Engagement Survey Frequency.* https://workforcescience.com/learn/articles/less-means-more-employee-engagement-survey-frequency/.

Zielinski, A., Middleton, S.E., Tokarchuk, L.N. and Wang, X. (2013, May). Social media text mining and network analysis for decision support in natural crisis management. In *ISCRAM.*

Chapter 9

An Introduction to the Utilization and Application of Text Analysis

Dane Luke Wagner, MS

Over the last 100 years, technology has advanced at an exponential rate and the result of that advancement has been its widespread adoption and use by individuals and companies alike. Studies show that as of the writing of this chapter: over 97% of Americans own a cellphone with 85% of those being considered "smartphones" (Pew Research Center 2021), more than 300 billion emails are sent worldwide daily, and the average office employee sends approximately 40 but receives near 100 work emails per day (Živković 2022), 85% of B2C companies and 91% of B2B companies host blogs (Byers 2021), and there are over 150 million daily users of business focused IM services such as Microsoft teams and Slack (Živković 2021). Though these numbers increase every year, many behavioral research and data collection methods when studying interpersonal relationships rely on outdated techniques that are obtrusive and may not capture accurate data due to having the participants know they are being studied, affecting how the participants work, relying on survey data, and numerous other issues (Meyerson 1990, Kazdin 1979). Thankfully the increased adoption of technology-based communication in the workplace has opened a plethora of previously untapped data to researchers. Some estimates place almost 80% of business data as unstructured with the majority of that unstructured data being text (Gandomi and Haider 2015). To utilize that text data for entrepreneurial research, researchers can rely on text analysis, a more modern, unobtrusive, and accurate method of measuring behaviors and interactions in the workplace.

Blackhawke Behavior Science, Florida International University, 3609 Monserrate Street, Coral Gables, FL, 33134, 305-926-6779.
Email: dane@blackhawke.io

At the start of the 20th century, the scientific community saw an increased interest in qualitative data analysis; psychoanalysis had taken the world by storm, but critics were finding the methods to be highly subjective and difficult to perform for multiple subjects. Thus, by the 1950's a new field of qualitative analysis emerged: Natural Language Processing (NLP). This branch of analysis sought to understand the rules behind the use of spoken and written language and accurately extract the meaning of various excerpts using those rules. The progress in this field was difficult as the sheer amount of information needed, and the variability within that information, made a set of effective standardized rules almost impossible to write down. The advent of machine learning algorithms in the 1980s finally made it possible, and since then, the growth in the field has been rapidly expanding into the field of Artificial Intelligence (AI). The success that the scientific community has seen with things such as online and virtual assistants has only been possible due to the progress made with NLP, but due to this tight interweaving of NLP and AI in the modern world, analysts and scientists have had to make a distinction between what it is today and its origins.

The strength of text analysis is its ability to comb through large amounts of qualitative text data and consistently provide the appropriate insights based on the researcher's needs. The methods can be applied to any open-ended survey responses, feedback, or review in addition to live communication data within companies and organizations.

The contemporary definition of NLP is as a branch of AI, specifically teaching machines to understand and replicate/manipulate speech (Lutkevich 2021). This is distinct from text analysis which seeks to gather insights and meaningful information from text without consideration for grammatical structures or semantics. Text analysis more closely parallels the earlier aspects of NLP that were more human driven. This is not to say text analysis is entirely reliant on human abilities, researchers still rely heavily on statistical software such as R and Python to do complex computations and create visualizations but unlike NLP, researchers must make their own choices for every step making the insights more personal and flexible to the researcher's needs.

The purpose of this chapter is to explore the value of engaging in text analytics from the perspective of startups and entrepreneurial research, prime the reader to understand the major factors involved in text analysis, and outline general steps to completing several aspects of text analysis.

Current Uses of Text Analysis

The most common use for text analysis is for marketing insights. Companies talk to consumers and consumers talk to companies as well as one another. Companies also communicate with investors and general society through their products and actions, all of which lead to or directly generate text. It is well established, through numerous academic papers and market research, that how consumers talk about the company and products affect other consumers' behaviors and thus, sales (e.g., Eliashberg et al. 2000, Reichheld and Teal 1996). By scraping social media posts, reviews, or even comment sections on a company's website, the researcher can gain knowledge on how people see the company and its products as well as *who* sees them that way (Humphreys and Wang, 2018, Kern et al. 2016, Opoku et al. 2006, Homburg et al. 2015, Packard and Berger 2021). The catch is that a large amount of communication content is inaccessible by the researcher and instead they must use "moments" of consumer-generated data or quantitative summaries (Chintagunta et al. 2010, Godes et al. 2005, Liu 2006). This area of text analysis begins with sentiment analysis (Liu 2011, Pang and Lee 2008) and continues through to topic analysis, both of which are covered in more depth later in this chapter, as well as Chapter 8.

Text Analysis for Understanding the Entrepreneurial Ecosystem

Currently, the majority of text analytics research is conducted by computer science or marketing groups, however, there is a growing area of research on utilizing text analysis and NLP in unique ways that impact companies at all stages of their development. There have been uses for studying how global integrations affect team learning (Zellmer-Bruhn and Gibson 2006), intentions to leave (Felps et al. 2009), how manager characteristics affect innovation (Heyden et al. 2018), and impression management (Wilhelmy et al. 2016). Currently, the author and his team are contributing to that research by using a type of topic modeling based on open response questions to craft a more accurate picture of contemporary startup cultures (Wagner et al. 2022). This area of research is dynamic and constantly pushes new insights while challenging old ideas based on larger established organizations. In addition, with the recent release of tools using GPT-3/4 (OpenAI 2023), startups are creating a unique collection of text data in the form of transcripts between the users and the GPT-3/4 powered AI. These transcripts could offer insights into creative ideation, problem solving, strategy, and direction, for example.

Improving the Startup Employee Experience

Though text analysis is most prevalent in marketing; product experience, customer experience, and even *employee* experience can be improved using those same techniques. As an example, it is useful to use when collecting help desk inquiries in that it can compile all the submissions into a report that conveys the most common issues or help monitor how effective the fixes to previous issues were by reading trends. In addition, scripts using keyword searches can route help requests to the correct department, reducing the need for active monitoring of inquiries and customer frustration. If startups offer an employee feedback portal, they can quickly identify pain points within their company's framework and decide which ones are the most imperative to address. A company's employees are some of its most powerful stakeholders having a direct impact on the functions of the company. Ensuring issues are identified and diagnosed quickly and effectively can increase many organizational metrics including employee engagement, perceptions of organizational commitment, satisfaction, and performance (Andrew and Sofian 2012, Chung 1997, Gardner et al. 2001).

Regarding talent management, text analysis can assist in boosting the accuracy of year end performance reviews and monitoring the success of coaching interventions. Often these processes are riddled with biases such as confirmation bias (only seeking out confirmatory evidence after deciding), availability bias (making decisions from easily recalled information), presumed associations (overestimating how related two events are), and hindsight bias (overestimating the degree to which an outcome was predicted) (Sadler-Smith and Shefy 2004, Viswesvaran et al. 1996, 2005). The less subjective the steps required for ratings, the more accurate the ratings should become. In this case, management and coaches would be required to log critical incidents and initiatives regularly (at least monthly, but more often would be ideal). This creates a collection of documents that can be analyzed to give a better picture of trends, recurring issues or successes, and even insights into the scope of impact for the critical incidents; thus, providing a more accurate base to build off when conducting year end performance reviews and building future developmental interventions. As a note, researchers should be aware that this process does not remove subjectivity in its entirety as the researcher will need to make decisions as to what to focus on and what parts of the text are valuable.

Team Communication

Teams are an integral part of the entrepreneurial ecosystem as well as contemporary workplaces in general. The vast majority of startups (approximately 90%) are led and founded by teams, rather than by individuals (Beckman 2006, Cooney 2005, Kamm et al. 1990, Lechler 2001, Reynolds 1997, West 2007). Strategically structured and supported teams provide benefits ranging from increased productivity to increased satisfaction and external judgments (Campion et al. 1996, Guzzo and Shea 1992, Peeters et al. 2006). In addition, they provide effective means to solving problems in complex environments, of which startups are certainly included. Specifically, in the entrepreneurial context, there is a unique importance regarding the behaviors of leadership teams. These teams oversee designing and creating policies and procedures, solidifying social norms that will affect the future development of their company, likely beyond the tenure of the team itself (Johnson 2007, Mischel 1977, Staw 1991). For teams to achieve these desired outcomes they must engage in effective communication (Foushee 1984, Lingard et al. 2004, Sasou and Reason 1999, Sutcliffe et al. 2004). Thus, analyzing a team's communication can provide insights into their coordination, situational awareness, knowledge, learning, workloads, and perceptions of stress.

With the common integrations of communication and productivity applications such as Google Workplace, Microsoft 365, and Slack, most team communications are easily accessible in text form. These documents can be collated into a corpus of relevant communications allowing a researcher to engage in a kind of simple version of *Verbal Protocol Analysis* (VPA), focusing only on explicit content. VPA is a process tracing technique pioneered by Ericsson and Simon (1984) that aims to measure cognitive factors that lead to various performance outcomes. Using text analysis techniques akin to topic modeling, the researcher can conduct *content analysis* (measuring the frequency particular variables are mentioned) (Stemler

2000), *duration/ frequency analysis* (measuring the length and number of time(s) a particular variable was brought up and referred to or a particular team member was interacted with) (Gröchenig 2001), and *sequential analysis* (recording the order of member interactions, phrases referring to flagged variables, or even specific words) (Gottman et al., Wald 2004). These methods can reveal if a team is effective by assessing whether team members followed statements of uncertainty, planning, or actions with responses or acknowledgements (Cannon-Bowers and Salas 1998), if a team is experiencing high task urgency, if they include members with high experience and effective leadership (Xiao et al. 2003), the frequency of communication during high difficulty tasks, information elaboration, and their level of engagement in strategy formation (Bunderson and Sutcliffe 2003, Hoch and Kozlowski 2014, Marks et al. 2001, Marlow et al. 2017, Mosier and Chidester 1991, Orasanu 1990).

Another important aspect of team communication that such research can uncover is if patterns in communication have changed. This necessitates the building of communication models using algorithms that utilize quantified past relationships between terms, topics, and team members to predict future communication patterns. Such model techniques include *Markov analysis* (which is common but is also unrealistically strict when using strings of items making it very sensitive to "noise" in the data), *lag sequential modeling* (finds dependencies between items that are separated by one or more other items making this method more robust to noise), and even *Fourier analysis modeling* (converts top view patterns to sinusoidal functions making this one of the methods most robust to noise, but also the least detailed) (Gottman and Bakeman 1986, Kemeny and Snell 1960, Sanderson and Fisher 1994, Watt and VanLear 1996).

Organizational Culture

Culture is a topic that has been studied for centuries by various fields of expertise. The majority of those studies have historically focused on macro culture (e.g., observations of people at the world, nation, and general occupation) but as time and understanding of complex systems increase more attention has been paid to micro and meso culture (observations of smaller groups of people and their interactions). Within the entrepreneurial ecosystem, micro and meso cultures are of particular importance as a startup often consists of only a handful of people. While developing a startup one may be introduced to phrases such as "learning culture", "caring culture", "safety culture", and more, being told that they should be building one or all of them. Regardless of the culture one desires to foster, one needs to be able to effectively measure and identify the culture as it develops. Numerous theories and rules have been applied to the idea of culture but even when those are agreed upon there seems to be disagreement on how to actually *measure* culture (Scott et al. 2003) With improved access to communication data between people, text analysis has become one of the easier methods to identify and classify culture at a micro, meso, and macro level.

The main text analytical technique used by researchers to achieve culture classification and identification is topic modeling. Through LSA or LDA (both of which are covered in more depth later in this chapter) researchers can take live text data from within the company and gain real insights into the culture *actually* practiced rather than the culture a company *wants* to practice. Researchers can identify and classify cultural elements that are unknown to those in the company itself and depending on the calibration of the model, even pick up key social actors that champion the culture (Bail 2014). These are things that have not been truly feasible in the past.

Talent Acquisition

As startups grow, they find the need to hire more people and manage them throughout the talent pipeline. Here there are uses of text analysis at the recruitment stage: scraping resumes, cover letters or web pages to see if keywords are present, or to measure the applicants' qualifications and characteristics (Campion et al. 2016, Sen et al. 2012, Yi et al. 2007, Zhao et al. 2015). Companies like LinkedIn, Monster.com, and Indeed already employ these features to assist recruiters by recommending the "optimal" candidate, but if a startup is receiving a large number of direct applications, using a keyword search and extractor, or employing *named entity recognition* (NER) can save them large amounts of time and effort. The keyword search and extraction are as simple as if–then statements (e.g., if a number of the keywords or specific ones are present, then return the applicant ID). This can filter out unwanted applicants in seconds, even if there are hundreds of applications.

Often with keyword searches and extraction, the documents need a degree of pretreatment. Such pretreatment may involve eliminating images, converting tables to text blocks, unifying fonts, and even stemming the text. These keywords should be developed with the help of SMEs and current high performing talent alongside an in-depth work analysis (see Morelli, this volume). It is important when using this method to recognize that each skill can be described using multiple words, symbols, or spellings, some words may define multiple notions (Word Sense Disambiguation [WSD]), and that some skills and experiences are transferable and present even if not clearly identified by the applicant. When using keyword searches, extraction, or research for talent acquisition purposes, there will be a degree of range restriction that may eliminate some unique high-performance talent. For this reason, it is only advisable to use this strategy when filtering large numbers of applicants. As a note, this text analysis method alone is not enough to base selection decisions. Even if a single candidate has the highest density of keywords present, it does not mean they are qualified for the position. Using this method can, however, make the applicant pool more manageable for hiring managers. Studies have shown repeatedly that the selection process that results in the lowest adverse impact and highest performers is a battery consisting of a standardized interview, a personality measure, and a high-fidelity performance assessment such as a work sample (Sackett and Ellingson 1997, Schmidt and Hunter 1998, Bobko Roth and Potosky 1999).

To discover the ideal applicants for a position, it can be useful to create *a basic text classifier* that can identify if a part of the resume is referring to the name of a company, the name of a university, a job description, and a job position. The classifier can then be paired with a text extractor to form corpora that collect all of the applicant's information into those aforementioned groups. Then stop words (common words that provide no contextual value to the research goal) and phrase delimiters (punctuation like "," and ".") are removed. The final step would be to run a frequency count on the words left in these groups looking at who the people are applying to the company. This can inform recruitment strategies.

Key Concepts of Text Analytics

The Process

Text analysis follows four general phases: (1) preprocessing the textual data, (2) utilizing an analytical technique, (3) converting the results of that analytical technique to a quantifiable insight, and (4) reviewing the validity of that insight.

Preprocessing is important because text tends to be unstructured and "untidy." Machines treat any variation of the input as a separate instance meaning that two terms that are grammatically the same but written differently (e.g., "1" versus "one") will be treated as two different terms. To structure the data into a manageable form that can handle analytic processing, researchers often tokenize, clean, adjust spelling, remove stop words, and stemming and lemmatization.

There are several basic analytical techniques used after the preprocessing as covered in the first half of the chapter such as keyword extraction, sentiment analysis, topic analysis, and relation analysis (sequential analysis).

Though the analytical techniques are easily replicable and consistent, the outcomes can vary dramatically based on preprocessing decisions the researcher makes, in addition, context may lead to the interpretation of those analyses one way rather than another. It is important whenever running text analysis to record one's decisions and rationale while progressing through those four steps. Note and record keeping are vital to ensuring the validity of the insights.

Tokenization

Tokenization is breaking the text into specific units (e.g., words or sentences). This is often determined by locating delimiters (e.g., spaces, commas, periods). Tokenization can often be handled by pre-generated algorithms, but there are some edge cases of which the researcher should be aware. If, for example, a word is determined by spaces or periods, some tokens may not make sense. A common case of this occurring is with named entities that are frequently expressed as initialisms (e.g., U.S.A.). In such cases, the tokenization would return "U", "S" and "A" as separate tokens. Thankfully most modern text mining and analytics software and algorithms include procedures to prevent such accidents, but the researcher should still understand the corpora being used, predict potential instances of unusual tokens, and review the produced statistics for those instances.

Cleaning

Cleaning is a multistep process that rearranges and modifies text into a format that machines can analyze with minimum error. It seeks to normalize text, remove unnecessary symbols and text, and minimize variance between grammatically similar words. For this chapter, the removal of stop words and stemming and lemmatization are described in the sections following this one, but they are functionally part of the pre-processing step of text analytics and often bundled together when referencing "cleaning".

Normalization of text involves standardizing cases, standardizing spaces, removing unicode, translation, typo correction, and number unification. Each of these parts will require different parameters, and different levels of strictness, or may not even occur at all depending on the research question. The majority of programs and packages offer buttons, selections, or single lines of code that achieve each of these steps with little effort from the researcher but understanding the basic rules behind each step is paramount for accurate insights.

Standardizing cases is often as simple as converting all uppercase characters to lowercase characters. Example:

"Hello, My name is Rayyan and i have a question about Companyname's return polecy? I need to return 2 items: https://www.companyname.com/myaccount/order-history?ref_=54321_abc."

would be converted to:

"hello, my name is rayyan and i have a question about companyname's return polecy? i need to return 2 items: https://www.companyname.com/myaccount/order-history?ref_=54321_abc."

In the output, not a lot has changed, but what *has* changed is significant. There is now no formatting information identifying when certain terms are proper nouns versus common nouns. Without knowledge of possible words being used going into the analysis, it may prove difficult to gain insights from the future output. However, this step prevents words with accidentally capitalized letters or words in all uppercase from being treated as a separate term in the document.

Standardizing spaces involves the elimination of excess spaces. For the majority of programs, this automatically changes any sequence of spaces where $n > 1$, to $n = 1$. There is rarely a reason to preserve sequences of multiple spaces after tokenization, but the option is available within many algorithms. Continuing with the previous example, upon standardizing spaces the output would be:

"hello, my name is rayyan and i have a question about companyname's return polecy? i need to return 2 items: https://www.companyname.com/myaccount/order-history?ref_=54321_abc."

The necessity of this action lies in that most programs read spaces as characters and so having variable spaces between terms may affect how the program reads the data, introducing errors in the output.

Removing unicode is often simply referred to as "removing punctuation," but with the prevalence of emojis, URLs, and "@" in contemporary text data. It should be noted that only removing punctuation could result in outputs that make little sense (e.g., the emoji "face with tears of joy" is converted to "U+1F602" by most programs and thus without punctuation may just be "U1F602" in an output). Removing unicode from the running example, the new output would be:

"hello my name is rayyan and i have a question about companynames return polecy i need to return 2 items."

The elimination of unicode removes inflection data but again eliminates the likelihood that the program may flag the same term as two unique terms just because one occurred at the end of a sentence. Some studies have suggested that emojis provide important information when interpreting textual language specifically regarding sentiment and may be valuable to maintain within the text (Barbieri et al. 2016, Weissman and Tanner 2018) but there are ways to extract such information if desired (Aoki and Uchida 2011).

Translation and *Typo Correction* is the identification of words that do not exist in the language being researched followed by the replacement of the correct words. Often this requires a library of "incorrect" words linked to the "correct" words. Then a script is run to read through the text data and when one of the "incorrect" words from the library is flagged, they are replaced with the "correct" word linked to it. These libraries can be built manually with smaller data sets and are often recommended for industries with particularly unique words that may not be in common dictionaries such as the medical field. Be mindful that in some cases the "incorrect" words may be found *within* valid terms, so setting the "incorrect" words to be bounded on both ends is a good practice to instill when performing such strategies. However, when it comes to translation or more general typo correction, dictionary-based algorithms may prove more useful. This chapter goes in depth about one, Hunspell, in the later section about stemming and lemmatization. In addition, it may be beneficial to ensure the use of a spell checker during the collection of data to minimize the sparsity of this step. If typo correction was applied to the example the output would read:

"hello my name is rayyan and i have a question about companynames return policy i need to return 2 items"

There's an important note to be made about this step: exceptions may need to be made for names of companies and organizations. It is not uncommon for companies to alter the spellings of words and then use them as their name. When building the library or using a dictionary-based algorithm, the researcher may need to add a list of terms that are to be ignored. Again, there is a necessity of keeping the research question in mind and familiarizing oneself with the data sources and content before processing or pre-processing it.

Number Unification is not wholly different from the previous steps, as the term unification is ultimately making two terms identical by substitution. Here, however, some considerations need to be kept in mind: Do numbers add any value to the information? How much of the "noise" is numerical? What kind of numbers should be expected in the dataset? How are the numbers formatted? These questions affect the depth of unification required. If there are addresses, phone numbers, or identification numbers, they need to be standardized into an easily referenceable format. This means that special care will be taken during the tokenization step and even before to identify these strings and ensure they are kept together and formatted the same as similar occurrences (is there an apartment number in the address? Ensure that it keeps it with the street address for easy queries). If instead, the numbers add meaning to the message (e.g., number of occurrences, number of items, times, etc.) the researcher should choose to unify them with the library substitution method described before. When applied to the example, changing Arabic numbers to spelled words, the following results:

"hello my name is rayyan and i have a question about companynames return policy i need to return two items."

Edge cases to be aware of are the substitution of letters or words with numbers. This is common in numeronyms and leetspeak. In numeronyms, common words are substituted with a number or alpha-numerical combination that is usually phonetically similar ("4" instead of "for", "K9" instead of "canine", "sk8tr" instead of "skater") but sometimes have a different relationship ("143" instead of "I love you", or "i18n" instead of "internationalization"). Leetspeak is the substitution of glyphically similar numbers and letters ("h3110" instead of "hello" or "f4k3" instead of "fake"). Understanding the culture and common communication practices of those contributing to the data set will inform the researcher on how to handle numbers and expected "aberrant terms".

Stop words

Stop words, as defined earlier in this chapter, are common words that provide no contextual value to the research goal. Examples of stop words are "and", "the", "of", and "as," but hundreds have been identified over the years. Francis and Kucera (1982) found that the top ten most frequently used words in the English language account for between 20 and 30 percent of the terms in any document. Stop words quickly clutter frequency counts and affect any analysis based on them. There are two basic ways to eliminate stop words: using a "static" library or using a "dynamic" library. Static libraries are provided by programs and packages and are likely to contain 100–200 common stop words. Dynamic libraries are often generated using the term frequency-inverse document frequency (TF-IDF) values (covered in the later TF-IDF section) but other methods have been developed (Lo et al. 2005). Both ways check the words in the libraries with the words in the corpora and substitute " " for them. This means for later computations tokens with the value of "0" in the term matrix may need to be deleted.

Some research has shown that using a dynamic library or both a dynamic *and* static library leads to better results when conducting analysis (Saif et al. 2014, Lo et al. 2005, Zaman et al. 2011). Part of the reason for this is each corpus may come from unique contributors with unique purposes. Many domain specific words that may not be very common throughout the English language as a whole, but are common for the contributors of the data. Such instances are most common in specialized fields but may include slang and other terms that are common due to world events or cultural shifts. With stop words, there is no "one size fits all" approach and this step may require several iterations before the right balance is found.

Stemming and Lemmatization

For grammatical reasons, a document will often utilize different forms of a word. Changing the tense, denoting ownership and plurals, using it as a noun, verb, adjective, adverb, etc. The same word and concept may be written in several different ways. Since each sequence of symbols is considered unique by text analysis algorithms, there is often a need to unify these various word forms into a single one. This would lead to more accurate insights when considering analyses built atop frequency counts.

Stemming is the act of mapping words to stems (basic forms of words) usually by removing suffixes. Stemming helps minimize the amount of variance between forms and reduce dimensionality and sparsity but will usually only look to remove common word endings such as "-s", "-es", "-al", "-ure", "-ible", and "-ily". Due to this strict rule-based approach, it is possible some stems produced will not make sense or be "real" words at all.

> Example: "Sleeping", "Sleeper", "Sleepy", "Sleepily", and "Sleepiest" could all be stemmed to "Sleep".
> "Features", "Biologically", "Variation", and "Visible" might be stemmed to "Featur", "Biolog", "Vari", and "Vis".

Several stemming algorithms have been developed over the years (and more are being worked on every day), most packages and software will decide which algorithm is run for the user, but it is important for the researcher to study the rules that define those algorithms and choose the ones which they are most comfortable. One of the more common stemming algorithms which is empirically effective on English words is Porter's algorithm (Porter 1980). Here the rules are not just focused on the word being stemmed but also the result of the stemming, taking into consideration the number of syllables, and the ending of the output word. Another algorithm is the Hunspell algorithm. Developed in the early 2000s as a robust spell checker for the Hungarian language, it has gained popularity since its 2016 release as a versatile set of linguistics algorithms applicable to multiple languages. Unlike the more rule-based algorithms like Porter's algorithm, Hunspell falls under a different group referred to as *dictionary-based* and thus performs Lemmatization.

Lemmatization uses vocabulary and morphological analysis to return the *lemma* or base dictionary form. Being the de facto spell checker for LibreOffice, Google Chrome, macOS, Opera, and more, Hunspell has access to multiple dictionaries in multiple languages and utilizes word stems to speed up its processing. Because its design was for spell checking non-English languages, each stem is a complete and actual word, but it can produce multiple stems for the groups of word forms meaning that the output needs to be managed differently than traditional non-dictionary based stemming algorithms. There are also occasionally irrelevant words produced alongside the relevant ones and often the words are not reduced as low as desired (particularly when using the English language).

> Example: "Numbers" is stemmed to "Numb" and "Number".
> "Sleeping", "Sleeper", "Sleepy", "Sleepily", and "Sleepiest" would be stemmed to "sleep", "sleep", "sleepy", "sleepily", and "sleepy" respectively.

The struggle comes from the fact that language is complex and a language like English, pulls words and etymology from various other languages making it difficult to apply a single set of rules effectively. Some stem completion algorithms seek to unify stems by completing them based on either the highest frequency, the shortest, the longest, or the first term that shares the stem in the corpus. Often such heuristics are in the control of the researcher, and so the research question should always be kept in mind when performing such actions. In addition, some research has begun to show that for certain applications, such as topic modeling, stemming does not help and can actually lower performance when training models for text (Schofield and Mimo 2016).

Collocations and Co-occurrences

Collocations, also known as N-gram statistics, are terms that are frequently found in immediate succession to each other. These collocated terms are expressed as 2-gram (bigrams, two words next to each other), 3-gram (tri-grams, three words next to each other), etc.

As an example, let's use the following text:

> "Text analysis is an exciting field that can provide many insights. For text analysis to provide many insights though, the researcher needs to understand the appropriate techniques and when to apply them."
>
> "Text analysis" is a 2-gram and "provide many insights" is a 3-gram both with the frequency of two.

Often N-gram analysis is involved in prediction models such as Markov Analysis, organizational communication theories, and information theories as well as building a more robust frequency analysis.

Co-occurrences are when terms are commonly present within the same units of text, whether documents, pages, or sentences. These terms are not required to be directly sequential but must have some degree of semantic proximity depending on the scope of the analysis.

Co-occurrences become important when using NER, topic analysis, relation analysis, and building document networks. Both of these statistics, collocations and co-occurrences, are reported as correlations and probabilities with a general rule being that the higher their statistic the higher their frequency.

Term Frequency—Inverse Document Frequency

Term Frequency—Inverse Document Frequency (TF-IDF) measures the relative importance of a term to a document by multiplying how often the term appears within a particular document (the term's frequency) by how rare it is in the entire corpus (the inverse document frequency). This calculation is the backbone to generating keywords for documents, topic analysis, similarity analysis, and vectorization for machine learning purposes. TF-IDF helps correct frequency counts for words such as "is", "the", "of", etc. (stop words) that are highly frequent but do not add inherent value to the context of the documents. This can be a valuable improvement over the classical *Bag of Words (BoW)* approach which simply finds the TF of every word without consideration of importance.

The output values for TF-IDF will range from 0–1 with 1 being the "most important" term and 0 being the "least important" term (often these are the words identified as stop words). When performing the TF-IDF calculation manually the result will often be above 1, this is because programs will apply regularization to the data set using either L1 or L2 norms. Not all programs or packages use the same norm so it is important to identify the rules used to best fit the data being analyzed. The L1 norm utilizes rectilinear distance (taxicab distance) while the L2 norm utilizes squared Euclidean distance. The contemporary mindset then is that L1 is more robust to outliers, but L2 is more stable to "horizontal adjustments". Currently, the L2 norm is the most common standard for programs and often there's no reason to alter it, but edge cases exist and it is valuable to be aware of the alternative.

Topic Analysis

Topic analysis uses a degree of machine learning to flag the subjects (i.e., topics) of each text excerpt. The goal is not just to find the topics within a corpus but also the *strength* of those topics. Two main assumptions are made throughout topic analysis: each document is made up of multiple topics, and each topic is made up of multiple words. The most common topic analysis techniques are *latent semantic analysis* (LSA) and *latent dirichlet allocation* (LDA) but many more exist requiring both supervised and unsupervised machine learning. Some popular alternatives such as lda2Vec lend themselves better to NLP than text analysis and thus are beyond the scope of this chapter. Each technique provides slightly different benefits whether being quicker, more refined, or less complex.

The backbone of *latent semantic analysis* is word frequency counts. The assumption this technique makes is that the same approximate distribution of words will occur in similar documents. To identify the highest value words, the researcher runs the TF-IDF and produces a *document-term matrix* that will show the highest value terms versus the documents in the corpus being considered. After putting in the hyperparameter T (the number of topics the researcher hopes to find) this matrix is then broken into USV matrices by an algorithm: the U matrix being a document-topic matrix, V being a term-topic matrix, and S being a diagonal of the single values from the Document-Term matrix. This division helps eliminate the sparsity and noise within the corpus and allows the researcher to evaluate the quality of the topics by looking at the U matrix, and the quality of the terms *within* those topics by looking at the V matrix.

LSA is quick to perform, and highly efficient specifically with large amounts of text, but it fails to identify what the topics are, will sometimes apply terms in a way that seems arbitrary at best, and can be difficult to visualize effectively.

Like LSA, *latent dirichlet allocation* is based on the distributional hypothesis, assuming that similar topics use similar words. LDA uses Bayesian statistics, locating n-grams in the corpus and assigning probability values to them that describe how likely they are to belong to a particular topic. Unlike LSA, LDA assumes that the terms within each topic are made up as dirichlet distributions. Like LSA, this technique utilizes a hyperparameter that reflects the number of topics to be detected, but there are two additional hyperparameters: alpha and beta. The alpha hyperparameter controls the number of topics assigned to each document while the beta controls the number of words used to model those topics. With all of these hyperparameters, the larger the value set by the researcher, the higher the affected parameter will be. The output is a vector containing the values of the coverage per topic.

Again similar to LSA, this technique is effective on large corpora and will not provide labels for what the identified topics are, however, this method is more flexible and can easily generalize to new documents as the algorithm is "trained" on the original corpus. Depending on the hardware used to run this technique, it may take longer than LSA as the algorithm is more complex, but it is still relatively quick to run.

Knowledge Graphs

A *knowledge graph (KG)* is a graphical representation of the relationships between concepts, entities, relationships, and events. Often a visualization of a knowledge model and a byproduct of relational analysis, a KG seeks to describe how words found in a document, or corpus, relate to each other using formal semantic descriptions.

The recommended format is *resource description framework (RDF)* statements. RDF statements are structured in three positional statements (triples):

SUBJECT -> PREDICATE (OF VERB) -> OBJECT.

In this way, one can take a term from a document (e.g., "U.S.A") and not only define it (predicate: "type of", object: "country") but describe why it is mentioned in the corpus by its relation to other terms in that corpus ("Company A" -> "located in" -> "U.S.A."). The reasons for the recommendation of RDF statements are: standardization (it is a fundamental structure through the W3C community for the Semantic Web Stack making it easy to use and understand by logicians, data management professionals, web researchers, and more), expressivity (allows for a fluent and thorough representation of numerous data types and levels), and interoperability (utilizing a formal semantic structure of three parts creates unique identifiers for each pathway allowing both humans and computers to interpret, serialize, access and unambiguously manage the data).

KGs are used to enrich the understanding of the text and are common when building ontological and semantic research, creating more robust predictions and inferences, and developing ML algorithms and AI for deeper NLP applications. An added bonus is that the process creates semantic metadata that helps future research into similar data groups.

Use Cases

Sentiment analysis uses pre-loaded libraries of words and phrases which are each flagged as either "positive", "negative" or "neutral". This library is then compared to the text generated from emails, chat rooms, or scraped from social media such as twitter. When matches are found between the words in the library and the words in the studied documents, they are then linked to the appropriate sentiment. Using the email example, this specific basic sentiment analysis will provide the overall percentage of emails moving within the company that are positive, the percent that are negative, and the percent that are neutral. This insight is beneficial as a quick and easy replacement for employee polls. It can be run in minutes and discovers how people generally feel within the company. But often, that is not quite enough. What specifically do they feel that way about? Is it the organization as a whole, or is it a specific product/task? Is it the physical location? What about specific people that run the company? The answer to these questions is topic analysis and classification.

Topic analysis and *classification* use a degree of machine learning, often including *Named Entity Recognition* (NER), to flag the subjects of each text excerpt. This means in the example, each email will have not just the sentiment flagged, but also the subject each sentiment is towards. A researcher can choose whether to create a library themselves to look for specific subjects or allow algorithms to detect the subject based on sentence structure. Both are common practices, though the researcher can always look for those specific subjects after the algorithm has flagged them.

Using such techniques, as well as including other variables such as location data, time of posting, and more, the researcher can discover what aspects of the employee experience are leading to the most positive sentiments, what the employee's top complaints are, as well as when they are performing specific tasks, the duration of the projects, and how different levels of the company are being affected by certain variables. In addition to those useful insights, a researcher can still use this information to monitor internal and external perspectives of the company's brand, learn about views of other competitors (are any of them becoming attractive to employees?), understand trends and how the company's culture has evolved over time, as well as identify potential PR crises before they grow out of hand (Lee and Bradlow 2011, Netzer et al. 2012). Running regular aspect-based sentiment analysis on various sources can provide a company with a competitive edge by boosting one's awareness of their most important stakeholders' needs.

Regarding attracting talent, whether the company is relying on a strategy focused on content marketing, search engine optimization, social media marketing, account based marketing, email marketing, video marketing, a mix of all the above, or even something different entirely: the end goal is usually converted to a successful hire. At the very least these strategies need to result in a flat return on investment or else they are not likely to be considered effective. Text analysis does not just empower the researcher to study the engagement with the marketing strategies, but also to understand the hiring process and inform the researcher of the roadblocks in that process.

Hiring teams can put a lot of time and effort into finding the right talent and so minimizing the number of calls and leads that fail to be converted to an employee is a boon for any company. Text analysis can help warm up calls by scraping potential hires' social media, file hosting sites, or even personal websites and qualifying them with their summary descriptions. It is common for hiring staff to have a persona or profile of the type of hire they are trying to attract. By creating a library that is populated by terms associated with that persona or profile, the researcher can start the process by simply performing a keyword search that returns potential hires that feature any or all of those keywords in their descriptions. In addition, because the potential hire may not be describing themselves using the same words the organization chose, returning a frequency count of the terms used may provide a quick and surprisingly thorough look into the individual, enabling hiring professionals to save a large amount of time performing due diligence before the call.

Prospecting is not the only area of hiring that benefits from the use of text analytics. Reviewing transcripts of past hiring conversations can provide valuable insights into the process and how it can be improved. If, for instance, a startup is losing talent that are all members of a specific demographic, a researcher could employ aspect-based sentiment analysis to the hiring transcripts to find the most frequent questions, terms, or responses when engaging with that demographic as well as how the sentiment changed throughout the conversation. This can inform future hiring teams on more effective processes with that demographic, ensuring avoiding common roadblocks and building on what sparks their interest.

The exciting part for startups and entrepreneurship researchers is that these techniques do not just apply to potential hires, but other key stakeholders like customers and investors as well. Just like with customers, each conversation with a new investor is functionally a sales conversation. Utilizing text analysis to learn how investors describe themselves, what their portfolio looks like, and what attracts certain investors or is steering other investors away, can be worth upward of hundreds of thousands of dollars.

Tips for Beginners

To perform text analysis techniques, interested researchers do not necessarily need to know how to code, but it helps significantly and enables the researcher a higher degree of control over the parameters of the techniques. Most data scientists use Python to run their analyses but there are many, myself included, that have adopted the R programming language to meet their needs. Python is a more generalized language with support for all areas of production, because of this the syntax has been designed to be easy to learn but the libraries and packages developed for Python may be overwhelming. As a whole, Python is so vastly used because it is easily integrated within apps and web services. Conversely, R is a slightly younger programming language specifically designed for statistical processes. R is well suited and more powerful for analyses and visualizations that are not built into parts of other products. That being said, R is constantly being developed and updated with new packages that enable new interactions and processes.

Common Python libraries for text analysis are:

- NLTK (https://www.nltk.org/)
    - Steep learning curve but can cover Entity Extraction, Tokenization, Parsing, Stemming, and more. This Library is more specifically built for NLP though and may be better for more advanced users.
- TextBlob (https://textblob.readthedocs.io/en/dev/)
    - This library is built off the NLTK library, but is significantly more beginner friendly. Run Sentiment Analysis, Term Frequencies, Tokenization, Spelling Correction, Wordnets, and more.
- SpaCy (https://spacy.io/)
    - This has the same functionalities as the above libraries but is written in a variation of python making it faster when used with larger datasets. In addition, it has a wider language base allowing for more accurate analyses using languages other than English (if gathering non english language text another alternative may be Polyglot which functions with over 100 languages).

Common R libraries for text analysis are:

- Tidytext (https://cran.r-project.org/web/packages/tidytext/index.html)
    - This may be one of the easiest text analytics libraries to start with as it is built with the Tidyverse in mind (another popular library commonly used as an entry point for learners). Much like the Tidyverse, it is built with visualization in mind. Though it is simple, it is still powerful; capable of performing tasks ranging from tf-idf and n-gram statistics to sentiment analysis and topic modeling.
- Quanteda (https://cran.r-project.org/web/packages/quanteda/index.html)
    - Quanteda was created as an answer for all quantitative text research needs. There are a few things this library can't do regarding text analysis. This library provides versatile corpus management and matrix manipulation alongside the analysis techniques covered in this chapter. The visualizations of this package may be less attractive than that of Tidytext though.
- Tm (https://cran.r-project.org/web/packages/tm/index.html)
    - Tm stands for Text Mining. This library is more focused on data import, corpus management, preprocessing, meta-data management, and matrix creation. Thus Tm is a tool for only part of the text analysis process.
- Spacyr (https://cran.r-project.org/web/packages/spacyr/index.html)
    - Spacyr is a wrapper of the python library SpaCy (i.e., the functionalities of SpaCy from within your R IDE). An important note is that this library in particular integrates very well with the Tidytext and Quateda libraries.

If going the coding route, it may also be advantageous to learn other programming languages such as Java to bridge the gaps between raw data and the presentation of insights. However, learning these languages may be time consuming and prove incredibly tedious for some. In such cases, it is useful to look to alternatives built on Software as a Service (SaaS) application programming interfaces (APIs). This way all that is needed is providing the SaaS data and requesting specific insights. Many of these services provide integrations with tools already commonly used by business owners making the shift to utilizing text analysis painless as possible.

Common "No-Code" options.

- Google NLP (https://cloud.google.com/natural-language)
- Lexalytics (https://www.lexalytics.com/)
- Meaning cloud (https://www.meaningcloud.com/)
- Amazon Comprehend (https://aws.amazon.com/comprehend/)
- MonkeyLearn (https://monkeylearn.com/)

Each of the above options has its benefits and drawbacks that vary per the researcher's needs. For that reason, it is out of the scope of this chapter to go into detailed descriptions of each option. Instead, researchers ought to explore respective sites to determine adequacy.

Conclusion

As a whole, text analysis is a set of techniques that may prove a steep learning curve. However, the depth and scope of insights that are possible from such practices are well worth the effort. Text analysis can enable a startup to separate itself from the competition, specify its product to its customer's needs, and ensure the internals of the company are built to perform as well as possible.

References

Andrew, O.C. and Sofian, S. (2012). Individual factors and work outcomes of employee engagement. *Procedia-Social and Behavioral Sciences*, 40: 498–508. https://doi.org/10.1016/j.sbspro.2012.03.222.

Aoki, S. and Uchida, O. (2011, March). A method for automatically generating the emotional vectors of emoticons using weblog articles. *Proceedings of the 10th WSEAS International Conference on Applied Computer and Applied Computational Science*, 132–136.

Bail, C.A. (2014). The cultural environment: Measuring culture with big data. *Theory and Society*, 43(3-4), 465–482. doi:10.1007/s11186-014-9216-5.

Barbieri, F., Kruszewski, G., Ronzano, F. and Saggion, H. (2016, October). How cosmopolitan are emojis? Exploring emojis usage and meaning over different languages with distributional semantics. *Proceedings of the 24th ACM International Conference on Multimedia*, 531–535. https://doi.org/10.1145/2964284.2967278.

Beckman, C.M. (2006). The influence of founding team company affiliations on firm behavior. *Academy of Management Journal*, 49(4): 741–758. https://doi.org/10.5465/amj.2006.22083030.

Bobko, P., Roth, P.L. and Potosky, D. (1999). Derivation and implications of a meta-analytic matrix incorporating cognitive ability, alternative predictors, and job performance. *Personnel Psychology*, 52(3): 561–589. https://doi.org/10.1111/j.1744-6570.1999.tb00172.x.

Bunderson, J.S. and Sutcliffe, K.M. (2003). Management team learning orientation and business unit performance. *Journal of Applied Psychology*, 88(3): 552. https://doi.org/10.1037/0021-9010.88.3.552.

Byers, K. (2021, September). *106 Effective Content Marketing Examples: With Details*. https://growthbadger.com/resources/content-marketing-examples/.

Campion, M.C., Campion, M.A., Campion, E.D. and Reider, M.H. (2016). Initial investigation into computer scoring of candidate essays for personnel selection. *Journal of Applied Psychology*, 101(7): 958. https://doi.org/10.1037/apl0000108.

Campion, M.A., Papper, E.M. and Medsker, G.J. (1996). Relations between work team characteristics and effectiveness: A replication and extension. *Personnel Psychology*, 49(2): 429–452. https://doi.org/10.1111/j.1744-6570.1996.tb01806.x.

Cannon-Bowers, J.A. and Salas, E. (1998). Team performance and training in complex environments: Recent findings from applied research. *Current Directions in Psychological Science*, 7(3): 83–87. https://doi.org/10.1111/1467-8721.ep10773005.

Chintagunta, P.K., Gopinath, S. and Venkataraman, S. (2010). The effects of online user reviews on movie box office performance: Accounting for sequential rollout and aggregation across local markets. *Marketing Science*, 29(5): 944–957. https://doi.org/10.1287/mksc.1100.0572.

Chung, B.G. (1997). Collecting and using employee feedback: An effective way to understand customers' needs. *Cornell Hotel and Restaurant Administration Quarterly*, 38(5): 50–57. https://doi.org/10.1177/001088049703800535.

Cooney, T.M. (2005). What is an entrepreneurial team? *International Small Business Journal*, 23(3): 226–235. https://doi.org/10.1177/0266242605052131.

Eliashberg, J., Jonker, J.J., Sawhney, M.S. and Wierenga, B. (2000). MOVIEMOD: An implementable decision-support system for prerelease market evaluation of motion pictures. *Marketing Science*, 19(3): 226–243. https://doi.org/10.1287/mksc.19.3.226.11796.

Ericsson, K.A. and Simon, H.A. (1984). *Protocol Analysis: Verbal Reports as Data*. The MIT Press.

Felps, W., Mitchell, T.R., Hekman, D.R., Lee, T.W., Holtom, B.C. and Harman, W.S. (2009). Turnover contagion: How coworkers' job embeddedness and job search behaviors influence quitting. *Academy of Management Journal*, 52(3): 545–561. https://doi.org/10.5465/amj.2009.4133107.

Foushee, H.C. (1984). Dyads and triads at 35,000 feet: Factors affecting group process and aircrew performance. *American Psychologist*, 39(8): 885. https://doi.org/10.1037/0003-066X.39.8.885.

Francis, W.N. and Kucera, H. (1982). Frequency Analysis of English Usage: Lexicon and Grammar. *Boston: Houghtin Mifflin*.

Gandomi, A. and Haider, M. (2015). Beyond the hype: Big data concepts, methods, and analytics. *International Journal of Information Management*, 35(2): 137–144. https://doi.org/10.1016/j.ijinfomgt.2014.10.007.

Gardner, T.M., Moynihan, L.M., Park, H.J. and Wright, P.M. (2001). Beginning to unlock the black box in the HR firm performance relationship: The impact of HR practices on employee attitudes and employee outcomes. https://hdl.handle.net/1813/77407.

Godes, D., Mayzlin, D., Chen, Y., Das, S., Dellarocas, C., Pfeiffer, B., Libai, B., Sen, S., Shi, M. and Verlegh, P. (2005). The firm's management of social interactions. *Marketing Letters*, 16(3): 415–428. https://doi.org/10.1007/s11002-005-5902-4.

Gottman, J.M. and Bakeman, R. (1997). *Observing Interaction: An Introduction to Sequential Analysis*. United Kingdom: Cambridge University Press.

Gottman, J.M., Gottman, J.M. and Roy, A.K. (1990). *Sequential Analysis: A Guide for Behavioral Researchers*. Cambridge University Press. https://doi.org/10.1017/CBO9780511529696.

Gröchenig, K. (2001). *Foundations of Time-frequency Analysis*. Springer Science & Business Media.

Guzzo, R.A. and Shea, G.P. (1992). Group performance and intergroup relations in organizations. pp. 269–313. *In*: M.D. Dunnette and L.M. Hough (eds.). *Handbook of Industrial and Organizational Psychology*. Consulting Psychologists Press.

Heyden, M.L., Sidhu, J.S. and Volberda, H.W. (2018). The conjoint influence of top and middle management characteristics on management innovation. *Journal of Management*, 44(4): 1505–1529. https://doi.org/10.1177/0149206315614373.

Hoch, J.E. and Kozlowski, S.W. (2014). Leading virtual teams: Hierarchical leadership, structural supports, and shared team leadership. *Journal of Applied Psychology*, 99(3): 390. https://doi.org/10.1037/a0030264.

Homburg, C., Ehm, L. and Artz, M. (2015). Measuring and managing consumer sentiment in an online community environment. *Journal of Marketing Research*, 52(5): 629–641. https://doi.org/10.1509/jmr.11.0448.

Humphreys, A. and Wang, R.J.H. (2018). Automated text analysis for consumer research. *Journal of Consumer Research*, 44(6): 1274–1306. https://doi.org/10.1093/jcr/ucx104.

Johnson, T.E., Lee, Y., Lee, M., O'Connor, D.L., Khalil, M.K. and Huang, X. (2007). Measuring sharedness of team-related knowledge: Design and validation of a shared mental model instrument. *Human Resource Development International*, 10(4): 437–454.

Kamm, J.B., Shuman, J.C., Seeger, J.A. and Nurick, A.J. (1990). Entrepreneurial teams in new venture creation: A research agenda. *Entrepreneurship Theory and Practice*, 14(4): 7–17. https://doi.org/10.1177/104225879001400403.

Kazdin, A.E. (1979). Unobtrusive measures in behavioral assessment. *Journal of Applied Behavior Analysis*, 12(4): 713–724. https://doi.org/10.1901/jaba.1979.12-713.

Kemeny, J.G. and Snell, J.L. (1960). *Finite Markov Chains*. D. van Nostrand Co. Ltd.

Kern, M.L., Park, G., Eichstaedt, J.C., Schwartz, H.A., Sap, M., Smith, L.K. and Ungar, L.H. (2016). Gaining insights from social media language: Methodologies and challenges. *Psychological Methods*, 21(4): 507.

Lechler, T. (2001). Social interaction: A determinant of entrepreneurial team venture success. *Small Business Economics*, 16(4): 263–278. https://doi.org/10.1023/A:1011167519304.

Lee, T.Y. and Bradlow, E.T. (2011). Automated marketing research using online customer reviews. *Journal of Marketing Research*, 48(5): 881–894. https://doi.org/10.1509/jmkr.48.5.881.

Lingard, L., Espin, S., Whyte, S., Regehr, G., Baker, G.R., Reznick, R., Bohnen, J., Orser, B., Doran, D. and Grober, E. (2004). Communication failures in the operating room: An observational classification of recurrent types and effects. *BMJ Quality & Safety*, 13(5): 330–334. http://dx.doi.org/10.1136/qshc.2003.008425.

Liu, B. (2011). Opinion mining and sentiment analysis. *Web Data Mining*, 459–526. Springer, Berlin, Heidelberg. https://doi.org/10.1007/978-3-642-19460-3_11.

Liu, Y. (2006). Word of mouth for movies: Its dynamics and impact on box office revenue. *Journal of Marketing*, 70(3): 74–89. https://doi.org/10.1509/jmkg.70.3.074.

Lo, R.T.W., He, B. and Ounis, I. (2005, January). Automatically building a stopword list for an information retrieval system. *Journal on Digital Information Management: Special Issue on the 5th Dutch-Belgian Information Retrieval Workshop (DIR)*, 5: 17–24.

Lutkevich, B. (2021, March). *What is Natural Language Processing?* Tech Target. https://www.techtarget.com/searchenterpriseai/definition/natural-language-processing-NLP#:~:text=Natural%20language%20processing%20(NLP)%20is,referred%20to%20as%20natural%20language.

Marks, M.A., Mathieu, J.E. and Zaccaro, S.J. (2001). A temporally based framework and taxonomy of team processes. *Academy of Management Review*, 26(3): 356–376. https://doi.org/10.5465/amr.2001.4845785.

Marlow, S.L., Lacerenza, C.N. and Salas, E. (2017). Communication in virtual teams: A conceptual framework and research agenda. *Human Resource Management Review*, 27(4): 575–589. https://doi.org/10.1016/j.hrmr.2016.12.005.

Meyerson, D.E. (1990). Uncovering socially undesirable emotions: Experiences of ambiguity in organizations. *American Behavioral Scientist*, 33(3): 296–307. https://doi.org/10.1177/0002764290033003004.

Mischel, W. (1977). On the future of personality measurement. *American Psychologist*, 32(4): 246. https://doi.org/10.1037/0003-066X.32.4.246.

Mosier, K.L. and Chidester, T.R. (1991). Situation assessment and situation awareness in a team setting. *Designing for Everyone*, 798–800.

Netzer, O., Feldman, R., Goldenberg, J. and Fresko, M. (2012). Mine your own business: Market-structure surveillance through text mining. *Marketing Science*, 31(3): 521–543. https://doi.org/10.1287/mksc.1120.0713.

OpenAI. (2023, January). *ChatGPT.* https://openai.com/blog/authors/openai/.

Opoku, R., Abratt, R. and Pitt, L. (2006). Communicating brand personality: Are the websites doing the talking for the top South African business schools?. *Journal of Brand Management*, 14(1): 20–39. https://doi.org/10.1057/palgrave.bm.2550052.

Orasanu, J. (1990, July). Shared mental models and crew decision making. In *12th Annual Conf. CSS Pod*, 1066–1066. Psychology Press.

Packard, G. and Berger, J. (2021). How concrete language shapes customer satisfaction. *Journal of Consumer Research*, 47(5): 787–806. https://doi.org/10.1093/jcr/ucaa038.

Pang, B. and Lee, L. (2008). Opinion mining and sentiment analysis. *Foundations and Trends® in Information Retrieval*, 2(1-2): 1–135. http://dx.doi.org/10.1561/1500000011.

Peeters, M.A., Van Tuijl, H.F., Rutte, C.G. and Reymen, I.M. (2006). Personality and team performance: A meta-analysis. *European Journal of Personality*, 20(5): 377–396. https://doi.org/10.1002/per.588.

Pew Research Center. (2021, April). *Mobile Fact Sheet.* https://www.pewresearch.org/internet/fact-sheet/mobile/.

Porter, M.F. (1980). An algorithm for suffix stripping. *Program: Electronic Library and Information Systems*, 14(3): 130–137. https://doi.org/10.1108/eb046814.

The Radicati Group. (2021, June). *Email statistics report 2021-2025.* https://www.radicati.com/wp/wp-content/uploads/2021/Email_Statistics_Report,_2021-2025_Executive_Summary.html.

Reichheld, F.F., Teal, T. and Smith, D.K. (1996). *The Loyalty Effect*, 1(3): 78–84. Boston, MA: Harvard Business School Press.

Reynolds, P.D. (1997). Who starts new firms? Preliminary explorations of firms-in-gestation. *Small Business Economics*, 9(5): 449–462. https://doi.org/10.1023/A:1007935726528.

Sackett, P.R. and Ellingson, J.E. (1997). The effects of forming multi-predictor composites on group differences and adverse impact. *Personnel Psychology*, 50(3): 707–721. https://doi.org/10.1111/j.1744-6570.1997.tb00711.x.

Sadler-Smith, E. and Shefy, E. (2004). The intuitive executive: Understanding and applying 'gut feel'in decision-making. *Academy of Management Perspectives*, 18(4): 76–91. https://doi.org/10.5465/ame.2004.15268692.

Saif, H., Fernández, M., He, Y. and Alani, H. (2014). On stopwords, filtering and data sparsity for sentiment analysis of twitter. *LREC 2014, Ninth International Conference on Language Resources and Evaluation. Proceedings*, pp. 810–817. http://lrec2014.lrec-conf.org/en/.

Sanderson, P.M. and Fisher, C. (1994). Exploratory sequential data analysis: Foundations. *Human–Computer Interaction*, 9(3-4): 251–317. https://doi.org/10.1080/07370024.1994.9667208.

Sasou, K. and Reason, J. (1999). Team errors: Definition and taxonomy. *Reliability Engineering & System Safety*, 65(1): 1–9. https://doi.org/10.1016/S0951-8320(98)00074-X.

Schmidt, F.L. and Hunter, J.E. (1998). The validity and utility of selection methods in personnel psychology: Practical and theoretical implications of 85 years of research findings. *Psychological Bulletin*, 124(2): 262. https://doi.org/10.1037/0033-2909.124.2.262.

Schofield, A. and Mimno, D. (2016). Comparing apples to apple: The effects of stemmers on topic models. *Transactions of the Association for Computational Linguistics*, 4: 287–300. https://doi.org/10.1162/tacl_a_00099.

Scott, T., Mannion, R., Davies, H. and Marshall, M. (2003). The quantitative measurement of organizational culture in health care: A review of the available instruments. *Health Services Research*, 38(3): 923–945. https://doi.org/10.1111/1475-6773.00154.

Sen, A., Das, A., Ghosh, K. and Ghosh, S. (2012, October). Screener: A system for extracting education related information from resumes using text based information extraction system. In *Proceedings of the International Conference on Computer and Software Modeling, Cochin, India*, 20–21.

Staw, B.M. (1991). Dressing up like an organization: When psychological theories can explain organizational action. *Journal of Management*, 17(4): 805–819. https://doi.org/10.1177/014920639101700412.

Stemler, S. (2000). An overview of content analysis. *Practical Assessment, Research, and Evaluation*, 7(1): 17. https://doi.org/10.7275/z6fm-2e34.

Sutcliffe, K.M., Lewton, E. and Rosenthal, M.M. (2004). Communication failures: An insidious contributor to medical mishaps. *Academic Medicine*, 79(2): 186–194.

Viswesvaran, C., Ones, D.S. and Schmidt, F.L. (1996). Comparative analysis of the reliability of job performance ratings. *Journal of Applied Psychology*, 81(5): 557. https://doi.org/10.1037/0021-9010.81.5.557.

Viswesvaran, C., Schmidt, F.L. and Ones, D.S. (2005). Is there a general factor in ratings of job performance? A meta-analytic framework for disentangling substantive and error influences. *Journal of Applied Psychology*, 90(1): 108. https://doi.org/10.1037/0021-9010.90.1.108.

Wagner et al. (2022). [Unpublished raw data on using LDA to create a taxonomy of start up cultures]. Blackhawke Behavior Science.

Wald, A. (2004). *Sequential analysis*. Courier Corporation.

Watt, J.H. and VanLear, A.C. (1996). *Dynamic Patterns in Communication Processes*. Thousand Oaks, CA: Sage Publications.

Weissman, B. and Tanner, D. (2018). A strong wink between verbal and emoji-based irony: How the brain processes ironic emojis during language comprehension. *PloS One*, 13(8): e0201727. https://doi.org/10.1371/journal.pone.0201727.

West III, G.P. (2007). Collective cognition: When entrepreneurial teams, not individuals, make decisions. *Entrepreneurship Theory and Practice*, 31(1): 77–102. https://doi.org/10.1111/j.1540-6520.2007.00164.x.

Wilhelmy, A., Kleinmann, M., König, C.J., Melchers, K.G. and Truxillo, D.M. (2016). How and why do interviewers try to make impressions on applicants? A qualitative study. *Journal of Applied Psychology*, 101(3): 313. https://doi.org/10.1037/apl0000046.

Xiao, Y., Seagull, F.J., Mackenzie, C., Ziegert, J. and Klein, K.J. (2003, October). Team communication patterns as measures of team processes: Exploring the effects of task urgency and shared team experience. In *Proceedings of the Human Factors and Ergonomics Society Annual Meeting*, 47(12): 1502–1506. Sage CA: Los Angeles, CA: SAGE Publications. https://doi.org/10.1177/154193120304701228.

Yi, X., Allan, J. and Croft, W.B. (2007, July). Matching resumes and jobs based on relevance models. In *Proceedings of the 30th Annual International ACM SIGIR Conference on Research and Development in Information Retrieval*, 809–810. https://doi.org/10.1145/1277741.1277920.

Zaman, A.N.K., Matsakis, P. and Brown, C. (2011, September). Evaluation of stop word lists in text retrieval using Latent Semantic Indexing. In *2011 Sixth International Conference on Digital Information Management*, 133–136. IEEE. https://doi.org/10.1109/ICDIM.2011.6093315.

Zellmer-Bruhn, M. and Gibson, C. (2006). Multinational organization context: Implications for team learning and performance. *Academy of Management Journal*, 49(3): 501–518. https://doi.org/10.5465/amj.2006.21794668.

Zhao, M., Javed, F., Jacob, F. and McNair, M. (2015, March). SKILL: A system for skill identification and normalization. In *Twenty-Seventh IAAI Conference*.

Živković, M. (2022, June 17). *Slack vs microsoft teams: How do the tools compare?* Chanty. https://www.chanty.com/blog/slack-vs-microsoft-teams/#:~:text=vs%20Microsoft%20Teams.-,Which%20tool%20is%20ahead%20of%20the%20game%3F,advantage%20of%20the%20free%20platform.

Chapter 10

Promoting Well-Being and Innovation in Startups

The Role of the Social Environment

*Lindsey M Freier** and *Ian M Hughes*

Startups and scaleups face a unique set of challenges and opportunities. In the midst of building and testing a minimally viable product at a breakneck pace, pitching to investors, building a brand, selling, and maintaining a resourceful mindset, an essential aspect of a thriving organization can slip through the cracks: building a positive social environment. While the consequences may not be immediately obvious, this oversight comes at a substantial cost to founders (e.g., a struggling startup), employees (e.g., lowered morale and health), and nascent organizations (e.g., the financial cost of employee turnover and poor performance). A positive social environment, in contrast, is the foundation of a thriving company.

WeWork, once denoted as a "unicorn" startup, is a quintessential example of the importance of the social environment. Founded in 2010 by Adam Neumann, WeWork had a vision for more dynamic, collaborative workspaces. They quickly partnered with businesses across the United States to "elevate" the way work was done (Sherman 2019). In 2017, the Japanese multinational conglomerate holding company SoftBank invested $4.4. billion into the company, raising their value to around $20 billion (the company would peak at $47 billion; Konrad 2017, Pietsch 2020). The following years, however, proved to be the downfall of WeWork, with hundreds of employees decrying the toxic culture of the company, including the

[1] Department of Psychology, Bowling Green State University, 822 E Merry Ave, Bowling Green, 43402, 720-480-5928.

[2] Department of Psychology, Bowling Green State University, 822 E Merry Ave, Bowling Green, 43402, 3077576573,

Email: ihughes@bgsu.edu

* Corresponding author: lfreier@bgsu.edu

filing of multiple discrimination-related lawsuits (Burden and Clarey 2020). Indeed, even the company's Director of Culture, Ruby Anaya—who proclaimed the company had an "entitled, frat-boy culture"—stated that she had been fired after reporting to HR that she had been sexually assaulted by another employee (Sherman 2019). By May 2020, fresh off another scandal surrounding unsafe working conditions (i.e., formaldehyde in phone booths, Sandler 2019), the company's value had plummeted to $2.9 billion (Pietsch 2020). In the aftermath, WeWork was forced to work to restore the damage to its reputation resulting from its poor treatment of employees and other scandals. While this story exists at the extreme end of the spectrum, it demonstrates the power of the social environment in startups, and the dire consequences of a negative environment (see also: Uber, Isaac 2017).

One of the essential features of a thriving organization, and the focus of this chapter, is a positive social environment. For this chapter, a positive social environment is defined as a work environment characterized by positive exchanges and relationships (e.g., friendships, social support), and a lack of negative exchanges and interactions (e.g., abuse, bullying, sexual harassment). Research suggests that a positive work environment is integral to the long-term flourishing of any startup or scaleup. Indeed, a workplace where organizational members feel respected, valued, and included is one where employees are healthier (e.g., Panaccio and Vandenberghe 2009), more productive (e.g., Kurtessis et al. 2017), and more innovative (e.g., Yu and Frenkel 2013). Conversely, stressful environments, such as those perceived as uncivil, chaotic, or toxic, contribute to particularly harmful outcomes, such as job dissatisfaction and the intent to quit (e.g., Paulin and Griffin 2017). For this reason, leaders of startups ought to invest in the creation of a supportive and inclusive social environment.

How, though, is a positive social environment fostered? This chapter is focused on the social environment of the workplace, which includes the social exchanges (i.e., exchanges whose outcomes are contingent on the actions of all parties, e.g., emotional social support) and interactions (i.e., briefer exchanges whose outcomes are driven by one party, e.g., rudeness) of organization members, and the relationships in which these interactions and exchanges take place (e.g., friendships). This chapter also discusses the important role that the social environment plays in supporting employee well-being and innovation and offers recommendations for cultivating aspects of a positive social environment, while curtailing features of a negative social environment. The chapter concludes with actionable recommendations for developing a positive social environment, such as the role of measurement and analytics.

Building a Positive Environment Early

Symbolic interactionism, though broad, can be applied to explain and predict organizational dynamics. *Symbolic interactionism* clarifies the particular importance of establishing a positive social environment early in the lifecycle of an organization. Symbolic interactionism is a micro theory focused on the construction of societies, wherein society is a web of communication between people. The guiding principle of symbolic interactionism is that people act toward things (e.g., situations, people) based on assigned meanings, and that things obtain meaning through communication. Communication contains symbols (e.g., artifacts that elicit feeling and action), which help to guide and constrain behavior in future interactions. Over time, repeated styles of action and communication lead to the development of behavioral and communicatory norms (Stryker and Vryan 2006). Norms are very powerful, and shape the behavior of those who occupy particular environments—newcomers, tenured occupants, and everyone in between (e.g., Miller and Prentice 2016).

Exchanges, such as social support cycles, and interactions, such as incivility (e.g., someone using an improper or rude tone), each contains symbols that people use to assess their workplace and the situations they encounter within it. Social support cycles, for instance, signal to others that they are in a workplace where psychosocial resources such as compassion and kindness are often exchanged, whereas incivility signals that the workplace permits rudeness or aggression. Over time, people will modify their behavior to comply with the communicatory symbols they perceive, helping to create social norms within the workplace (Stryker and Vryan 2006). As noted by Schneider (1987), "it is the kind of persons in environments who determine the kinds of human environments they are" (p. 439); the way people interact with each other largely determines the environment that will be perceived.

In line with the symbolic interactionist frame (Stryker and Vryan 2006), founders and startup leaders are in a unique position to construct an environment from the "ground up"—putting an emphasis on positive social exchanges and interactions and working to prevent or eliminate negative ones. Indeed, once an organizational culture is in place, it can be incredibly difficult, costly, and time-consuming to change (De Witte and van Muijen 1999). Thus, startup founders should capitalize on the opportunity to create a desirable, collaborative, pleasant culture from the beginning, as a positive social environment offers a variety of advantages to new organizations.

Well-Being and Innovation as Mechanisms for Startup Survival

While there are many benefits to a positive social environment, this chapter focuses on well-being and innovation as two outcomes that are key to startup success. Beginning with the former, employee well-being is related to employee productivity, retention, and business-unit level profitability (e.g., Krekel et al. 2019, Wright and Bonnett 2007). As discussed in the previous section, social exchanges and interactions are of great importance for the formation of workplace environments. Organizational research positions a favorable working environment as a cornerstone for employee health and well-being. For example, empirical work from Burns and Machin (2012) found that a positive work environment directly contributed to positive morale and affect. Conversely, when one perceives their working environment to be uncivil or hostile, they are more likely to experience a bevy of negative outcomes, such as reduced health and life satisfaction and increased psychological distress and burnout (Miner-Rubino and Cortina 2004, Sloan 2012, Spector and Bruk-Lee 2007).

A positive social environment fosters continued creativity and innovation as a startup matures. Innovation is important for sustained organizational success (e.g., Spender et al. 2017) — Fortune 500 companies like IBM list creativity as a top organizational priority (Perry-Smith and Shalley 2014). Innovation and startups are theoretically and practically intertwined: startups bring new, disruptive, or inventive ideas to the market and then develop them over time into economically sustainable and scalable enterprises (Spender et al. 2017). For example, Airbnb (valued at over $110 billion, Forbes 2021), launched in 2007, helped pioneer the peer-to-peer lodging model, competing with large conglomerates such as Hyatt and Marriott that have dominated the market for decades. To compete with Airbnb and adapt to the changing industry, these lodging firms are attempting to emulate aspects of Airbnb's business model (Zach et al. 2020). In other words, innovation is essential to business survival. Innovation, the "successful implementation of creative ideas", and its precursor, creativity, the "production of novel and useful ideas by an individual or small group" (Amabile and Pratt 2016, p. 158) can be either positively or negatively impacted by the work environment in which they take place. Thus, a positive social environment — through its effect on innovation and creativity — is essential to survival.

Creativity and innovation are inherently social processes, and the knowledge and perspectives of multiple organizational members are often necessary to develop and implement creative ideas (Amabile and Pratt 2016). Thus, the quality of innovations is tied to the quality of interpersonal exchanges and relationships in the workplace (e.g., Muñoz-Doyague and Nieto 2012, Ozer and Zhang 2022). A positive social environment shares many of the same features as an environment that fosters creativity, including psychological safety, group cohesion, frequent communication, and the integration of diverse perspectives (Hulsheger et al. 2009) and functions as an effective foundation for building an environment conducive to creative ideas and outcomes. In addition, as mentioned previously, positive social environments more frequently induce positive emotional states (e.g., Jolly et al. 2021), which have been consistently linked to creativity (Davis 2009). Finally, a positive social environment helps retain top creative talent by increasing well-being, which helps to embed employees within the startup (e.g., Park and Min 2020). Employees that are more embeded within their organizations are less like to leave (Jiang et al. 2012), which allows them to continue to contribute to innovation. The following sections discuss key exchanges, interactions, and relationships, the ways in which these aspects of the social environment, both positive and negative, influence worker well-being and innovation, and how they can be promoted or discouraged.

The Role of Social Support and Informal Relationships in the Social Environment

As discussed in the above section, the social environment of a startup or scaleup is an important feature of a thriving organization, but what exactly does a desirable social environment entail? In a positive social environment, organizational members feel supported, comfortable asking for help when necessary, able to collaborate and form connections, and safe to take necessary risks. A positive environment is marked by trust, open communication, equity, and inclusion. While an entire book could be written on any one of these topics, this section will focus on the exchanges and relationships that form the foundation of a positive environment and have been the focus of a substantial body of research: social support and informal workplace relationships.

Social Support

Social support is defined as the "psychological or material resources that are provided to a focal individual by partners in some form of social relationship" (Jolly et al. 2021, p. 229). Although there are many types of social relationships (e.g., friends, family, coworkers), this chapter focuses on social support in the workplace. Social support takes a variety of forms such as talking a coworker through a difficult work problem, explaining an unfamiliar task, engaging in a discussion of an emotional event, or providing encouragement on a difficult day.

Social support is generally divided into two categories: instrumental and emotional (Mathieu et al. 2019). Instrumental support is oriented toward work tasks, while emotional support is oriented toward psychological or interpersonal difficulties. Importantly, all social support behaviors provide resources (i.e., facilitate goal attainment; Halbesleben et al. 2014) to the receiving party. Instrumental support resources may take the form of conveying knowledge, or time spent helping with a work task. In contrast, the resources shared in emotional support are more social or emotional in nature, such as empathy, warmth, and understanding. Although instrumental and emotional support are conceptually distinct and the categories are helpful for understanding different behaviors and resources, they tend to be linked in interpersonal interactions (Mathieu et al. 2019) and both are essential aspects of a positive work environment.

Establishing Social Support Norms: Reciprocity and Spirals

As symbolic interaction theory states, norms are built over time through repeated patterns of behaviors (Stryker and Vryan 2006) and play an important role in shaping ongoing behaviors. Two aspects of social support are particularly beneficial for establishing norms and each plays a unique role in allowing social support to proliferate throughout the organization: reciprocity norms and resource gain spirals.

Reciprocity is a social norm across human societies, with the basic principle being that actions ought to be returned in kind (Cropanzano and Mitchell 2005). Thus, a positive social support behavior from a coworker is likely to be returned by the recipient (Bowling et al. 2005). This return need not occur immediately, or be of the same kind (e.g., instrumental to instrumental) and instead may depend on perceived need. For example, imagine Deandra, an experienced programmer, helping Alex, who is new to the team, with troubleshooting a difficult line of code (instrumental support). While Alex may not be knowledgeable enough to reciprocate by helping Deandra with a coding problem, Alex will be more likely to offer a listening ear and kind words (emotional support) when Deandra seems to be having a difficult day a week later. Deandra, in turn, is more likely to help Alex next time a difficulty arises, regardless of whether it requires emotional or instrumental support. Indeed, if these exchanges continue, they form the basis for a high-quality social relationship (Cropanzano and Mitchell 2005), in addition to the formal working relationship required by Deandra and Alex's roles. Social relationships, which are discussed in the next section, provide further benefits to both startups and their members.

Resource gain spirals also provide a key avenue through which positive norms are established. Resources must often be invested to gain further resources (Hobfoll 2001), so the more resources available to an individual, the more they have available to invest, and the more that can be further gained. This reciprocal process is referred to as a resource gain spiral. For example, when Deandra helps Alex to solve a coding problem more quickly, Alex can instead use that time (a resource), to invest in generating further resources, such as by taking the time to learn a new programing skill (a resource). Alex can then invest the new skill by using it to solve a different problem more quickly (gaining more time) or teaching it to Deandra. The latter is an example of an interindividual resource spiral, in which resource investments by each exchange partner benefit the other in an ongoing positive cycle (Halbesleben and Wheeler 2015). Further, the spiral need not be limited to just the two exchange partners but can also extend throughout both partners' work networks.

Together, these two aspects of social support feed off one another and can turn a seemingly small offer of help into a substantial resource for those within and beyond the focal pair. They also allow early efforts towards facilitating social support to establish norms quickly, because one positive action precipitates future positive actions. Thus, the benefits of one act of social support continue beyond the original resources and recipient. These mechanisms also function, often on a long-term scale, within informal social relationships in the workplace.

Informal Social Relationships in the Workplace

Exchanges such as social support often result in the development of workplace relationships, which are defined as long-term patterns of exchanges between two members directed toward common purposes (Ferris et al. 2009). In the modern world of work where collaboration is often essential, relationships form the basis for a productive startup. While workplace relationships include both those formally designated by the organizational structure, and those that grow informally, this chapter focuses on the latter as key differentiators of a positive work environment.

Workplace Friendships

One of the most common informal workplace relationships is friendship. Many startup founders are friends before they begin working together, providing direct benefits to the leadership team, such as improved cooperation and decision-making outcomes (Francis et al. 2004, Rank and Tuschke 2010). In addition, this early basis for the social environment, if leveraged appropriately, can form a solid foundation for a workplace in which friendships are the norm. If founders model positive social interactions and relationships, and encourage such behavior throughout the startup, this approach is likely to become pervasive in the culture of the organization as it grows.

Workplace friendships are differentiated from other workplace relationships by four core features: voluntariness (i.e., chosen freely), informality (i.e., lack of standardized expectations for role behavior), communal norms (i.e., the understanding that support will be given based on needs of the other party), and socioemotional goals (i.e., the primary purpose of the relationship being interpersonal well-being; Pillemer and Rothbard 2018). Friendships are also holistic, meaning that friends see one another as whole individuals, including both work and non-work aspects (Morrison and Cooper-Thomas 2016). Friendships are formed from a combination of mutual attraction, where each member is interested in engaging with the other, and proximity (Nahemow and Lawton 1975, Sias and Cahill 1998).

Ferris et al. (2009) propose a four-stage model of relationship development, in which the quality and progression of the relationship are contingent on met expectations. In the first stage, initial interaction, parties often meet and begin interacting because of formal role requirements (e.g., task interdependence; Zhang et al. 2021) or physical proximity in the workspace (e.g., adjoining desks). Instrumentality (i.e., the perceived potential value of the relationship), positive affective reactions, and a sense of respect are the key dimensions of the initial stage. The second stage is characterized by the development and expansion of roles, growing trust and self-disclosure (e.g., Gibson 2018), early offerings of social support, such as those Deandra and Alex provide in the previous section, and a continued importance of instrumentality. In the third stage, as a relationship matures and is seen as a value in and of itself, the importance of instrumentality decreases substantially and is replaced by an emphasis on positive affect, trust, mutual commitment, respect, and the offering of support to meet the needs of the other party. Flexibility and adaptability also become increasingly important as the relationship persists through changing external conditions (e.g., one party moves to a different department). In the fourth and final stage, mutual loyalty and increased interpersonal commitment become defining aspects of the relationship, in addition to trust, respect, positive affect, and support. The characteristics that define the third and fourth stages, in

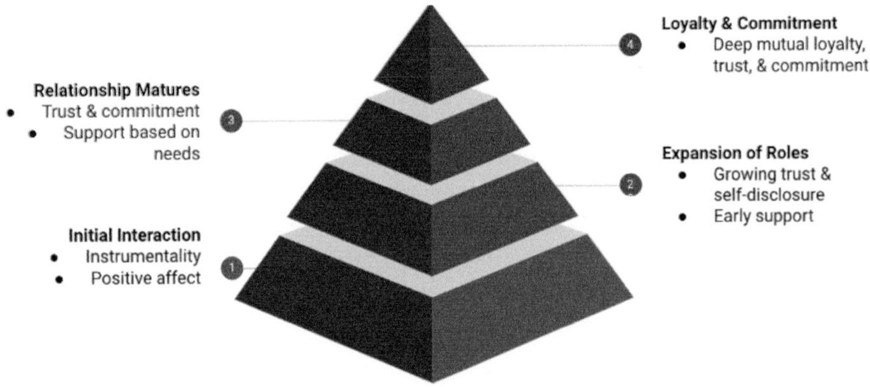

Figure 1. Stages of friendship development.

particular, form the basis for the benefits of friendships to relationship members, and the startups in which they work.

Romantic Workplace Relationships

Another common, though somewhat controversial type of informal workplace relationship, is a romantic relationship. A workplace romance is defined as a "consensual relationship between two members of the same organization that entails mutual sexual attraction" (Pierce and Aguinis 2009, p. 447). Although romances can, and do, occur within supervisor-subordinate relationships, this chapter focuses on those between peers. Though the instinctive response of many organizations is to discourage workplace romances, romantic relationships can contribute positively to the social environment through many of the same mechanisms of friendships. For example, interacting with their significant other in the workplace can increase each partner's positive emotions, increasing their friendliness and pro-social behavior toward other co-workers throughout the day (Biggs et al. 2012). As discussed below, romances and friendships often have similar benefits. It should be noted, however, that while romances can have a positive influence on the work environment, there can also be drawbacks, such as legal concerns and sexual harassment. Thoughtful workplace policies (see Pierce and Aguinis 2009 for a review and recommendations), however, can mitigate risks without reducing the broader benefits of romantic relationships.

Table 1. Definitions of positive social exchanges and relationships.

Positive Social Exchange	Definition
Emotional Social Support	The provision of psychological resources, e.g., listening empathetically to a colleague discussing a personal problem.
Instrumental Social Support	The provision of work-related resources, e.g., helping a colleague to solve a difficult work problem.
Workplace Friendship	A voluntary, informal, and holistic relationship between colleagues that is guided by communal norms, for the purpose of meeting socioemotional goals.
Workplace Romantic Relationship	A consensual relationship between two members of the same organization that entails mutual sexual attraction.

Benefits of Social Support and Informal Relationships

Features of a positive social environment, such as social support and informal relationships offer benefits for both organizational members and the organizations themselves. Startups that work to build a positive social environment from the beginning will reap the rewards throughout the life cycle of the organization because of the strength and persistence of organizational norms (Miller and Prentice 2016).

Well-Being

Social support and informal social relationships provide a variety of benefits for employee well-being. Social support has a positive impact on wellbeing, both directly, and by mitigating the negative effects of other stressors (e.g., work overload, conflicting expectations, ambiguity) in the workplace (Cohen and Wills 1985), the latter of which is referred to as the 'buffering hypothesis'. The benefits of social support for wellbeing have been widely studied, and meta-analyses (e.g., Mathieu et al. 2019, Viswesvaran et al. 1999) suggest that social support is consistently related to reduced burnout (see later section for a more in-depth discussion of burnout), negative physical symptoms (e.g., headaches), and voluntary turnover, and increased job satisfaction, commitment, and task performance. The buffering hypothesis has also been widely supported in that social support reduces perceived stressors and resulting negative outcomes. The type of social support may also be important for some outcomes. Emotional social support may be particularly beneficial for negative physical symptoms, while instrumental social support is particularly beneficial for job satisfaction and performance (Mathieu et al. 2019).

Social relationships also provide a variety of benefits for wellbeing through increased social support, positive affect (e.g., Methot et al. 2016), and the satisfaction of needs (i.e., autonomy, competence, and relatedness; Quinn 2017). Empirical studies have established relationships between informal workplace relationships and job satisfaction, commitment, health, wellbeing, and reduced voluntary turnover (e.g., Nielsen et al. 2000, Morrison 2005, Winstead et al. 1995, Zhang et al. 2021). Romantic relationships in the workplace, which are sometimes perceived negatively, have also been linked to positive outcomes such as psychological wellbeing, affective commitment, and job involvement (Khan et al. 2017, 2018, 2022). The positive well-being outcomes described above also provide avenues through which both friendships and romantic relationships have been linked to improved job performance and work effort (e.g., Jung and Yoon 2020, Khan et al. 2017, 2022).

Innovation

Social relationships often form the basis for creativity and innovation in the workplace as a positive social environment contains many of the necessary ingredients for innovation. Although social support exchanges may have similar benefits for creativity and should be investigated in future research, friendships have been the focus of most studies linking the social environment to creativity. First, friendships increase psychological safety (Cao and Zhang 2020, Helmy and Wiwoho 2020), which is defined as the shared belief that the workplace is safe for interpersonal risks (Edmondson 1999). Psychological safety is important for innovation; as novel ideas may be perceived as risky (Mueller et al. 2012). Friendships also increase knowledge sharing (Helmy and Wiwoho 2020), cooperation, and positive communication (Jehn and Shah 1997), all of which have been shown to positively impact creativity (Hulsheger et al. 2009). Friendships also reduce the negative effects of faultlines within groups (Ren et al. 2015), which may otherwise function as a barrier to innovation.

Similarly, friendships play an important role in linking together different groups within the workplace and maximizing the benefits of the diverse knowledge and backgrounds that organizational members bring to the table. For example, intercultural friendships and romantic relationships increase creativity and workplace innovation (Lu et al. 2017). Friendships can also link together separate organizations and disparate functions and teams within a single organization: when individuals move through different roles, teams, or organizations, strong friendship ties may remain even though there is no longer a formal link. These informal links increase information sharing, communication, and the crossing of functional or project boundaries (Yakubovich and Burg 2019), thus improving innovation requiring multiple teams or divergent backgrounds.

Fostering a Positive Social Environment

Fostering a positive social environment requires a multifaceted approach. Five recommendations are provided below to guide the development of a strategy for fostering a positive work environment. Each of the five recommendations should be applied based on the overall strategy and values of the startup and its founders. The recommendations will be most effective if customized in a careful and thoughtful way.

Recommendation 1: Establish Values

The values of a startup should be established early and should reflect the founders' desires for the organization both externally and internally. These values should be explicitly stated, as in a mission statement or guiding principles document, and referred to by all employees as a guide for behavior. The latter is essential as principles that are seen as window dressing will not have a positive effect on the work environment. In contrast, well established and lived values can have a strong influence on behavior (Gonzalez-Roma and Peiro 2014). Emphasis on a positive social environment should be included in stated values. What the exact language looks like will be different for every startup, but could include words such as 'respect', 'collaboration', or 'support'. The below sections provide suggestions for ensuring that stated values become lived values.

Recommendation 2: Lead by Example

The behaviors of startup founders and leaders establish early norms for the organization. Startup founders should build and engage a leadership team that not only brings in important skills and competencies but shares the values of founders and lives them out. This should be a focus for leaders at all levels of the startup as it grows into a larger organization. Just as organizational founders set the tone for the organization (Schneider 1987), lower-level leaders establish norms for their immediate work groups based on the norms set by their superiors. Thus, leaders at all levels should be encouraged to model behaviors related to a positive social environment and support their team members in providing social support and establishing informal relationships.

Recommendation 3: Provide Opportunities

A positive social environment will not result from established values and leaders who model positive behaviors if organizational members do not have the opportunity to engage in behaviors reflective of a positive social environment themselves. To truly establish norms, employees must have the time, autonomy, and space for forming relationships and providing social support. Socializing should be encouraged (Holt-Lunstad 2018) not only verbally and with the modeling of behaviors, but also through the physical work environment (see Khazanchi et al. 2018 for a thoughtful overview).

Special consideration should also be taken for remote workers, for whom forming a relationship and receiving necessary support can be more difficult (Beauregard et al. 2019). When possible, it is beneficial for the social environment for employees to work remotely only part time (e.g., coming into the office 2–3 days per week; ten Brummelhuis et al. 2010), as it allows for the formation of relationships face-to-face, and increases perceptions of support (Collins et al. 2016). If part-time in-person work is not possible, norms of socializing virtually should be established. This can take a variety of forms, from encouraging employees to pause to eat lunch together virtually, to setting aside time at the beginning of a small-group meeting to catch up, to hosting virtual happy hours at the end of the week.

Recommendation 4: Create Thoughtful Policies and Work Systems

Formal policies and work systems should also correspond to the desired positive work environment. For example, high commitment work systems which are defined as, "human resource management practices (such as employee participation, internal promotion, team rewards, profit sharing, extensive training and benefits, and job security) that signal commitment to the employees" (Xiao and Tsui 2007, p. 2) have been found to increase employee wellbeing by increasing friendships (Zhang et al. 2021). Formal policies, such as those surrounding romantic relationships, should be designed with a positive work environment in mind. Policies should balance the reduction of risk and maximization of benefits (Pierce and Aguinis 2009). Finally, the quality of the work environment and success of any interventions should be measured frequently to demonstrate success and identify areas of opportunity. Measurement can take a variety of forms, from annual organizational culture surveys to small pulse surveys for specific work groups.

Recommendation 5: Reward and Reinforce Desired Behaviors

Finally, establish a system to reward desired behaviors. Rewards can take a variety of forms, from a simple thank you from a colleague, to a shout-out in a meeting, to an organization wide recognition system, or the inclusion of prosocial behavior in performance reviews. Employees' direct leaders play a particularly important role as they often have the most direct effect on formal reward decisions and contribute to the establishment of reward norms for the teams they lead.

The Role of Negative Exchanges and Interactions in the Social Environment

In addition to fostering and encouraging positive workplace exchanges and interactions, startup founders and leaders should be mindful of the potential negative exchanges and interactions that can take place at work. According to symbolic interactionism (Stryker and Vryan 2006), negative exchanges and interactions also work to establish cultural and communicatory norms—in this instance, toxic ones. Curtailing such negative exchanges and interactions is of great importance to startups, as there can be substantial negative consequences if toxic norms are established (e.g., WeWork). In this section, a variety of negative exchanges and interactions that contribute to workplace dynamics are highlighted. In addition, the impacts of these various exchanges and interactions on employee well-being and innovation, as well as actionable recommendations for curtailing them, are described.

Types of Negative Exchanges and Interactions

Incivility

Incivility, commonly defined as "low-intensity deviant behavior with ambiguous intent to harm the target, in violation of workplace norms for mutual respect" (Andersson and Pearson 1999, p. 457), is an almost ubiquitous workplace interaction. Porath and Pearson (2010) report that only *1% of employees* surveyed in their research reported having *not* witnessed incivility. An example of incivility is illustrative: Rashad is walking to his desk at his job and says hello to a coworker in the hallway, who ignores him and keeps walking. This would be considered an act of incivility, as the intent to harm him was unclear. Did the person not hear him? Are they choosing to ignore him? Such rumination is what makes incivility particularly harmful to those routinely exposed. While most would agree that having such an interaction once or twice would be tolerable, an environment where employees routinely experience such behavior can be catastrophic if left unattended (see Yao et al. 2021).

Unhelpful Help

Although social support is often a net positive for employees (see the sections above), social support can, if not provided appropriately, function as a stressor. This is referred to as unhelpful workplace social support (UWSS), defined by Gray et al. (2020) as "any action taken by a supervisor and/or colleague that the recipient believes was intended to benefit him or her but is perceived as unhelpful or harmful" (p. 376). Incivility and UWSS are alike in that they are both subtle exchanges, and UWSS can come across as ambiguous with intent to harm, but UWSS—unlike incivility—still represents the provision, or attempted provision, of some sort of resource. Uncivil exchanges do not make attempts at such provisions. UWSS can take several forms; for instance, one could be rude while providing support to another (critical social support), provide incomplete or unclear support (partial social support), or provide social support when it is not wanted by the recipient (imposing social support). Perhaps a simpler way to conceptualize UWSS is by using logic from Beehr and colleagues (2010), who describe social support as stressful if it meets any of the following three categories: (1) the social support makes negative aspects of the environment more salient, (2) the social support is not wanted, or (3) the social support makes the recipient feel inadequate or incompetent. Empirical research on this form of social support is nascent, though recent work (e.g., Gray et al. 2020, Hughes et al. 2022) has helped to illustrate the disruptive, deleterious nature of UWSS.

Abusive Behavior

More overt negative exchanges are also a point of concern for organizations. These explicit exchanges typically take the form of abusive behaviors toward others. Verbal harassment or hostile teasing, use of ethnic or racial slurs, threats of physical violence, obscene gestures (e.g., the finger), or nastiness to subordinates or customers, all constitute abuse. These exchanges, perhaps intuitively, leave little to no ambiguity in their intent to harm and are of higher intensity, separating them from uncivil exchanges. Organizational scholars have assessed the effects of abuse from a variety of sources, including peer-to-peer abuse (e.g., Berry et al. 2007, Spector et al. 2006), as well as supervisor-to-subordinate abuse (e.g., Mackey et al. 2017, Martinko et al. 2013).

Bullying

Bullying, although less frequent than other forms of mistreatment (e.g., 2–17% of employees report experiencing bullying, Hansen et al. 2006), represents another social exchange of concern for organizational dynamics. Although bullying is related to incivility and abuse, it is distinct from these behavioral families. Those who engage in bullying at work *may* leverage behaviors related to incivility or abuse against others (typically against those in lesser positions of power, Glasø et al. 2011), but bullying is defined by the frequency and target of the behaviors rather than by the behaviors themselves. Hauge and colleagues (2010) define bullying as "[when] an employee faces repeated and prolonged exposure to various forms of predominately psychological mistreatment" (p. 427). Indeed, other researchers (e.g., Hansen et al. 2006) also note the systematic and targeted nature of bullying.

Sexual Harassment

While workplace relationships, such as friendships and romantic relationships, have many benefits for startups and their members, there are serious consequences when boundaries are crossed (Elsesser and Peplau 2006). Sexual harassment is a family of behaviors that is, unfortunately, all too common in organizations. In fact, research suggests that one in every *two* women will experience some form of sexual harassment during her working life (e.g., Fitzgerald 1994). Although there are a variety of ways to conceptualize sexual harassment, Fitzgerald and colleagues' (1995) three category model is most clear. The model includes (1) gender harassment, which refers to "crude, verbal, and symbolic behaviors that convey hostile attitudes" (e.g., making fun of someone who is female for possessing "overtly" feminine qualities), (2) unwanted sexual attention, which refers to "sexual attention that is unwanted and/or unreciprocated by the target", and (3) sexual coercion, which refers to "subtle or explicit efforts to make job rewards contingent on sexual cooperation"

Table 2. Definitions of negative social exchanges and interactions.

Negative Social Exchange	Definition
Incivility	Low intensity behavior that is ambiguous in its intent to harm others, e.g., rolling eyes at a colleague, ignoring a colleague.
Unhelpful help	Behavior that a recipient believes was intended to be helpful but is not; help that makes negative environmental factors more salient, is unwanted, or makes the recipient feel incompetent.
Bullying	Frequent and targeted mistreatment of a focal employee, which can take subtle (incivility) and explicit (abuse) forms.
Abuse	Harmful behavior that more explicitly breaches social norms, e.g., threatening others, verbal harassment.
Sexual Harassment	Gender harassment, unwanted sexual attention, sexual coercion.

(Lapierre et al. 2005, p. 156). Sexual harassment has long been a point of interest for organizational scholars (e.g., Loy and Stewart 1984, Maypole and Skaine 1983), and is a family of behaviors that founders of startups should work to prevent to ensure the longevity and success of their organizations.

Implications for Well-Being and Innovation

Many scholars (e.g., Nixon et al. 2021) have noted that negative social exchanges and interactions tend to be interrelated. Bullying, for instance, may consist of targeted incivility, abuse, sexual harassment, and UWSS. In addition, negative exchanges, like incivility, can result in negative spirals (Andersson and Pearson 1999), just as positive exchanges, such as social support, can result in positive spirals: small uncivil exchanges have the potential to escalate over time into more heated conflicts. Although they are conceptually distinct, because these exchanges are closely tied together, they tend to negatively impact well-being and innovation in similar ways.

Well-Being and Burnout

It is perhaps intuitive that the focal negative exchanges elaborated upon thus far would all work to erode employee well-being. Indeed, no one desires mistreatment or disrespect. The Golden Rule, or the ethic of reciprocity (for a deeper dive into reciprocity as a psychological construct, see the previous section on reciprocity and spirals), permeates all societies and works as a reminder to treat others in positive ways. The negative exchanges and interactions highlighted thus far all contravene this rule. They also contribute to lowered well-being in the form of burnout and negative emotions, which, in turn, impact behavior and overall health.

Burnout is typically viewed through the lens of Maslach and Jackson's (1981) three-factor model. Burnout consists of three different elements: emotional exhaustion, which refers to cognitive, physical, or emotional fatigue, cynicism (or depersonalization) which concerns feelings of being skeptical about the value of one's occupation, and reduced personal accomplishment, where one feels a sense of inefficacy—downplaying achievements and ability (Leiter et al. 2014, Maslach and Jackson 1981, Schaufeli et al. 2009). Burnout is a psychological construct of incredible organizational importance, as it contributes to massive revenue loss for organizations. Indeed, the American Institute of Stress reports that U.S. businesses lose up to *300 billion dollars* yearly because of stress- and burnout-related issues (Heckman 2019).

Each of the negative exchanges and interactions highlighted above can increase burnout. Incivility is thought to contribute to burnout primarily through rumination. That is, those who experience incivility have to spend time and precious cognitive and emotional resources attempting to unpack the meaning of the behaviors they are experiencing (Andersson and Pearson 1999, Cortina et al. 2001). For example, someone who is ignored after greeting someone in the hallway will likely expend resources determining whether they were intentionally or accidentally ignored. UWSS can result in similar ruminations (e.g., "did they provide improper information/bad advice on purpose, or did they think they were being helpful?"). Certain UWSS exchanges also leave recipients in a place where they must expend resources to fix issues resulting from improper support (e.g., someone providing "help" by doing a task incorrectly). Finally, bullying, experienced abuse, and sexual harassment all work to generate negative emotions such as anger, frustration, and shame, which in turn contribute to burnout (e.g., Laschinger and Fida 2014, Mathews et al. 2019, Wu and Hu 2009).

There are also other mechanisms through which these exchanges and interactions negatively affect well-being. For example, workplace bullying contributes to physiological symptoms, such as lower morning salivary cortisol levels (Hansen et al. 2006)—which are associated with disorders such as chronic fatigue (e.g., Strickland et al. 1998). Similarly, sexual harassment from men contributes to cardiovascular symptoms, such as increased diastolic and systolic blood pressure in women, particularly when they blame themselves for the harassment they receive (Schneider et al. 2001).

These exchanges can have ramifications for behavioral outcomes as well, namely retaliation, resulting from reciprocity norms, and displaced aggression. Experienced bullying (e.g., Fox and Stallworth 2005), UWSS (Hughes et al. 2022), abuse (e.g., Zhang et al. 2019), incivility (e.g., Welbourne and Sariol 2016), and sexual harassment (e.g., Anderson et al. 2022) are all related to perpetrated workplace deviance, such as being hostile to others and stealing time/property from one's organization. These kinds of behaviors are more formally known as counterproductive work behaviors (CWB), or voluntary workplace behaviors that intend to harm the organization, its members, or its stakeholders (Marcus et al. 2016). These behavioral reactions to negative social exchanges and interactions are often discussed as forms of displaced aggression (e.g., Spector and Fox 2002). Indeed, it is estimated that, in the United States alone, 2.8 million productive workdays are lost due to absenteeism (Ones and Dilchert 2013), and theft and fraud cost businesses up to $50 billion annually (Coffin 2003). Moreover, anywhere from 33 to 75 percent of all employees engage in CWB at some point in their career (Robinson and Bennett 1995)—often in response to the exchanges and interactions highlighted thus far. Thus, it is imperative that these negative social exchanges and interactions be given the proper level of attention and care.

Innovation

Although the relation between negative exchanges and interactions and innovation has not been researched as extensively as well-being, emerging research suggests these behaviors have important implications for employee and workforce innovation. Beginning with incivility, recent research from Motro et al. (2021) positioned incivility as detrimental to team creativity. The authors suggested that incivility would reduce positive affect within the team, thus reducing creativity. Interestingly, through three experiments, the authors found that incivility from women—but not men—reduced team creativity. The authors suggested that this may be due to men getting a "free pass" to engage in incivility, as it is more in line with expectations for their ("masculine") behavior. Research from Hur et al. (2016) took a more employee-oriented approach to the incivility–creativity relation. The authors found that incivility from colleagues worked to engender burnout (as mentioned in the previous section) which, in turn, reduced employee intrinsic motivation. As a result of this motivation decrease, individual employee creativity decreased. The extant literature, though nascent, seems to position colleague incivility as a net negative for creative performance (e.g., Zhan et al. 2019).

Similarly, behaviors such as abuse or bullying work to reduce employee creative performance. Starting with the former, researchers, such as Jiang and colleagues (2019), have found that abusive behavior erodes one's perceived creative self-efficacy, contributing to lower creative performance ratings. It has also been proposed from a theoretical perspective that bullying would work to stifle creativity on both an individual and team level (Creasy and Carnes 2017) and Mathisen et al. (2008) found that both being subjected to and observing bullying was negatively related to creative performance, though more research is needed Unfortunately, there has been no research to date examining the impacts of UWSS or sexual harassment on creative performance at work. Given the interrelatedness and similar mechanisms of these types of interactions, it is possible that UWSS and sexual harassment also share similar relationships with creativity and innovation. Psychological and behavioral dynamics are complex and complicated, however, and empirical research is needed to further investigate potential relations with creativity.

Reducing the Prevalence and Impact of Negative Social Exchanges and Interactions

Incivility, UWSS, abuse, bullying, and sexual harassment have substantial effects on well-being and innovation. Thus, it is important to deter negative exchanges and interactions whenever possible and reduce impacts when they do occur. To that end, four recommendations are provided below. As with fostering a positive environment, a multifaceted approach should be taken, and the recommendations provided should be considered within the broader context of the startup and its strategy.

Build Intolerance to these Exchanges into Company Culture

As the symbolic interactionist frame would suggest (Stryker and Vryan 2006), environmental norms can be difficult to change once they are in place—hence this chapter's heavy emphasis on promoting positive exchanges and interactions and curtailing negative ones. As discussed in the above section on establishing positive values, perhaps the most straightforward way to help craft an environment where these negative exchanges and interactions are frowned upon is by incorporating stances against such behaviors into organizational culture. For example, materials such as a document(s) clearly elaborating what behaviors are tolerated (positive exchanges and interactions) and which are not (negative exchanges and interactions) should be included in onboarding materials for new employees. Overly punitive language in such materials should be avoided, but it is nonetheless important that there is a clear message that incivility, bullying, sexual harassment, and the like will not be tolerated.

Establish Communication Channels

Having a clear, direct, and easy to access to a channel of communication for employees who are experiencing mistreatment is *paramount* for ensuring deviant acts, such as incivility or sexual harassment, do not become rife in the workplace. Indeed, many perpetrators of such behavior operate on the assumption that they will be able to get away with what they do (Cortina et al. 2013), and that they have power over their victims (e.g., Glasø et al. 2011). Founders and leaders should be sure that a Human Resources Officer, or other entity with authority to handle interpersonal disputes, can be easily reached by employees. Too often, employees feel that reporting such behavior (e.g., sexual harassment) is a muddy, frustrating process (Perry et al. 2009). Providing employees with a clear and accessible channel for reporting instances of mistreatment—and not punishing them for using it (see WeWork)—is paramount for ensuring such behaviors do not become a pattern.

Provide Training

Training employees to be mindful of such behaviors and their impacts can go a long way to ensuring that they do not occur in the workplace. Perhaps the exchange that has received the most focus in terms of training development and effectiveness is sexual harassment. Scholars (e.g., Perry et al. 2009) suggest that sexual harassment training should be made mandatory, and should be tailored, at least to some degree, to the individual receiving the training. For example, supervisors should receive training on what is or is not acceptable behavior when interacting with subordinates versus peers. In addition, scholars recommend that periodic voluntary training sessions on sexual harassment be available for those interested in attending. A great deal of information is disseminated informally throughout organizations (Oh et al. 2004), and those interested in attending sexual harassment training are more likely to retain and share this information with others, resulting in a more conscious workforce. Voluntary training on other negative exchanges and interactions can also be made available. For example, training on UWSS could emphasize the three-factor model (i.e., that is, avoid offering unwanted help, communicating a feeling of intellectual dominance over recipients, or making stressful stimuli more salient), and discuss methods for identifying when social support may be helpful or unhelpful to the recipient.

Address Negative Exchanges and Interactions in the Moment

As previously mentioned, perpetrators of negative exchanges and interactions often operate under the assumption that their behavior will go unpunished (Cortina et al. 2013). It is imperative, then, that founders and other leaders politely address negative exchanges (particularly incivility, abuse, sexual harassment, or what appears to be bullying) *when they are witnessed*. Directly addressing problematic behavior when it occurs helps prevent such behavior from being perpetrated by the same individuals in the future.

Summary

Norms of engaging in social support and building informal workplace relationships are foundational to a positive environment and should be actively encouraged by startup founders and leaders. Reciprocity norms (i.e., obligations to return in kind) and gain spirals (i.e., the cycle of resource gain resulting from investing existing resources) facilitate the spread of benefits beyond the exchange or relationship partners to the broader social environment. They extend and amplify the effects of even small interpersonal interactions, such as the offer of help to a colleague. In contrast, negative exchanges and interactions such as incivility, UWSS, abuse, bullying, and sexual harassment erode the social environment and should be discouraged and promptly addressed.

Practical Implications

- A positive social environment has a variety of benefits, including increases in employee wellbeing and innovation through mechanisms such as boosted positive emotions, support, and communication. These outcomes are essential for startup success and building a thriving organization in the long-term.
- Fostering a positive social environment should be an intentional effort including establishing values early, engaging leaders in efforts, providing employees with opportunities to engage in behaviors that form a positive social environment, creating thoughtful policies and work systems, and rewarding desired behaviors.
- Negative interactions and exchanges, however, have negative effects on employee wellbeing, such as increased burnout, negative emotions, and health concerns. The existing literature also suggests that they reduce creativity and innovation, but more research is needed.
- Negative exchanges and interactions can be addressed by building a culture intolerant to such behaviors, providing easy access to confidential communication channels, offering training, and delivering immediate feedback when negative behaviors are witnessed.

The Importance of Measurement and Analytics

Measurement and analytics, though not the focus of this chapter, should be noted for the critical role they play in ensuring the success of any initiative, including fostering a positive work environment. To maximize the benefits of a positive social environment and assess the effectiveness of the provided recommendations for the unique environment of an individual startup, a rigorous analytics process should be employed. Empirical measurement provides a more objective assessment of the current state of the social environment, helps identify areas of opportunity, creates a comparison point for the results of implemented changes, and provides the means for data-driven decision-making. Social exchanges, interactions, and relationships may be more difficult to measure than organizational metrics such as revenue or headcount, but they can be effectively assessed through the careful development of psychological scales utilizing the perceptions of organizational members. For example, social relationships may be measured through the mutual perceptions of both members of the relationship, or a social network analysis. In contrast, the pervasiveness of incivility, abuse, or bullying may be best measured by asking each individual the frequency with which they have experienced certain negative behaviors from other employees. Although a detailed discussion of measurement development and data analysis is beyond the scope of this chapter, a strong measurement and analysis process is essential for supporting a positive social environment.

Future Research Suggestions

Research has worked to extensively map out the nomological network of the social exchanges and interactions covered here in this chapter (e.g., Jolly et al. 2021, Yao et al. 2021). More research is needed, however, to investigate direct relations between certain aspects of the social environment and creativity and entrepreneurship. The social dynamics unique to startups and their implications for facilitating a positive social environment should also be studied further. Below are points of consideration for those looking to build on the extant body of research.

- Explore the direct benefits of social support, and the direct consequences of negative exchanges and interactions, such as UWSS and sexual harassment, on employee creative performance.
- Explore the unique impact startup dynamics have on the influence of these exchanges and interactions, both positive and negative.
- Explore best practices for establishing positive norms in startups and scaleups specifically.

Conclusion

To conclude, the social environment of a startup, though often overlooked for more salient factors such as profitability, is key for ensuring the long-term viability of a startup. To foster a positive social environment, founders and leaders should focus on promoting positive social exchanges and interactions, such as social support, and curtailing negative social exchanges and interactions, such as incivility and sexual harassment. A positive social environment holds many benefits, such as increases in employee well-being and innovation. This chapter provides relevant information on key exchanges and interactions, as well as practical recommendations for how to promote or prevent them.

References

Amabile, T.M. and Pratt, M.G. (2016). The dynamic componential model of creativity and innovation in organizations: Making progress, making meaning. *Research in Organizational Behavior*, 36: 157–183. https://doi.org/10.1016/j.riob.2016.10.001.

Anderson, E., Tomeh, D.H., Sackett, P.R. and McGue, M. (2022). Relationships between targeting recipients of OCB and CWB based on gender and engaging in sexual harassment. *PsyArXiv*. https://doi.org/10.31234/osf.io/3hgz8.

Andersson, L.M. and Pearson, C.M. (1999). Tit for tat? The spiraling effect of incivility in the workplace. *The Academy of Management Review*, 24(3): 452–471. https://doi.org/10.2307/259136.

Beauregard, T. Alexandra and Basile, K.A. and Canónico, E. (2019). Telework: Outcomes and facilitators for employees. *In*: R.N. Landers (ed.). *The Cambridge Handbook of Technology and Employee Behavior.* Cambridge, UK: Cambridge University Press.

Beehr, T.A., Bowling, N.A. and Bennett, M.M. (2010). Occupational stress and failures of social support: When helping hurts. *Journal of Occupational Health Psychology*, 15(1): 45. https://doi.org/10.1037/a0018234.

Berry, C.M., Ones, D.S. and Sackett, P.R. (2007). Interpersonal deviance, organizational deviance, and their common correlates: A review and meta-analysis. *Journal of Applied Psychology*, 92(2): 410–424. https://doi.org/10.1037/0021-9010.92.2.410.

Biggs, D., Matthewman, L. and Fultz, C. (2012). Romantic relationships in organisational settings: Attitudes on workplace romance in the UK and USA. *Gender in Management: An International Journal*, 27(4): 271–285. https://doi.org/10.1108/17542411211244803.

Bowling, N.A., Beehr, T.A. and Swader, W.M. (2005). Giving and receiving social support at work: The roles of personality and reciprocity. *Journal of Vocational Behavior*, 67(3): 476–489. https://doi.org/10.1016/j.jvb.2004.08.004.

Burden, L. and Clarey, K. (2020, July 15). *We Work Hit with 3 Suits Alleging Race Discrimination, Sexual Harassment.* HR Dive. https://www.hrdive.com/news/wework-hit-with-3-suits-alleging-race-discrimination-sexual-harassment/581507/.

Burns, R.A. and Machin, M.A. (2012). Moving beyond the pleasure principle: Within and between-occasion effects of employee eudaimonia within a school organizational climate context. *Journal of Vocational Behavior*, 80(1): 118–128. https://doi.org/10.1016/j.jvb.2011.04.007.

Cao, F. and Zhang, H. (2020). Workplace friendship, psychological safety and innovative behavior in China: a moderated-mediation model. *Chinese Management Studies*, 14(3): 661–676. https://doi.org/10.1108/CMS-09-2019-0334.

Coffin, B. (2003). Breaking the silence on white collar crime. *Risk Management*, 50(9): 8–9.

Collins, A.M., Hislop, D. and Cartwright, S. (2016). Social support in the workplace between teleworkers, office-based colleagues and supervisors. *New Technology, Work and Employment*, 31(2): 161–75. https://doi.org/10.1111/ntwe.12065.

Cohen, S. and Wills, T.A. (1985). Stress, social support, and the buffering hypothesis. *Psychological Bulletin*, 98(2): 310. https://doi.org/10.1037/0033-2909.98.2.310.

Cortina, L.M., Kabat-Farr, D., Leskinen, E.A., Huerta, M. and Magley, V.J. (2013). Selective incivility as modern discrimination in organizations: Evidence and impact. *Journal of Management*, 39(6): 1579–1605. https://doi.org/10.1177/0149206311418835.

Cortina, L.M., Magley, V.J., Williams, J.H. and Langhout, R.D. (2001). Incivility in the workplace: Incidence and impact. *Journal of Occupational Health Psychology*, 6(1): 64. https://doi.org/10.1037/1076-8998.6.1.64.

Creasy, T. and Carnes, A. (2017). The effects of workplace bullying on team learning, innovation and project success as mediated through virtual and traditional team dynamics. *International Journal of Project Management*, 35(6): 964–977. https://doi.org/10.1016/j.ijproman.2017.04.005.

Cropanzano, R. and Mitchell, M.S. (2005). Social exchange theory: An interdisciplinary review. *Journal of Management*, 31(6): 874–900. https://doi.org/10.1177/0149206305279602.

Davis, M.A. (2009). Understanding the relationship between mood and creativity: A meta-analysis. *Organizational Behavior and Human Decision Processes*, 108(1): 25–38. https://doi.org/10.1016/j.obhdp.2008.04.001.

De Witte, K. and van Muijen, J.J. (1999). Organizational culture. *European Journal of Work and Organizational Psychology*, 8(4): 497–502. https://doi.org/10.1080/135943299398122.

Edmondson, A. (1999). Psychological safety and learning behavior in work teams. *Administrative Science Quarterly*, 44(2): 350–383. https://doi.org/10.2307/2666999.

Elsesser, K. and Peplau, L.A. (2006). The glass partition: Obstacles to cross-sex friendships at work. *Human Relations*, 59(8): 1077–1100. https://doi.org/10.1177/0018726706068783.

Ferris, G.R., Liden, R.C., Munyon, T.P., Summers, J.K., Basik, K.J. and Buckley, M.R. (2009). Relationships at work: Toward a multidimensional conceptualization of dyadic work relationships. *Journal of Management*, 35(6): 1379–1403. https://doi.org/10.1177/0149206309344741.

Fitzgerald, L.F. (1994). Sexual harassment: Violence against women in the workplace. *American Psychologist*, 48(10): 1070–1076. https://doi.org/10.1037/0003-066X.48.10.1070.

Fitzgerald, L.F., Gelfand, M.J. and Drasgow, F. (1995). Measuring sexual harassment: Theoretical and psychometric advances. *Basic and Applied Social Psychology*, 17(4): 425–445. https://doi.org/10.1207/s15324834basp1704_2.

Forbes. (2021). Why Airbnb Stock Isn't the Best Travel Recovery Play. *Forbes*. Retrieved from: https://www.forbes.com/sites/greatspeculations/2021/04/07/why-airbnb-stock-isnt-the-best-travel-recovery-play/?sh=57e10a638b6b.

Fox, S. and Stallworth, L.E. (2005). Racial/ethnic bullying: Exploring links between bullying and racism in the US workplace. *Journal of Vocational Behavior*, 66(3): 438–456. https://doi.org/10.1016/j.jvb.2004.01.002.

Francis, D.H., Huang, L.Y. and Carraher, S. (2004). Top management teams and friendship: Results from the USA and Taiwan. *International Journal of Family Business*, 1: 73–86.

Gibson, K.R. (2018). Can I tell you something? How disruptive self-disclosure changes who "we" are. *Academy of Management Review*, 43(4): 570–589. https://doi.org/10.5465/amr.2016.0317.

Glasø, L., Vie, T.L., Holmdal, G.R. and Einarsen, S. (2011). An application of affective events theory to workplace bullying: The role of emotions, trait anxiety, and trait anger. *European Psychologist*, 16(3): 198–208. https://doi.org/10.1027/1016-9040/a000026.

González-Romá, V. and Peiró, J.M. (2014). Climate and culture strength. pp. 496–531. *In*: B. Schneider and K.M. Barbera (eds.). *The Oxford Handbook of Organizational Climate and Culture*. Oxford University Press.

Gray, C.E., Spector, P.E., Lacey, K.N., Young, B.G. and Taylor, M.R. (2020). Helping may be harming: Unintended negative consequences of providing social support. *Work & Stress*, 34(4): 359–385. https://doi.org/10.1080/02678373.2019.1695294.

Halbesleben, J.R.B., Neveu, J.-P., Paustian-Underdahl, S.C. and Westman, M. (2014). Getting to the "COR": Understanding the role of resources in conservation of resources theory. *Journal of Management*, 40(5): 1334–1364. https://doi.org/10.1177/0149206314527130.

Halbesleben, J.R. and Wheeler, A.R. (2015). To invest or not? The role of coworker support and trust in daily reciprocal gain spirals of helping behavior. *Journal of Management*, 41(6): 1628–1650. https://doi.org/10.1177/0149206312455246.

Hansen, Å.M., Hogh, A., Persson, R., Karlson, B., Garde, A.H. and Ørbæk, P. (2006). Bullying at work, health outcomes, and physiological stress response. *Journal of Psychosomatic Research*, 60(1): 63–72. https://doi.org/10.1016/j.jpsychores.2005.06.078.

Hauge, L.J., Skogstad, A. and Einarsen, S. (2010). The relative impact of workplace bullying as a social stressor at work. *Scandinavian Journal of Psychology*, 51(5): 426–433. https://doi.org/10.1111/j.1467-9450.2010.00813.x.

Heckman, W. (2019, September 25). *42 Worrying Workplace Stress Statistics*. The American Institute of Stress. https://www.stress.org/42-worrying-workplace-stress-statistics.

Helmy, I. and Wiwoho, G. (2020). Workplace friendship and innovative work behaviour: Investigating the mediation model in Indonesian SMEs context. *International Journal of Innovation, Creativity, and Change*, 13(10): 973–993.

Hobfoll, S.E. (2001). The influence of culture, community, and the nested-self in the stress process: Advancing conservation of resources theory. *Applied Psychology*, 50(3): 337–421. https://doi.org/10.1111/1464-0597.00062.

Holt-Lunstad, J. (2018). Fostering social connection in the workplace. *American Journal of Health Promotion*, 32(5): 1307–1312. https://doi.org/10.1177/0890117118776735a.

Hughes, I.M., Freier, L.M. and Barratt, C.L. (2022). "Your help isn't helping me!" Unhelpful workplace social support, strain, and the role of individual differences. *Occupational Health Science*. https://doi.org/10.1007/s41542-022-00115-x.

Hülsheger, U.R., Anderson, N. and Salgado, J.F. (2009). Team-level predictors of innovation at work: A comprehensive meta-analysis spanning three decades of research. *Journal of Applied Psychology*, 94(5): 1128–1145. https://doi.org/10.1037/a0015978.

Hur, W.-M., Moon, T. and Jun, J.K. (2016). The effect of workplace incivility on service employee creativity: The mediating role of emotional exhaustion and intrinsic motivation. *Journal of Services Marketing*, 30(3): 302–315. https://doi.org/10.1108/JSM-10-2014-0342.

Isaac, M. (2017, February 23). *Inside Uber's Aggressive, Unrestrained Workplace Culture*. The New York Times. https://www.nytimes.com/2017/02/22/technology/uber-workplace-culture.html.

Jehn, K.A. and Shah, P.P. (1997). Interpersonal relationships and task performance: An examination of mediation processes in friendship and acquaintance groups. *Journal of Personality and Social Psychology*, 72(4): 775–790. https://doi.org/10.1037/0022-3514.72.4.775.

Jiang, W., Gu, Q. and Tang, T.L.-P. (2019). Do victims of supervisor bullying suffer from poor creativity? Social cognitive and social comparison perspectives. *Journal of Business Ethics*, 157(3): 865–884. https://doi.org/10.1007/s10551-017-3660-x.

Jiang, K., Liu, D., McKay, P.F., Lee, T.W. and Mitchell, T.R. (2012). When and how is job embeddedness predictive of turnover? A meta-analytic investigation. *Journal of Applied Psychology*, 97(5): 1077–1096. https://doi.org/10.1037/a0028610.

Jolly, P.M., Kong, D.T. and Kim, K.Y. (2021). Social support at work: An integrative review. *Journal of Organizational Behavior*, 42(2): 229–251. https://doi.org/10.1002/job.2485.

Jung, H.S. and Yoon, H.H. (2020). How does workplace romance influence employee performance in the hospitality industry?. *Sustainability*, 12(13): 5478, 1–13. https://doi.org/10.3390/su12135478.

Khan, M.A.S., Du, J., Anwar, F., Qalati, S.A., Waqas, M. and Iqbal, S. (2022). Relationship between workplace romance, job involvement, and work effort: moderating roles of gender and workplace romance types. *Current Psychology*, 1–15. https://doi.org/10.1007/s12144-021-02665-9.

Khan, M.A.S., Jianguo, D., Hameed, A.A., Mushtaq, T.U. and Usman, M. (2018). Affective commitment foci as parallel mediators of the relationship between workplace romance and employee job performance: a cross-cultural comparison of the People's Republic of China and Pakistan. *Psychology Research and Behavior Management*, 11: 267–278. https://doi.org/10.2147/PRBM.S168542.

Khan, M.A.S., Jianguo, D., Usman, M. and Ahmad, M.I. (2017). Moderated mediation model of interrelations between workplace romance, wellbeing, and employee performance. *Frontiers in Psychology*, 8: 2158, 1–13. https://doi.org/10.3389/fpsyg.2017.02158.

Khazanchi, S., Sprinkle, T.A., Masterson, S.S. and Tong, N. (2018). A spatial model of work relationships: The relationship-building and relationship-straining effects of workspace design. *Academy of Management Review*, 43(4): 590–609. https://doi.org/10.5465/amr.2016.0240.

Konrad, A. (2017). *We Work Confirms Massive $4.4 Billion Investment from SoftBank and its Vision Fund.* Forbes. https://www.forbes.com/sites/alexkonrad/2017/08/24/wework-confirms-massive-4-4-billion-investment-from-softbank-and-its-vision-fund/?sh=5d3a70675b3c.

Krekel, C., Ward, G. and De Neve, J.E. (2019). Employee wellbeing, productivity, and firm performance. *Saïd Business School WP*, 3–43.

Kurtessis, J.N., Eisenberger, R., Ford, M.T., Buffardi, L.C., Stewart, K.A. and Adis, C.S. (2017). Perceived organizational support: A meta-analytic evaluation of organizational support theory. *Journal of Management*, 43(6): 1854–1884. https://doi.org/10.1177/0149206315575554.

Lapierre, L., Spector, P. and Leck, J. (2005). Sexual versus nonsexual workplace aggression and victims' overall job satisfaction: A meta-analysis. *Journal of Occupational Health Psychology*, 10: 155–169. https://doi.org/10.1037/1076-8998.10.2.155.

Laschinger, H.K.S. and Fida, R. (2014). A time-lagged analysis of the effect of authentic leadership on workplace bullying, burnout, and occupational turnover intentions. *European Journal of Work and Organizational Psychology*, 23(5): 739–753. https://doi.org/10.1080/1359432X.2013.804646.

Leiter, M.P., Bakker, A.B. and Maslach, C. (2014). *Burnout at Work: A Psychological Perspective*. Psychology Press.

Loy, P.H. and Stewart, L.P. (1984). The extent and effects of the sexual harassment of working women. *Sociological Focus*, 17(1): 31–43. https://doi.org/10.1080/00380237.1984.10570460.

Lu, J.G., Hafenbrack, A.C., Eastwick, P.W., Wang, D.J., Maddux, W.W. and Galinsky, A.D. (2017). "Going out" of the box: Close intercultural friendships and romantic relationships spark creativity, workplace innovation, and entrepreneurship. *Journal of Applied Psychology*, 102(7): 1091–1108. https://doi.org/10.1037/apl0000212.

Mackey, J.D., Frieder, R.E., Brees, J.R. and Martinko, M.J. (2017). Abusive supervision: A meta-analysis and empirical review. *Journal of Management*, 43(6): 1940–1965. https://doi.org/10.1177/0149206315573997.

Marcus, B., Taylor, O.A., Hastings, S.E., Sturm, A. and Weigelt, O. (2016). The structure of counterproductive work behavior: A review, a structural meta-analysis, and a primary study. *Journal of Management*, 42(1): 203–233. https://doi.org/10.1177/0149206313503019.

Martinko, M.J., Harvey, P., Brees, J.R. and Mackey, J. (2013). A review of abusive supervision research. *Journal of Organizational Behavior*, 34(S1): S120–S137. https://doi.org/10.1002/job.1888.

Maslach, C. and Jackson, S.E. (1981). The measurement of experienced burnout. *Journal of Organizational Behavior*, 2(2): 99–113. https://doi.org/10.1002/job.4030020205.

Mathews, E., Hammarlund, R., Kullar, R., Mulligan, L., Le, T., Lauve, S., Nzodom, C. and Crapanzano, K. (2019). Sexual harassment in the house of medicine and correlations to burnout: A cross-sectional survey. *Ochsner Journal*, 19(4): 329–339. https://doi.org/10.31486/toj.19.0019.

Mathieu, M., Eschleman, K.J. and Cheng, D. (2019). Meta-analytic and multiwave comparison of emotional support and instrumental support in the workplace. *Journal of Occupational Health Psychology*, 24(3): 387–409. https://doi.org/10.1037/ocp0000135.

Mathisen, G.E., Einarsen, S. and Mykletun, R. (2008). The occurrences and correlates of bullying and harassment in the restaurant sector. *Scandinavian Journal of Psychology*, 49(1): 59–68. https://doi.org/10.1111/j.1467-9450.2007.00602.x.

Maypole, D.E. and Skaine, R. (1983). Sexual harassment in the workplace. *Social Work*, 28(5): 385–390. https://doi.org/10.1093/sw/28.5.385.

Methot, J.R., Lepine, J.A., Podsakoff, N.P. and Christian, J.S. (2016). Are workplace friendships a mixed blessing? Exploring tradeoffs of multiplex relationships and their associations with job performance. *Personnel Psychology*, 69(2): 311–355. https://doi.org/10.1111/peps.12109.

Miller, D.T. and Prentice, D.A. (2016). Changing norms to change behavior. *Annual Review of Psychology*, 67(1): 339–361. https://doi.org/10.1146/annurev-psych-010814-015013.

Miner-Rubino, K. and Cortina, L.M. (2004). Working in a context of hostility toward women: Implications for employees' well-being. *Journal of Occupational Health Psychology*, 9(2): 107–122. https://doi.org/10.1037/1076-8998.9.2.107.

Morrison, R.L. (2005). *Informal Relationships in the Workplace: Associations with Job Satisfaction, Organisational Commitment and Turnover Intentions* (Doctoral dissertation, Massey University).

Morrison, R.L. and Cooper-Thomas, H.D. (2016). Friendship among coworkers. *In*: M. Hojjat and A. Moyer (eds.). *The Psychology of Friendship*. New York: Oxford University Press.

Motro, D., Spoelma, T.M. and Ellis, A.P.J. (2021). Incivility and creativity in teams: Examining the role of perpetrator gender. *Journal of Applied Psychology*, 106(4): 560–581. https://doi.org/10.1037/apl0000757.

Mueller, J.S., Melwani, S. and Goncalo, J.A. (2012). The bias against creativity: Why people desire but reject creative ideas. *Psychological Science*, 23(1): 13–17. https://doi.org/10.1177/0956797611421018.

Muñoz-Doyague, M.F. and Nieto, M. (2012). Individual creativity performance and the quality of interpersonal relationships. *Industrial Management & Data Systems*, 112(1): 125–145. https://doi.org/10.1108/02635571211193671.

Nahemow, L. and Lawton, M.P. (1975). Similarity and propinquity in friendship formation. *Journal of Personality and Social Psychology*, 32(2): 205–213. https://doi.org/10.1037/0022-3514.32.2.205.

Nielsen, I.K., Jex, S.M. and Adams, G.A. (2000). Development and validation of scores on a two-dimensional workplace friendship scale. *Educational and Psychological Measurement*, 60(4): 628–643. https://doi.org/10.1177/00131640021970655.

Nixon, A.E., Arvan, M. and Spector, P.E. (2021). Will the real mistreatment please stand up? Examining the assumptions and measurement of bullying and incivility. *Work & Stress*, 35(4): 398–422. https://doi.org/10.1080/02678373.2021.1891584.

Oh, H., Chung, M.-H. and Labianca, G. (2004). Group social capital and group effectiveness: The role of informal socializing ties. *The Academy of Management Journal*, 47(6): 860–875. https://doi.org/10.2307/20159627.

Ones, D. and Dilchert, S. (2013). Counterproductive work behaviors: Concepts, measurement, and nomological network. *APA Handbook of Testing and Assessment in Psychology*, 643–659.

Ozer, M. and Zhang, G. (2022). Interpersonal relationships and creativity at work: A network building perspective. *Journal of Product Innovation Management*, 39(3): 312–333. https://doi.org/10.1111/jpim.12575

Panaccio, A. and Vandenberghe, C. (2009). Perceived organizational support, organizational commitment and psychological well-being: A longitudinal study. *Journal of Vocational Behavior*, 75(2): 224–236. https://doi.org/10.1016/j.jvb.2009.06.002.

Park, J. and Min, H. (Kelly). (2020). Turnover intention in the hospitality industry: A meta-analysis. *International Journal of Hospitality Management*, 90: 102599. https://doi.org/10.1016/j.ijhm.2020.102599.

Paulin, D. and Griffin, B. (2017). Team incivility climate scale: Development and validation of the team-level incivility climate construct. *Group & Organization Management*, 42(3): 315–345. https://doi.org/10.1177/1059601115622100.

Perry, E.L., Kulik, C.T. and Field, M.P. (2009). Sexual harassment training: Recommendations to address gaps between the practitioner and research literatures. *Human Resource Management*, 48(5): 817–837. https://doi.org/10.1002/hrm.20316.

Perry-Smith, J.E. and Shalley, C.E. (2014). A social composition view of team creativity: The role of member nationality-heterogeneous ties outside of the team. *Organization Science*, 25(5): 1434–1452. https://doi.org/10.1287/orsc.2014.0912.

Pierce, C.A. and Aguinis, H. (2009). Moving beyond a legal-centric approach to managing workplace romances: Organizationally sensible recommendations for HR leaders. *Human Resource Management*, 48(3): 447–464. https://doi.org/10.1002/hrm.20289.

Pietsch, B. (2020). *We Work's Valuation has Fallen from $47 Billion Last Year to $2.9 Billion.* Business Insider. https://www.businessinsider.com/wework-valuation-falls-47-billion-to-less-than-3-billion-2020-5.

Pillemer, J. and Rothbard, N.P. (2018). Friends without benefits: Understanding the dark sides of workplace friendship. *Academy of Management Review*, 43(4): 635–660. https://doi.org/10.5465/amr.2016.0309.

Porath, C.L. and Pearson, C.M. (2010). The cost of bad behavior. *Organizational Dynamics*, 39(1): 64–71. https://doi.org/10.1016/j.orgdyn.2009.10.006.

Quinn, R.W. (2017). Energizing others in work connections: How positive relationships create a context for self-discovery and self-actualization. pp. 73–90. *In*: J.E. Dutton and B.R. Ragins (eds.). *Exploring Positive Relationships at Work.* Psychology Press.

Rank, O.N. and Tuschke, A. (2010). Perceived influence and friendship as antecedents of cooperation in top management teams: a network approach. *Business Research*, 3(2): 151–171.

Ren, H., Gray, B. and Harrison, D.A. (2015). Triggering faultline effects in teams: The importance of bridging friendship ties and breaching animosity ties. *Organization Science*, 26(2): 390–404. https://doi.org/10.1287/orsc.2014.0944.

Robinson, S.L. and Bennett, R.J. (1995). A typology of deviant workplace behaviors: A multidimensional scaling study. *Academy of Management Journal*, 38(2): 555–572. https://doi.org/10.5465/256693.

Sandler, R. (2019). *We Work Removes in-office Phone Booths due to Formaldehyde Contamination.* Forbes. https://www.forbes.com/sites/rachelsandler/2019/10/14/wework-removes-in-office-phone-booths-due-to-formaldehyde-contamination/?sh=4a0ef52540a7.

Schaufeli, W.B., Leiter, M.P. and Maslach, C. (2009). Burnout: 35 years of research and practice. *Career Development International*, 14(3): 204–220. https://doi.org/10.1108/13620430910966406.

Schneider, B. (1987). The people make the place. *Personnel Psychology*, 40(3): 437–453. https://doi.org/10.1111/j.1744-6570.1987.tb00609.x.

Schneider, K.T., Tomaka, J. and Palacios, R. (2001). Women's cognitive, affective, and physiological reactions to a male coworker's sexist behavior. *Journal of Applied Social Psychology*, 31(10): 1995–2018. https://doi.org/10.1111/j.1559-1816.2001.tb00161.x.

Sherman, G. (2019). *"You Don't Bring Bad News to the Cult Leader": Inside the Fall of We Work. Vanity Fair*. https://www.vanityfair.com/news/2019/11/inside-the-fall-of-wework.

Sias, P.M. and Cahill, D.J. (1998). From coworkers to friends: The development of peer friendships in the workplace. *Western Journal of Communication*, 62(3): 273–299. https://doi.org/10.1080/10570319809374611.

Sloan, M.M. (2012). Unfair treatment in the workplace and worker well-being: The role of coworker support in a service work environment. *Work and Occupations*, 39(1): 3–34. https://doi.org/10.1177/0730888411406555.

Spector, P.E. and Bruk-Lee, V. (2007). Conflict, health, and well-being. In *The Psychology of Conflict and Conflict Management in Organizations*. Psychology Press.

Spector, P.E. and Fox, S. (2002). An emotion-centered model of voluntary work behavior: Some parallels between counterproductive work behavior and organizational citizenship behavior. *Human Resource Management Review*, 12(2): 269–292. https://doi.org/10.1016/S1053-4822(02)00049-9.

Spector, P.E., Fox, S., Penney, L.M., Bruursema, K., Goh, A. and Kessler, S. (2006). The dimensionality of counterproductivity: Are all counterproductive behaviors created equal? *Journal of Vocational Behavior*, 68(3): 446–460. https://doi.org/10.1016/j.jvb.2005.10.005.

Spender, J.C., Corvello, V., Grimaldi, M. and Rippa, P. (2017). Startups and open innovation: A review of the literature. *European Journal of Innovation Management*, 20(1): 4–30. https://doi.org/10.1108/EJIM-12-2015-0131.

Strickland, P., Morriss, R., Wearden, A. and Deakin, B. (1998). A comparison of salivary cortisol in chronic fatigue syndrome, community depression and healthy controls. *Journal of Affective Disorders*, 47(1): 191–194. https://doi.org/10.1016/S0165-0327(97)00134-1.

Stryker, S. and Vryan, K.D. (2006). The symbolic interactionist frame. In *Handbook of Social Psychology* (pp. 3–28). Springer.

ten Brummelhuis, L., Haar, J. and van der Lippe, T. (2010). Collegiality under pressure? The effects of family demands and flexible work arrangements in the Netherlands. *International Journal of Human Resource Management*, 21(15): 2831–47. https://doi.org/10.1080/09585192.2010.528666.

Viswesvaran, C., Sanchez, J.I. and Fisher, J. (1999). The role of social support in the process of work stress: A meta-analysis. *Journal of Vocational Behavior*, 54(2): 314–334. https://doi.org/10.1006/jvbe.1998.1661.

Welbourne, J. and Sariol, A. (2016). When does incivility lead to counterproductive work behavior? Roles of job involvement, task interdependence, and gender. *Journal of Occupational Health Psychology*, 22. https://doi.org/10.1037/ocp0000029.

Winstead, B.A., Derlega, V.J., Montgomery, M.J. and Pilkington, C. (1995). The quality of friendships at work and job satisfaction. *Journal of Social and Personal Relationships*, 12(2): 199–215. https://doi.org/10.1177/0265407595122003.

Wright, T.A. and Bonett, D.G. (2007). Job satisfaction and psychological well-being as nonadditive predictors of workplace turnover. *Journal of Management*, 33(2), 141–160. https://doi.org/10.1177/0149206306297582.

Wu, T.Y. and Hu, C. (2009). Abusive supervision and employee emotional exhaustion: Dispositional antecedents and boundaries. *Group & Organization Management*, 34(2): 143–169. https://doi.org/10.1177/1059601108331217.

Xiao, Z. and Tsui, A.S. (2007). When brokers may not work: The cultural contingency of social capital in Chinese high-tech firms. *Administrative Science Quarterly*, 52(1): 1–31. https://doi.org/10.2189/asqu.52.1.1.

Yakubovich, V. and Burg, R. (2019). Friendship by assignment? From formal interdependence to informal relations in organizations. *Human Relations*, 72(6): 1013–1038. https://doi.org/10.1177/0018726718789479.

Yao, J., Lim, S., Guo, C.Y., Ou, A. and Ng, J. (2021). Experienced incivility in the workplace: A meta-analytical review of its construct validity and nomological network. *Journal of Applied Psychology*. https://doi.org/10.1037/apl0000870.

Yu, C. and Frenkel, S.J. (2013). Explaining task performance and creativity from perceived organizational support theory: Which mechanisms are more important? *Journal of Organizational Behavior*, 34(8): 1165–1181. https://doi.org/10.1002/job.1844.

Zach, F.J., Nicolau, J.L. and Sharma, A. (2020). Disruptive innovation, innovation adoption and incumbent market value: The case of Airbnb. *Annals of Tourism Research*, 80: 102818. https://doi.org/10.1016/j.annals.2019.102818.

Zhan, X., Li, Z. and Luo, W. (2019). An identification-based model of workplace incivility and employee creativity: Evidence from China. *Asia Pacific Journal of Human Resources*, 57(4): 528–552. https://doi.org/10.1111/1744-7941.12204.

Zhang, Y., Sun, J.M., Shaffer, M.A. and Lin, C.H. (2021). High commitment work systems and employee well-being: The roles of workplace friendship and task interdependence. *Human Resource Management*, 61(4): 399–421. https://doi.org/10.1002/hrm.22093.

Zhang, Y., Liu, X., Xu, S., Yang, L.Q. and Bednall, T.C. (2019). Why abusive supervision impacts employee OCB and CWB: A meta-analytic review of competing mediating mechanisms. *Journal of Management*, 45(6): 2474–2497. https://doi.org/10.1177/0149206318823935.

Chapter 11

Understanding the Basics of Startup Development Organizations

Allison Piper Kimball

Startup companies often engage with startup development organizations (SDOs), a class of companies that provide critical support and building blocks to startups such as physical office and meeting space, access to mentoring and intellectual capital, and access to funding and investor capital. SDOs offer collaborative environments and community building in addition to physical, mentoring, and business services. The role and impact of each type of SDO individually and together within the broader startup incubation ecosystem (SUPIE) are not well understood. Do engagement with SDOs and their programs increase a company's odds of success and in turn, lead to higher returns for investors? It is difficult to answer this question without a clear understanding of SDOs. Understanding the effectiveness of SDOs is a critical need for entrepreneurs, investors, and researchers (Novotny et al. 2020).

Understanding the structure and objectives of an SDO is critical for entrepreneurs before deciding to join. Before applying to SDOs, entrepreneurs need to understand what the SDO offers and the price they will pay for this service, whether in time, cash burn, or equity (Richards 2021). Due to the relative newness and varieties of SDOs, limited information is available to compare program success across SDO types. There is little research on precisely what makes SDOs effective. Still, some studies provide insight into performance metrics, such as the number of SDO graduates who have had an "exit," defined as an acquisition or an IPO (Paluch 2021).

Understanding the type of funding, mentoring, and resources provided to a startup by an SDO is valuable for investors. SDOs can provide access to early-stage investment opportunities and help accelerate the progress of their portfolio companies. SDOs enlist investors and mentors to work with the companies throughout the program, providing an extended period to interact with the founders and get to

Business School Lausanne, 281-788-8450.
Email: alirpiper@gmail.com

know them before making an investment decision. Many SDOs culminate in "Demo Days," where startups present to potential investors and have an opportunity to engage interested investors in further dialogue. Some SDOs may invest directly in or form the startups themselves, leveraging expertise and capital to scale the opportunities quickly and provide investors an opportunity to invest in diversified startup funds.

Developing a solid conceptualization and understanding of SDOs is critical for science and practice. With inconsistent definitions, no clear boundaries between types of SDOs, and overlapping features, the study of SDO effectiveness is more complicated than it would appear. Precise definitions and clear relationships between the constructs and the measurement models are needed to evaluate SDO effectiveness (Mackenzie 2003). This chapter reviews the evolving landscape of SDOs and provides insight into research regarding their impact on the success of startups. This chapter intends to contribute to the literature by examining SDOs, their definition, and important nuances by introducing, defining, and describing SDOs so academics and practitioners can begin to advance their understanding of what makes an SDO effective.

Startup Development Organizations and the Link to Business Development Stage

Startup company development and growth can be broken into three fundamental phases: formation, validation, and growth, as depicted in Fig. 1. During the initial formation stage, an idea becomes an actionable concept and business model. The business model assumptions must then be validated in the next phase concerning operability, relevance, importance, market opportunity, and profitability. Companies that successfully navigate the first two phases then attempt entry to a growth phase, where the product or service is scaled for expansion, and the launch occurs. Startup companies tend to move through this development process from *formation* through *validation* (and ultimately *growth and scaling* if successful) at various speeds and over widely divergent time periods Startup Commons 2019).

SDOs support startup companies through this development process in different ways depending on their intentions and capabilities. Some SDOs merely provide a workspace to conduct the work of the various business stages. In contrast, others offer all the necessary financial, intellectual, and physical support to take an idea from *formation* through the *validation* stage to becoming an established, growing company with sustaining revenue. This transition to growth can be precarious and is a critical time for startups. Failure to reach and surpass a commercial inflection point,

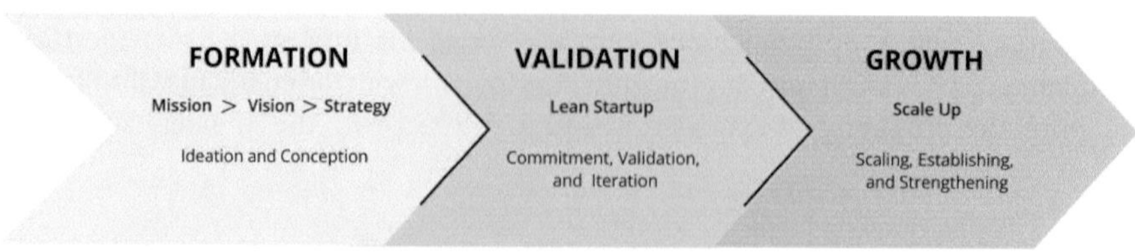

Figure 1. The successful startup progression.

where revenue begins to support expenses before running out of money, usually results in the end of the company.

For this reason, the transition to growth is often referred to as the "Valley of Death," as depicted in Fig. 2. Financial gaps, unsupportive ecosystems, infrastructure gaps, and lack of internal and external cooperation are common causes of startup demise during this stage (AlNatsheh et al. 2021). SDOs targeting these areas may assist startup survival. Some in the startup industry refer to startup companies that have passed the commercial inflection point and are entering or achieving consistent and significant growth as "scale-ups" (Isenberg and Onyemah 2016). Research is beginning to investigate factors important in achieving scale, as it is estimated that in Europe, for example, just half of one percent of startups become scale-ups (Reypens et al. 2020).

Startup Development Organizations are a relatively recent but quickly expanding phenomenon. The first business incubator was founded in New York state in 1959 (Calza et al. 2014, Mancuso 2022), and since then, thousands of SDOs spanning co-working spaces, incubators, and accelerators have promulgated worldwide. With the rapid expansion of SDOs, entrepreneurs have myriad options for physical, financial, and mentoring support, ranging from shared office and co-working spaces, and regional or functionally focused incubators, to joining with a venture or startup studio or a studio-accelerator hybrid. According to the International Business Innovation Association (IBIA), as of their most recent data in 2018, there were over 3,200 co-working spaces, 2,300 "economic development organizations", 1,400 incubators, 1,100 small business development offices, 500 accelerators, 442 makerspaces, 118 women's business centers, and 30 "super hubs" in the USA alone (IBIA 2018, Schuenke 2022). These various models offer different services with different goals and objectives.

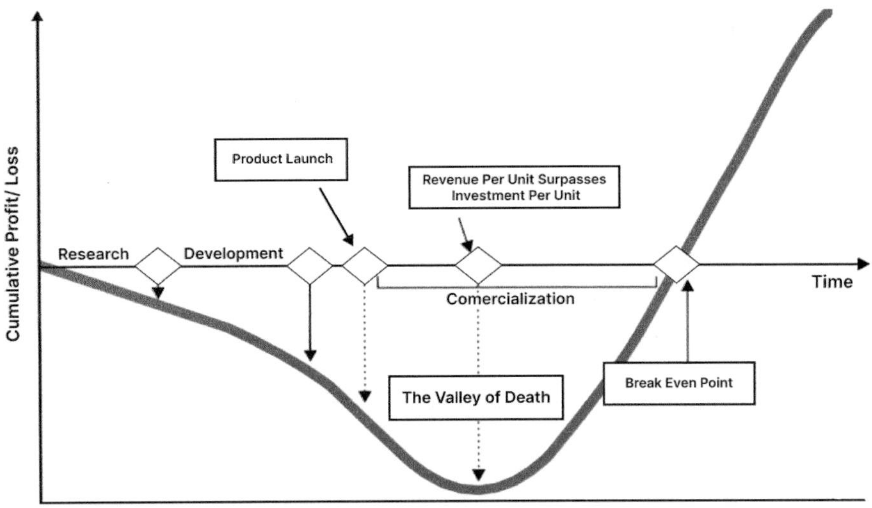

Figure 2. The valley of death.

Lack of Standardization in SDO Terminology

First, the reader should be aware that there is a general lack of terminology and performance metrics standardization in both the SDO industry and the scientific literature. There is no consensus as to what an SDO means, nor is there a taxonomy to organize the different types or features. Conceptualizations and taxonomies are usually the foundation of an area of research - without an agreed-upon, consistent definition and understanding of the concept, scientists cannot study it. So, from a research perspective, it's essential to conceptualize (i.e., define boundaries) and organize. This paper aims to help develop a clear conceptualization for SDOs that can serve as a foundation for a taxonomy.

For example, the term "incubator" is itself a specific type of SDO. Yet, the term is sometimes applied generically to various forms of business "incubation," including accelerators and startup studios. The words "venture studio" and "startup studio" are often used interchangeably, yet at other times differentiated by the formation of the company concept and source of funding. As a result, attempts to rank and evaluate SDOs result in inconsistencies across naming conventions and functional outcome metrics. The performance metrics used by the SDOs and third-party groups also vary widely, from startup-based feedback and net promoter scores to throughput-based metrics such as the number of companies that graduated or the number of companies funded. Yet other SDOs use outcome-based metrics such as the number of jobs created and the number of years after graduation the startup is still active. Some SDOs use investor-based success metrics such as the number of "unicorns" (companies valued at over $1B) or the number of exits, defined as initial public offerings or private sales. One Houston-based venture studio Director reported a recent rise in interest in diversity-based metrics by current and prospective investors. In addition, some SDOs originated to fill a social service function, such as creating jobs in urban or under-developed areas or sectors driving social change. These SDOs tend to focus on performance metrics tied to their missions.

Types of Startup Development Organizations

A summary of startup development organization types is presented in Figure 3. Types of Startup Development Organizations. Variations from these types exist, as discussed in the opening section, due to a lack of standardization in terminology and the continuing evolution of startup support offerings. While definitions vary and continue to evolve, as of the writing of this chapter, a review of literature and practice coalesces into an approximation of the following definitions. Further detail and nuances are discussed in the sections below.

Variations on and deviations from the summaries above exist, as discussed in the opening section, due to a lack of standardization in terminology and the continuing evolution of startup support offerings. Of particular note, across SDO types, the common element of providing physical space that enables founders, investors, and mentors to interact with each other provides a corollary benefit of "fortuitous collisions" in which participants make unorchestrated connections. While the COVID-19 pandemic that began in 2019 resulted in some models moving to a virtual

Types of Startup Development Organizations (SDOs)	
Incubators	Physical or virtual space with experts and programming to help founders to refine business ideas, create a minimum viable product, and build their companies.
Coworking Communities	Shared physical office space for founders to interact with others; not unique to startups.
Accelerators	Physical or virtual space with a time-bound program duration designed to help startup companies that already have a minimum viable product to promote their models and build their businesses to a specific next level in a short period of time, often with investment as part of the program, and an opportunity to be introduced to potential customers and present to future investors at the end.
Startup Studios/ Venture Studios	Founding and incubation of multiple startup ideas with the support of a core team of experienced functional experts and repeatable processes, generally with a majority investment in the startup until they are proven to a point appropriate to then spin out into separate companies or be disbanded.
Corporate Studios / Corporate Accelerators/ Corporate Venture Labs	Corporate funded (in whole or part) physical or virtual space to enable commercialization of ideas from within the corporation or external to the corporation.

Figure 3. Types of startup development organizations.

environment, the SDOs still endeavor to build a sense of community and provide forums for the participants to make personal connections.

The reader is encouraged to understand each of these SDO types as an important part of the entrepreneurial ecosystem rather than as an evolution from one kind to the next. While incubators were created before co-working communities, startup studios originated before co-working communities, and the models each continue to evolve creativly fashion, meeting the needs of the diverse investing and startup industry.

Incubators

A statement made 35 years ago seems to hold true today: "The task of defining what is meant by an incubator has become difficult since the original concept is being adapted to fit the needs of economic areas." (Kuratko and LaFollette 1987 p. 49). In their literature review, Hausberg and Korreck (2020, p. 161) present a table of over 15 definitions of an incubator. At its most basic, "a business incubator is a building with affordable industrial, commercial, and/or office space that offers shared services, assistance, and guidance that helps people start and grow businesses to create local jobs" (Mancuso 2022). Startup incubators support entrepreneurs at the earliest stages when they still need to refine their business ideas and build from the ground up (Richards 2021).

The nation's first business incubator was founded in 1959 in an 850,000-square-foot complex in Batavia, New York when the closing of the area's largest employer led to 20% unemployment. A local family purchased the property and rented it to multiple tenants while providing shared support services, assistance with raising capital, and business startup advice to spur economic growth in the area (Mancuso 2022). Since this beginning, business incubation has become a critical component of economic development ecosystems, providing business development support services in public, private, university, corporate, and governmental settings. Startup incubators historically supported local or regional economic development, and many incubators are affiliated with or housed within universities. Recent incubation innovation recognizes the value of concentrating founders, investors, and mentors by subject matter, resulting in incubators focusing on familiar subject areas such as low carbon technologies or sustainability.

Common Characteristics of Incubators

Incubators provide office space, essential services such as Wi-Fi and printing, access to mentors, and purposeful instructional and educational content. The educational content, if provided, can consist of formal training materials and classwork taught in a traditional classroom setting or accessible online at the startup's discretion; or can be special topics or unstructured talks designed to take advantage of a visiting expert or mentor. Incubators also offer networking opportunities with area investors and mentors. Incubators do not usually take equity in participating startups and do not typically provide direct investment capital. Still, they may provide funding in exchange for current or future equity or offer non-dilutive prize money. Most incubators do not require the startup to have a minimum viable product (MVP) or traction, as the purpose of the incubator is to help entrepreneurs put together these building blocks and get started. Incubators can be useful in helping businesses vet their ideas, validate the initial business problem, and refine their solution; in effect, preparing them for entering an accelerator, which typically requires the business to be beyond this initial stage (Richards 2021). In addition, some startups maintain key founding staff within incubator settings during and after participating in accelerators, providing them with needed resources outside the highly structured and time-constrained accelerator environment.

With the state of Texas being a microcosm of the growth of the industry, the Texas Business Incubator Association (2015) published a detailed study of Texas incubators in April 2015 with this disclaimer, "The Texas incubators/accelerators listings in this directory are representational, in large part because these entities are very dynamic: going in and out of existence and changing their names, points of contact, services, and focus." At the time, there were just 103 incubators across the state. By 2017 the list had nearly doubled to almost 200 SDOs. Examples of incubators include:

- *Greentown Labs*: Based in Boston, MA, and Houston, TX, Greentown Labs incubates companies focusing on climate action. Greentown Labs provides physical space and office amenities, hardware, and technology labs to support prototyping and hosts events and programs to connect entrepreneurs, investors, and corporations to achieve climate-impacting startup development (Greentownlabs.com).
- *Awesome, Inc*: Based in Lexington, KY, Awesome Inc is an incubator, co-working space, and accelerator facility that also runs specialized training to learn software development and coding (awesomeinc.com).

- *ETC Baltimore*: Based in Baltimore, Maryland, ETC (Emerging Technology Centers) Baltimore ranked in the top 5 global public incubators by UBI Global in 2019, providing co-working, incubation, and accelerator programs. Member companies gain access to affordable physical office space in a collaborative community, including access to seed capital, mentors, and potential partners. ETC Baltimore offers both incubation and acceleration programs (ETCBaltimore.com 2022, Meyer 2019).
- *Founder Institute (FI)*: Founded in 2009 by serial entrepreneurs Adeo Ressi and Jonathan Greechan, FI provides several startup programs supporting over 6,000 startups in 200+ cities and six continents. Its core program bills itself as a "pre-seed accelerator" as it supports qualifying founders to develop initial traction and receive funding. Applicants take personality profiles and pay an admission fee ($899 as of 2022 to the Houston program); once accepted and enrolled, participant companies ultimately pledge 2.5% of the company in warrants to be paid at successful exit back to the program and its mentors (www.fi.co 2022).

Co-Working Communities

An office structure providing physical amenities to multiple companies within a shared environment is a "co-working community." Not limited to startup companies, co-working spaces offer a community-like environment through an open structure and shared services that enable users to build and maintain networks with others. (Rese et al. 2022) Co-working communities can be a cost-effective way to provide office structure to a startup without the commitment and expense of a formal lease. Individuals or groups can contract for dedicated office space or for "pay as you go" daily use of desk or meeting room space. An advantage to joining a co-working space is that many are part of national and even global networks, allowing members a place to work even when traveling or if they have partners, collaborators, or employees in other physical locations.

Origin and Growth of Co-working Communities

Co-working as separate from business incubation is claimed to have originated in San Francisco's Mission District in 2005, where Bradley Neuberg started working two days a week at an open concept office and invited others to join him (Franco 2015). Neuberg was looking for a way to have a sense of community of a great office culture while working for himself (Upsuite 2021, Richards 2021, DiRisio 2021). The Cambridge Innovation Center (CIC) opened its doors in Cambridge, Massachusetts, in 1999 also with the concept of different companies operating under the same space, yet without the term or concept of "co-working" being common at the time (CIC 2022). Some argue that hackerspaces, which originated in 1995 and are covered later in this chapter, were a precursor to co-working spaces (DiRisio 2021). The model is now global; according to coworkingresources.org, as of May 2020 the United States had over 3,700 shared workspaces, India had 2,197 spaces, and the United Kingdom had 1,044 spaces, representing almost 35% of the global co-working spaces available at the time. According to the Global Coworking Growth Study in 2020, the number of people working in co-working spaces increased from its origin in 2005 to 1.9 million in 2019, with a projected doubling by 2024 (DiRisio 2020).

Some co-working communities specifically market to startup companies and provide educational and social opportunities. These co-working spaces targeting startups create environments and networks to help companies grow within a common field of interest or provide specially designed support services for startups, resembling incubators by including mentoring and programming for a fee, blurring the lines of distinction. Some offer opportunities for more advanced assistance, including the ability to present to potential investors at pitch sessions or participate in accelerators or incubators that office on-site.

Maker Spaces, Hacker Spaces, and Fab Labs

Some co-working communities are designed to encourage people to make, collaborate, learn, and share physical and technical applications, including building prototypes or minimum viable products. A makerspace is a collaborative workspace inside a school, library, or separate public/private facility for making, learning, exploring, and sharing that uses "high tech" to "no tech" tools. Maker spaces are also known as innovation hubs, design labs, Fab Labs, and hackerspaces (Hackerspaces 2017, acceleratingbiz.com). As of 2016, Popular Science reported almost 1,400 makerspaces worldwide, either active or planned (Lau and Peek 2016). Makerspaces offer high-end tools and equipment like 3-D printing equipment, laser cutters, sewing machines, CNC machines, super high-tech computing, specialized fabrication tools, construction equipment and tools, and biological lab facilities for founder ideation. Trained professionals are available to either build the prototypes or to enable founders to use the equipment themselves independently or with instruction. Many makerspaces offer technical skills classes and provide access to qualified technical experts.

Fab Labs take the makerspace concept and apply standardization. A Fab Lab has a network of uniform workshop spaces with set tools and standards (makerspaces.com; affordablecollegesonline.org). Hackerspaces tend to focus on computing, from the first "Hackerspace" in Berlin in 1995 when a group of computer programmers met to collaborate and code. In the early 2020s, hackerspaces continue to focus on electronics and computing. Examples of co-working spaces include:

- *WeWork*: Founded in 2010 in New York, WeWork is an example of a co-working community that targets tech startups and serves other startups and established companies with over 800 locations worldwide in their network. WeWork offers daily use and monthly plans for access to dedicated or flex space, which members book through a specially designed app (www.wework.com).
- *Galvanize*: A co-working community consisting of 8 campuses across the US, Galvanize provides co-working spaces and high-quality software engineering education through courses and boot camps. Galvanize positions itself as a tech ecosystem connecting founders and developers and markets itself as a great place to co-work to find employment in coding (Moss 2019, galvanize.com).

- *Capital Factory*: Based in Austin, Texas, Capital Factory specifically targets startups, providing an extensive support community designed for founders. The company advertises 81,000 square feet of co-working space, with 900 tech-focused events, 150+ of its self-proclaimed "top tech mentors in Texas," quarterly Founders Academy, a virtual reality lab, and "all the cold brew you could ever want" (capitalfactory.com).
- *The Cannon*: Based in Houston, Texas, The Cannon brings together startup entrepreneurs with service providers, corporate innovators, and an advisor and investor network specifically developed by The Cannon and offered as part of the value of the co-working arrangement. rinWith six area locations, companies can pay a monthly fee for access to desk space, meeting rooms, coffee bars, and office supplies at any of the locations. For higher levels of membership, additional services and amenities are available, including dedicated office space. Not only does the Cannon cultivate its network of investors and advisors, but it also provides local space for third-party accelerators and incubators, including Capital Factory and Kurio Collective, a Christian co-working community (www.thecannon.com).
- *SheSpace*: Based in Houston, Texas, SheSpace is a co-working community designed to support women-led startups, staffed, and managed entirely by women. SheSpace offers memberships and office space, access to podcast studios and other tools for supporting a new business, as well as seminars and networking events (shespacehtx.com).
- *East End Maker Hub and TXRX Labs*: East End Maker Hub is a 300,000 sq. foot industrial space and manufacturing center east of downtown Houston, Texas, with space for crafting, light and heavy fabrication, and manufacturing, as well as traditional office co-working space. Partnered with TXRX Labs, founders can partner with trained fabricators to develop prototypes, attend classes and take classes in various technical craft skills, including welding, woodworking, crafting, fabrication, etc. (https://eastendmakerhub.org).

Accelerators

Accelerators are SDOs that are designed to take early-stage companies with a Minimum Viable Product (MVP) and product-market fit (PMF) and transform them over a relatively short period into scalable entities ready for growth. Similar to incubators, accelerators provide basic, essential physical support for the duration of the program (although sometimes this was conducted virtually during the COVID-19 pandemic, and some hybrid models have continued). Accelerators typically provide business strategy mentoring, with access to subject matter experts relevant to the functional area of the accelerator or specific startup needs, including sales and marketing, fundraising, startup management, and business model development. In addition, accelerators often provide access or referrals to structural support, which includes recruiting and human resources, and payroll and legal support. In contrast with incubators, accelerators typically offer some financial investment in exchange for a percentage of equity and have a specific duration with a closing event or "Demo

Day" at which the startups can showcase their businesses to investors (Cohen 2013, Cohen et al. 2019).

Accelerators fundamentally differ from incubators by running "cohorts" or targeted programming to a set number of startup companies participating simultaneously in a specific time-bound program. Startup companies apply to the accelerator for acceptance into a cohort. Programs typically run for up to three to four months. Many accelerators have a geographical or functional focus, targeting a specific region or specific industry, such as healthcare technology (healthtech) or financial technology (fintech), or technology focusing on women's issues (femtech), among others. To be accepted into an accelerator, startup companies are typically expected to have a clear definition of the problem being solved, the way the company's product solves the problem, the size of the market for the solution (total addressable market or TAM), and a basic plan for the product or even an MVP created. A progression for a startup could be to develop its MVP, demonstrate product-market fit within an incubator system, and then apply to and be accepted into an accelerator program to continue its progression to viability. Greentown Labs reports that some companies use incubators as a home base, returning to the incubator in between time-bound accelerator programs.

Interestingly, some established companies considered accelerators at one time are moving to variations of the model through changes in funding or the fundamental offering to cohort companies. These variations are discussed more fully in the next section. Examples of accelerators include:

- *Y Combinator*: Credited with being the original accelerator, Y Combinator (YC), founded in 2005, has invested in over 3,000 companies worth $300B with notable graduates including Airbnb, DoorDash, Stripe, Instacart, Dropbox, and Coinbase. Investment is $500,000 on a post-money Simple Agreement for Future Equity (SAFE) which converts to 7% of the company's equity. Over the years, YC has grown to include a venture fund, training and services to help founders succeed throughout the life of their businesses, and specialized programs for later rounds of fundraising and development beyond the initial accelerator (www.ycombinator.com, Sarath 2022).
- *Techstars*: Techstars offers acceleration programs around the world. Its accelerators offer $20 K for 6% of an accepted company (to convert to equity with a raise of $250 K) and an optional $100 K convertible note which converts at a 20% discount to the agreed company valuation cap, which is between $3 M and $5 M (Techstars 2022). Its program runs for three months and has a defined agenda for participants, with networking events, mentoring programs, educational seminars, and cohort activities. Techstars also runs programs in partnership with corporations looking to encourage innovation in specific targeted functional or industry areas (techstars.com).
- *Mass Challenge*: Based in Boston, MA, with global reach both virtually and in person, with cohorts in several countries, MassChallenge accepts cohorts for 3-month programs concluding with a demo day/showcase. Fairly unusual in the accelerator space, MassChallenge operates as a non-profit and does

not take an equity position in the companies, instead cultivating sponsorships from local businesses and encouraging innovation in their communities (Masschallenge.com, Richards 2021).

Company Builders, Venture Studios, Startup Studios

The "Startup Studio" idea: leverage expertise, infrastructure, and funding to start multiple businesses. Successful entrepreneurs saw that founders were spending too much time securing funding and dealing with administrative hassles, taking away from the development of the business itself. These founders created a business model leveraging their prior success, putting their experience and capital earned from prior exits to work funding multiple new ventures and called it a startup or venture studio. The model has evolved since its inception in the mid-1990s, but in general these studios have common elements including moving forward those with building multiple startups simultaneously; providing a core team for infrastructure support; and testing a variety of ideas quickly and spinning out the successful ones, often hiring "founders" for the job. Frequently the ideas for the startups are developed within the studio and the studio holds a majority of the equity in the startup (Szigeti 2016), although as of 2022 these factors allow some classification into three general models, with variations on each (Anderson 2021).

- Formation Studios: models in which the studio starts the business, funds the business, and provides core functions and expertise.
- Early-Stage Incubator Studios: models in which externally started companies apply to the studio to receive funding and benefit from the studio's active support of critical functions and expertise.
- Technology Commercialization Studios: models in which the studio pairs founders and core functional expertise; funding may be provided, but variations on this model in the early 2020s may not provide significant (or any) direct funding, instead focusing on matchmaking of founders with technology and mentoring.

Figure 4 depicts three common "Startup Studio" models based on the studio's involvement in funding, ideation/concept creation, and mentoring/building core functions of the company: formation studios, early-stage incubator studios, and technology commercialization studios. As of 2017 there were over 100 startup studios worldwide as tracked by eFounders (Ferres 2017) and by 2019 that number had grown o=to over 300 as compiled by Enhance Ventures, a startup studio based in Dubai (Alhokail 2019).

A. "Formation startup studios"–develop ideas, test them, and build a team to develop companies, funded and supported by the studio:

Idealab, founded by serial entrepreneur Bill T. Gross, is credited with founding the "company builder" or "startup studio" (also referred to as "Venture studio," "Venture builder," and others) movement in 1996, followed in 1998 by London based Blenheim Chalcot (Alhokail 2019). In this model, the startup studio provides the

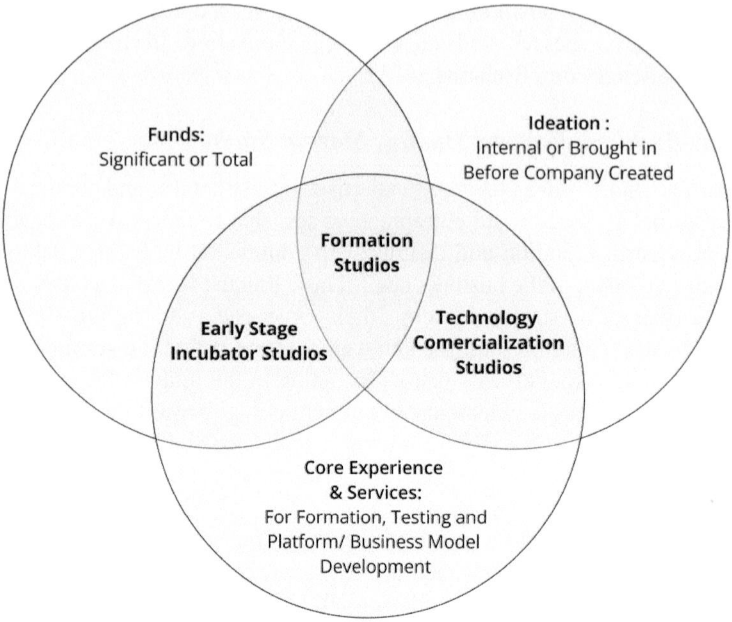

Figure 4. Startup studio models.

majority of initial funding, starting with and internally developed idea that is tested and validated within the studio. Ideas with traction are then supported by hiring a founding team to move the idea from testing into company formation. Instead of spending time raising funds and managing company administration, the founders focus on developing the business. As variations of the startup studio model evolved, differentiation led to calling this specific type of studio model "formation studios," as described by Vault Fund founder Sarah Anderson. In the formation studio model, founders are typically given minimal amounts of equity (5–10%) compared to if they were to raise funds and build the entire company by themselves (Alhokail et al. 2019, Baumann et al. 2018, Rajendran 2022). The resulting companies are primarily owned and controlled by the studio, even as they grow through subsequent raises.

The formation studio model expanded in the mid-2000s with Rocket Internet, eFounders, High Alpha, and others. Idealab started over 150 companies with more than 45 IPOs and acquisitions as of 2022 (idealab.com). Rocket Internet founded over 100 companies before 2017 by aggressively replicating existing business models using founders with local connections and insight. These studios offered a new vehicle for company creation and a corollary investment type, a venture fund designed to invest in the startup studios' companies, allowing outside investors to fund the studio model. For example, High Alpha's startup studio is funded by its venture fund, and the entity uses the name "Venture Studio" as a result. Examples of formation startup studios include:

- *High Alpha Venture Studio*: High Alpha's stated purpose is to combine company building with venture funding to conceive, launch, and scale B2B SaaS companies. The company has created over 30 companies since 2015. High Alpha

	Traditional Startup	Startups Created by Studios
Average Internal Rate of Return (IRR)	***21.3%*** (the best being 30%)	***53%***
Total Value to Paid In (TVPI)	***1.57***	***5.8%***
Time from Zero to Series A (months)	***56***	***25.2***
Time from Zero to Seed (months)	***36***	***10.7***
Time from Seed to Series A (months)	***20***	***14.5***

GSSN surveyed 258 startups created by studios, expertise, have skin in the game as a co-founder, and provide financial resources.

Figure 5. Startup studio to traditional startup comparisons.

hires founders through an application process and pairs them with internally generated ideas and support (highalpha.com).

- *eFounders*: With offices in Paris, New York, Berlin, and San Francisco, eFounders maintains a core team to support administration, accounting, finance, operations, and marketing, while hiring "founders" to lead the new companies as they launch. eFounders also currently has two other separately organized studios, one targeting FinTech startups and one targeting web3 startups (efounders.com).

B. "Early-stage incubator studios" -extend the startup studio idea to externally formed startups and leverage internal expertise and infrastructure to match technologies, entrepreneurs, and mentors:

While some startup studios fund and build companies from their ideas or ideas and IP sourced as raw technologies, others accept externally formed companies still in very early stages to benefit from leveraging internally sourced processes, business model playbooks, and infrastructure similar to that of the formation startup studios described above. In these "early-stage incubator" studios, the externally formed companies apply to join the startup studio for specific expertise such as software development and associated business process mentorship and funding (Anderson 2021). For example, some studios in this model bring in only business-to-business (B2B) companies that are software as a service (SaaS) platform companies. By limiting the cohort to similar company models, applying their expertise from earlier startup successes to move the startups forward much faster than if they were "learning from scratch." In this model, the studio typically has a lower ownership percentage than if it started and funded the company itself, leading to comparisons with the accelerator structure from the SDO perspective and the traditional venture capital structure from the financial investment perspective. Some early-stage incubator studios have accelerator-like models where the studio takes a relatively small percentage of equity (5–10%) in exchange for a fixed amount of funding and engagement over a short term in a cohort; others bring on the startups for an indeterminate or longer duration. Some studios offer the ability to continue engagement with the studio beyond an initial intensive period for a monthly fee or if the venture capital backing the studio continues investment in subsequent rounds. Examples of early-stage incubator studios include:

- *NineTwoThree Ventures*: Ninetwothree.co is a self-named "digital startup studio" formed by Andrew Amann and Pavel Kirillov, with a model that supports internally and externally developed ideas and various funding scenarios. The company developed a robust bench of software developers and project managers through experience working on application development for externally originated companies. These developers can build and scale new business projects quickly when brought from external sources, and they work on internally developed projects between external projects. NineTwoThree Ventures developed 8 internally developed startups by 2022, each led by a robust project manager who is typically transitioned into the company's CEO role when it spins out (ninetwothree.co).
- *Golden Section Studio*: This startup studio accepts applications on a rolling basis to its early-stage incubation program for B2B SaaS companies. Accepted startups benefit from the use of the studio's "playbooks," business strategies for a variety of common situations impacting B2B SaaS startups, and assigned venture associates who are entrepreneurs themselves with significant startup expertise. Funding includes a $250K convertible note investment and $250K of software development services; successful studio companies become eligible for investment consideration by Golden Section's technology venture fund (www.goldensection.com).
- *Softeq Venture Studio*: Based in Houston, Texas, Softeq Venture Studio runs four cohort programs per year and operates like an accelerator model, taking 6% of the company's equity for an investment of $125K and the use of its mentoring and programming for three months, with a demo day at the end of the program. Participant companies are not required to relocate to Houston for the cohort; the team can stay working remotely or wherever they have a corporate office and just travel to Houston one week each month for the three months of the program. Affiliated with Softeq software development company, cohort portfolio

companies have the opportunity to develop further relationships with Softeq for continued technical assistance. Softeq has competency in developing the software needed for hardware solutions, but applicants are not limited to having that business model. Softeq selects individual companies from its venture studio program for further investment (www.softeq.com/venture-studio).

- *Drukka Startup Studio*: Hungarian-based Drukka startup studio accepts a range of sophistication in startup companies, from founders with only an idea to early-stage businesses. Drukka bills itself as a "One-stop-shop" Startup Studio, providing startups with investment, mentoring, software development, and marketing services. Accepted founders progress toward a Demo Day like a traditional accelerator, where they have a chance to receive an investment from the Drukka Studio (https://drukka.hu/ 2022).

C. "Technology commercialization studios" – focus on business model formation by pairing technologies with founders in a studio environment:

Technology commercialization studios bring external entrepreneurs, opportunities, resources, and tools together to create a ready-to-launch enterprise (FedTech 2022) but may not provide funding. Promising technology or intellectual property (IP) from universities and national labs are paired with founders within the setting of experienced mentors to develop commercially viable business concepts. In these startup studios, common accelerator and incubator benefits like structural support and access to mentoring and education are combined with a hands-on approach to technology development and overall business strategy based on the expertise of the studio's leadership. Often the studio does not provide funding but provides mentors with information about how to apply for other governmental or non-profit funding once the company forms. Studios formed by corporations with this purpose are discussed in a later section.

Anderson (2021) describes the "commercialization studio" concept in the venture studio world, which licenses technology from universities or corporate labs to pair with founders and processes to develop commercially viable companies leveraging the technology. In this way, the studio assists in commercializing university and research ideas and can often work with a university's technology transfer office or even co-locate labs on campus. In these cases, the studio often funds most or all of the startup (vaultfund.com 2022). An example of a technology commercialization startup studio includes:

- *Homeland Security Startup Studio*: Operated by FedTech, a provider of US federal technology startup studios and accelerators, entrepreneur applicants are paired with technologies identified by the studio with the potential to support Homeland Security needs. External ideas and companies are not allowed in the program. Over seven-months, while still otherwise employed, paired teams spend 20 hours per week working through initial ideation and formation to identify commercial opportunities for the technologies. FedTech does not provide funding, nor does it take equity in the venture, but successful graduates of the program can receive assistance in securing future funding and negotiating technology licenses (Fedtech 2022).

In these three studio models, it is typical that once the startup achieves certain milestones in demonstrating product-market fit, revenue, or other business/industry-specific variables, the company will typically raise external funding or otherwise be "spun out" or exited from the studio. Many startup studios have venture funds associated with them, and some allow accredited investors to invest in the funds and thereby invest in the studio's portfolio. Once the startup raises outside capital or otherwise spins out from the studio, the fund associated with the startup studio may or may not participate in the next funding rounds, depending on its business model and prospects for the startup.

Investors have taken notice of venture studios, with studios reporting increasing funding from well-established individual and corporate VCs including Virgin Group, Peter Thiel, Jeff Bezos, and even Y-Combinator, the accelerator that now likens itself to being more of a seed fund. Enhance Ventures and the Global Studio Support Network describe the venture studio model as a new and separate investor asset class, from an investor perspective, as much as it is an SDO (Alhokail et al. 2019, Hochberg et al. 2015, Zasowski 2020).

Corporate Startup Studios, Accelerators, and Venture Labs

Large corporations with extensive experience in product and technology commercialization created startup studios for early-stage companies. Corporate-backed studios typically focus on specific niches or industries, called corporate startups, corporate accelerators, or venture labs. Startup companies apply to the corporate-backed studios to receive specific mentoring, technical support, and funding from successful companies with related customers, technical, or market needs; and the opportunity to build relationships that could lead to longer-term opportunities such as partnership or acquisition. In return, the corporations can identify technology or innovative ideas that may be useful for future acquisition or partnership at very early stages, staying ahead of the competition and accelerating innovation beyond what they develop internally. The structure of the corporate studio is often similar to that of an accelerator focused on a specific industry concerning the equity investment, the availability of physical space, and the time-bound nature of the offering, without a specific commitment to future funding or support after the program concludes. However, some corporate venture labs take large stakes in the startups and are therefore more aligned to traditional venture equity models in which the relationship becomes long-standing as the invested companies become portfolio companies with expectations for long-term support through exit.

Corporate venture labs have long partnered with university Technology Transfer Offices (TTO) to provide a method of commercializing ideas created and tested in the university lab setting. Spurred by the passage of the Bayh-Dole Act by the United States Congress in 1980 (Bayh–Dole Act or Patent and Trademark Law Amendments Act 35 USC § 200–212 (1980)), more than two-hundred fifty US universities established TTOs to support activity as of 2014. These TTOs work with student and corporate accelerators to increase the return on intellectual property developed at the university (Tseng and Raudensky 2014).

In a variation of the model, some independent accelerators and incubator companies will partner with corporate organizations to create the startup studio, rather than the corporation having to develop the competency of managing a startup studio or venture lab. For example, Techstars can host an accelerator for a corporation, focusing on emerging technology in a specific area of interest to the corporation. Key personnel from the corporation participate in the studio as mentors and facilitators and investments can be made at various levels.

In some instances, third-party startup studios and incubators will not host an official corporate startup studio. Instead, these third-parties serve as facilitators for a corporation's internal innovation efforts. This model is useful for corporations with no culture or history of rapid testing and deployment of new technologies but who desire to build these competencies. Corporations turn to these third-party providers for expertise about the entrepreneurial mindset and to enable their employees to operate outside their traditional systems.

- *Halliburton Labs*: Companies receive $100K in exchange for 5% equity to participate in the 12-month program, with no guarantee of further investment or assistance in additional raises. Participants receive dedicated laboratory working space and use of their physical research and testing facilities and may purchase monthly services such as office space, on-demand lab services, and raw materials procurement (not for commercial production) at Halliburton's preferred discounted rates (halliburtonlabs.com).
- *Chevron Studio*: Chevron Studio is a partnership between Chevron Technology Ventures and the National Renewable Energy Laboratory (NREL) to connect entrepreneurs with technologies from national labs and universities to commercialize concepts that support building a low-carbon future (https://nrel.smapply.io/res/p/chevron-studio/).
- *BAE Systems/FedTech*: BAE Systems partnered with FedTech to develop and execute the first BAE Systems Technology Accelerator: a 3-month program that will pair deep tech startups with BAE Systems research teams via a cohort-based accelerator program. The program ultimately seeks to deliver game-changing breakthroughs at speed and scale to BAE Systems' customers (https://www.fedtech.io/accelerators).
- *Incub & Co*: Based in Lausanne, Switzerland, Incub and Co, itself a startup company providing business incubation and co-working services, also partners with corporations to lead startup business ideation and formation, and conducts education sessions to assist the companies in fostering a culture of innovation (www.incub.co)
- *Techstars Norway*: One of several corporate partnership accelerators hosted by Techstars, Techstars Norway partners with Equinor and CapGemini to host an accelerator program focused on startups working across the energy transition within the critical areas of Energy Production, a Net-Zero Future, Digital, and Operational Enablers and Disruptors (https://www.techstars.com/accelerators/equinor-energy).

Assessing SDO Effectiveness

Investors and founders alike need ways to evaluate SDO effectiveness when making decisions for their businesses. For founders, time spent in engagement with an SDO cannot be retrieved, and with limited funding resources, it is critical to align their businesses with SDOs appropriate for their stage of development and resource needs. Likewise, investors benefit from understanding the relative expectations of each type of SDO, aligning their investment thesis with potential investment opportunities and finding ways to engage with promising startups.

Metrics used by SDOs include volume or operational activity, which don't necessarily equate to effectiveness, such as the number of program participants or the number of mentor hours or workshops provided. To truly assess the success of the SDO, participants could be tracked after graduation to compare outcomes, including years in operation, profitability, return of investment capital to investors, and other financial and operational metrics of the startups themselves against benchmarks (Batra 2021).

In addition to a lack of standardized metrics, challenges in evaluating the effectiveness of SDOs include the lack of suitable "control groups" (i.e., companies under similar circumstances targeting similar customer needs that do not engage with SDOs for assistance) against which to compare the incubated companies; the difficulty in controlling for the application and selection process across SDO types; the quality of the SDO leadership and management; and the variation in outcomes. SDO success is often dependent on hard to quantify factors such as the quality of the program and the strength of specific people working in leadership, management, teaching, networking, and mentoring roles; thoroughness of the vetting process to limit participants to those most likely for success (see Chapter 6 for more on the vetting process); and access to local financial and personnel resources to remove obstacles (Hackett and Dilts 2004a, 2004b).

In addition, the ability of the SDO to attract and hire qualified mentors and founders may be related to its business model and "cachet" or reputation. As Fowle (2017) found, reputation allows accelerators to attract more and better applicants, which is reinforced when the accelerator can signal its exclusivity. Developing a reputation for quality and desirability and demonstrating that alumni of the program are successful creates a virtuous cycle in which mentors and founders want to be associated with the program.

Performance of Incubators and Accelerator

While the scientific literature lacks performance measures from which these programs can be compared, there are efforts in practice to compare and rank SDOs, within and across categories.

Incubators and accelerators compared within their categories:

- UBI Global's World Rankings of Incubators and Accelerators 2019–2020: UBI compared and ranked the performance of business incubators and accelerators worldwide who were self-selected to participate in a benchmarking study. The

UBI ranking consists of metrics evaluating the SDOs themselves, not based on the ultimate success of their participants but rather based on factors about the SDOs along the dimensions of Value for Ecosystem, Value for Client Startups, and Value for Program with metrics like number of jobs created, services offered to the startup, mentoring hours offered, sponsorship attraction and number of out of state applicants, etc. (Meyer and Sowah 2019).

- Beta Boom 2021 comparison of accelerators: Beta Boom analyzed exit data from Crunchbase.com (with exits being "sale to a private company" or "IPO") and developed a ranking of the best startup accelerators with more than 100 investments in 2021. In this study, the top five accelerators achieved 22–40 exits with an exit rate of 19.3% to 24.7%. Techstars Boulder and Techstars Seattle topped the list (Beta Boom 2022).
- 2017 Literature Review of accelerator success factors: Michael Fowle at The Open University in the UK identified ten critical success factors for business accelerators based on his review with over 100 source references, which can be grouped into five areas: content (business expertise, product expertise, and the manner in which the founders acquire learning); context (location with respect to industry clusters and accessibility to the intended customers); community (local industry, cohort quality and access to innovation such as universities), cash (access to investors) and cachet (brand perception of the accelerator with quality and desirability driven by exclusivity and social reinforcement.) Interestingly, Fowle concludes that funding is most effective when it is a reward, not a guarantee by the accelerator. Funding as a reward is contrary to the standard practice of providing funding to all admitted (Fowle 2017).

Performance of Venture Studios and Startup Studios

In 2022, it was common to find SDOs using the name "Venture Studio" or "Startup Studio" with variations across the models, acting as formation studios, commercialization studios, accelerators, and early-stage incubators, without nomenclature for distinguishing them. An investor or potential founder would need to review the company carefully to understand its particular model, as it is often not evident through the name. Startup studios or venture studios compared within their category:

- Enhance Ventures 2019: In its December 2019 whitepaper, Enhance Ventures reported over 330 "venture" or "startup " studios were in existence. Despite the number of venture studios around the world, their tenure is relatively short, so it is still early to evaluate their exit potential (Alhokail et al. 2019).

Startup studio companies compared to accelerator companies:

- Comparison of Top Accelerator Companies and Startup Studio Companies: Attila Szigeti, entrepreneur and COO of Drukka Startup Studio, compared the 21 most successful studio companies with the 21 most successful accelerator startups using data from a commercially available database of investment metrics, Crunchbase (Szigeti 2015). He found that on average, accelerator companies raise more capital (more than double) and employ more people (~ 15% more) than startup studio companies. Considering accelerators are building competency in each startup rather than leveraging the competency of a group for multiple companies, this is not a surprising result. He also found that startup studio companies are more efficient and grow more quickly as evidenced by their 26% larger "MatterMark Growth Score" which is a measurement of how quickly a company gains traction (mattermark.com). This finding complements the assumption that using processes developed over time should eliminate some trial and error and enable a more efficient scale (Szigeti 2015).

- Global Startup Studio Network 2020: While still relatively young, the studio model seems to be showing signs of promise: the latest estimates by the Global Startup Studio Network state that startups created by venture studios offer an Average Internal Rate of Return of 53% as compared to 21.3% of a traditional startup with half the time from initiation to Series A round (Zasowski 2020).

As time goes on and more companies who participated in accelerators and incubators have exits and IPOs, more data is available about their return to investors as a success metric. Enhance Ventures reports in its 2019 study whitepaper that over 35% of Series A funding in the US went to accelerator graduates (Alhokail et al. 2019). In addition, more data is available to analyze the percentage of companies who graduate from an accelerator or incubator going on to have an exit, defined as a sale to private equity or IPO. As stated in all financial disclaimers, prior success does not guarantee future success; but evaluating the success metrics of an accelerator or startup studio may be a useful metric for investors vetting the quality of mentorship, networks, and financial assistance of startup company participants when conducting due diligence (Zasowski 2020).

SDO Factors for Entrepreneurs and Investors to Consider

Entrepreneurs considering joining an SDO should do so with a clear goal of the exercise in order to choose wisely. For entrepreneurs with an existing business idea, the critical path item is understanding what is needed to get to the next stage and determining if the SDO in question has the required services to help the company do that. Whether the entrepreneur is giving up equity, time, or both, the time with an SDO has a cost and, just like any other investment, should provide a clear ROI. For investors, these questions will also allow to evaluate SDO opportunities to connect them to companies that match their investment profiles and mentoring opportunities.

Theoretical Implications: The Role of the Startup Incubation Ecosystem

As startup support systems evolve with the infusion of more capital and the synergistic effects of serial entrepreneurial environments increase the sophistication of startup development, academic research has begun to take note of the importance

Table 1. Considerations when evaluating an SDO.

Factor	Questions to Evaluate
Business Model	Does the SDO have a sustainable model? Where does it receive its funding? If it is a non-profit, are sources of income sufficient to hire qualified staff and mentors? If for-profit, is it successful, and does it offer an opportunity for long-term engagement and ongoing support?
Experience and Expertise	Does the SDO have experience with companies in the same industry or similar business models? What is its track record with respect to exits, funding rounds, and revenue of the startups?
Mentors and Functional Leadership	Are the mentors experienced entrepreneurs themselves? How does the SDO screen and vet them before they are assigned to the startup? Does the SDO assist with securing talent and recruiting the right people to build the team?
Cost and Funding	How much funding will the SDO offer, and at what cost to the company? What is a reasonable range for similar services? Is there an ongoing fee to participate if the program has an end date, such as with an accelerator?
Peers and Network	Who are the other startups with which the entrepreneur will engage? Does the SDO have a strong alumni network? Are references available from other founders who participated in the program?

of these types of organizations (Novotny et al. 2020). This chapter contributes to the scientific literature by:

- Bridging the science-practice gap by illuminating the current state of the ecosystem
- Clarifying definitions and conceptualizations of SDOs
- Offering a starting point for a taxonomy of SDOs
- Providing insight into situational factors and characteristics that may affect performance
- Exploring methods of SDO performance measurement

Future Research

The role or influence of participating in one SDO may or may not be as crucial in startup success as the overall network engagement with the founders, mentors, and advisors of an SDO who have connections to other network players. There is an increasing understanding of the role of factors external to any individual SDO for the success of the startup, including quality of external networking relationships, availability of external sources of funding, access to human capital, and the cross-pollination of these factors when founders mix and mingle in cohorts after participating in prior activities and cohorts elsewhere.

- What is an SDO (agreed upon scientific definition)?
- What makes an SDO effective?

- Are startups that work with SDOs more successful?
- Which factors leading to a company's failure to exit the "valley of death" are best addressed through SDO participation?
- Does participation in an SDO enhance the likelihood of ROI for the investor?

Conclusion

Through understanding the structure and role of each SDO and its part of the SUPIE, today's investor can understand the relative progression of startup companies and can even choose to invest in a basket of companies at various development and potential risk levels, from incubators to startup studios, directly or in angel or venture funds. Founders can assess their company's development stage and resource needs and engage with the appropriate organization to deliver maximum support while minimizing costs in time and resources. The Angel Capital Association (ACA 2022) reports $950M invested by angels in 2021. According to ACA's Angel Report 2021, of the companies that received angel investment that exited or closed down in 2021, 30% failed (ACA 2021). How many of those companies and the investor funds they represent could have been turned into success stories if they received appropriate SDO support? As investment and founder activity increase, the importance of understanding SDOs and how they can support startups increases in parallel (ACA 2021, 2022). Further research into optimizing SDO effectiveness is critical to assist founders and investors in deploying capital wisely.

Acknowledgements

Thank you to everyone in the SDO community who shared your time with me to help pull this chapter together, to provide more context and information about this fast-growing industry. Through countless conversations, proofreading, and feedback, your contributions made all the difference. Special thanks to the following reviewers whose comments and suggestions on the manuscript were invaluable. Thank you all for your assistance and for everything you do to support entrepreneurship.

- Ganesh Nathan, Ph.D., Business School Lausanne
- Jason Ethier, Greentown Labs
- Payal Patel, Softeq Venture Studio
- Sarah Anderson, Vault Fund
- Stephanie Campbell, the Artemis Fund (and formerly MD of the Houston Angel Network)
- Sunny Shuoyang Zhang, Ph.D., Born Global Venture Fund

Special thanks go to Nikki Blacksmith, Ph.D., for including me in this valuable project and for providing expert guidance and editing, and for connecting me to Dane Wagner on the Blackhawke team for polishing up the tables and citations. It's been an amazing learning experience to be part of putting this book together and I've learned so much from you!

References

Alhokail, M., Celen, A. and Tilani, R. (2019). *Startup Studios: Innovating Innovation.* Enhance Ventures. https://www.enhance.online/#why_venture_builder.

Al Natsheh, A., Gbadegeshin, S.A., Ghafel, K., Mohammed, O., Koskela, A., Rimpiläinen, A., Tikkanen, J. and Kuoppala, A. (2021). The causes of valley of death: a literature review. *INTED2021 Proceedings*, 9289–9298.

Anderson, S. (2021, August 26). *What is a Venture Studio?* Vault fund. https://vaultfund.com/what-is-a-venture-studio/.

Angel Capital Association. (2021). *Angel Funders Report.* Retrieved from: https://www.angelcapitalassociation.org/angel-funders-report-2021/.

Angel Capital Association. (2022). *Angel Funders Report.* Retrieved from: https://www.angelcapitalassociation.org/angel-funders-report-2022/.

Batra, A. (2021, January 5). *What Incubators can Learn from 2020 and Prioritise for 2021.* Yourstory. https://yourstory.com/2021/01/incubators-learn-2020-prioritise-2021/amp.

Baumann, O., Bergenholtz, C., Frederiksen, L., Grant, R.M., Köhler, R., Preston, D.L. and Shane, S. (2018). Rocket internet: Organizing a startup factory. *Journal of Organization Design*, 7(1): 13. https://doi.org/10.1186/s41469-018-0037-2.

Beta Boom. (2022, May 17). *Beta Boom Pre-seed Fund Investing in Tech for Women and Black, Latinx, and Ethnic Minority Consumers*. Betaboom. https://betaboom.com/.

Calza, F., Dezi, L., Schiavone, F. and Simoni, M. (2014). The intellectual capital of business incubators. *Journal of Intellectual Capital*, 15(4): 597–610. https://doi.org/10.1108/JIC-07-2014-0086.

Cambridge Innovation Center. (2022, June 1). *CIC Cambridge Co-working.* Cic. https://cic.com/cambridge-coworking.

Cohen, S. (2013). What do accelerators do? Insights from incubators and angels. *Innovations: Technology, Governance, Globalization*, 8(3): 19–25.

Cohen, S., Fehder, D.C., Hochberg, Y.V. and Murray, F. (2019). The design of startup accelerators. *Research Policy*, 48(7): 1781–1797.

Di Risio, A. (2020, July 3). *Global Co-working Growth Study 2020: Market Size and Industry Trends.* Coworkingresources. https://www.coworkingresources.org/blog/key-figures-coworking-growth.

Di Risio, A. (2021). *The History of Coworking.* Coworking Resources. Retrieved from https://www.coworkingresources.org/blog/history-of-coworking retrieved 7/21/22.

ETC Baltimore. (2022, May 17). *Everything you Need and More.* Etc. https://www.etcbaltimore.com/.

FedTech. (2022, May 15). *Startup Studio Faq.* Fedtech. https://www.fedtech.io/startup-studio-faqs.

Fowle, M. (2017) *Critical Success Factors for Business Accelerators: A Theoretical Context.* British Academy of Management.

Franco, G. (2015). Interview with Bradley Neuberg, Creator of the Coworking Concept. Retrieved from: https://www.bcnewt.com/promociones/2015/02/we-interviewed-bradley-neuberg-creator-of-the-coworking-concept/.

Galvanize. (2022, May 17). *Entrepreneur Education.* Galvanize. https://www.galvanize.com/coworking-space.

Greentown Labs. (2022, May 17). *Greentown Labs Houston.* Greentownlabs. https://greentownlabs.com/houston/.

Hackerspaces. (2017, April 25). *Number of Makerspaces in the us by State.* Acceleratingbiz. https://acceleratingbiz.com/proof-point/number-of-makerspaces-in-the-us-by-state/.

Hackett, S.M. and Dilts, D.M. (2004a). A real options-driven theory of business incubation. *The Journal of Technology Transfer*, 29(1): 41–54.

Hackett, S.M. and Dilts, D.M. (2004b). A systematic review of business incubation research. *The Journal of Technology Transfer*, 29(1): 55–82.

Hausberg, J.P. and Korreck, S. (2020). Business incubators and accelerators: A co-citation analysis-based, systematic literature review. *Journal of Technology Transfer*, (45): 151–176. https://doi.org/10.1007/s10961-018-9651-y.

Hochberg, Y., Cohen, S. and Fehder, D. (2015, March 31). *The Top 20 Startup Accelerators in the U.S.* Harvard Business Review. https://hbr.org/2015/03/the-top-20-startup -1-accelerators-in-the-u-s.

Isenberg, D. and Onyemah, V. (2016). Fostering scale up ecosystems for regional economic growth. In *Global Entrepreneurship Congress*, 71–97.

International Business Innovation Association. (2018). *Number of US Entrepreneurship Centers by Type.* (Retrieved from: https://inbia.org/wp-content/uploads/2018/08/NumberofECsimage.jpg?x62369.

Kuratko, D.F. and LaFollette, W.R. (1987) Small business incubators for local economic development. *Economic Development Review*, 5(2): 49. https://hardeebusiness.com/entrepreneurs/small-business-incubators-for-local- economic-development.

Lau, N. and Peek, K. (2016, February 23). *By the Numbers, the Rise of the Makerspace*. Popular Science. https://www.popsci.com/rise-makerspace-by-numbers/.

MacKenzie, S.B. (2003). The dangers of poor construct conceptualization. *Journal of the Academy of Marketing Science*, 31(3): 323–326.

Mancuso Business Development Group. (2022, February 2). *Mancuso Business Development Group: What is a Business Incubator?* Mancusogroup. https://mancusogroup.com/what-is-a-business-incubator/.

Meyer, H. and Sowah, J. (2019). *Ubi Global World Rankings of Business Incubators and Accelerators 2019–2020.* UBI Global. https://doi.org/10.13140/RG.2.2.16066.53441.

Moss, W. (2019, October 10). *The Rise and Fall of Wework: What Actually Happened.* Wesmoss. https://www.wesmoss.com/news/the-rise-and-fall-of-wework-what-actually-happened/.

Novotny, A., Rasmussen, E., Clausen, T.H. and Wiklund, J. (eds.). (2020). *Research Handbook on Start-up Incubation Ecosystems*. Edward Elgar Publishing.

Office of the Governor Economic Development and Tourism. (2017, December). *Texas Startup Resource Directory.* The State of Texas. https://gov.texas.gov/uploads/files/business/ TXStartUpDirectory.pdf retrieved 3/8/22.

Paluch, S. (2021, January 21). *The Best Startup Accelerators: The Definitive List for 2021.* Betaboom. https://betaboom.com/blog/best-startup-accelerators/.

Rajendran, V. (2022, January 2). *The Secrets to Venture Studio Success.* 500. https://500.co/theglobalvc/the-secrets-to-venture-studio-model-success.

Rese, A., Görmar, L. and Herbig, A. (2022). Social networks in co-working spaces and individual coworker's creativity. *Review of Managerial Science*, 16(2): 391–428. https://doi.org/10.1007/s11846-021-00445-0.

Reypens, C., Rückert, D. and Delanote, J. (2020). *From Starting to Scaling: How to Foster Startup Growth in Europe*. European Investment Bank.

Richards, R. (2021, June 10) *Accelerators vs Incubators: How to Choose the Right One.* Masschallenge. https://masschallenge.org/article/accelerators-vs-incubators.

Schuenke, L. (2022). *Innovation Incubated: Awesome inc. Communications.* INBIA.

Sarath. (2022, May 13). Top *100 y Combinator Companies.* Eqvista. https://eqvista.com/top-100-y-combinator-companies/.

Startup Commons. (2019, May 17). *What is a Startup?* Startupcommons. https://www.startupcommons.org/what-is-a-startup.html.

Szigeti, A. (2015). *Startup Studio Trends 2015: A Detailed Look at 51 Startup Studios and 212 Portfolio Companies.* Attila Szigeti. https://aszig.gumroad.com/l/sssp1.

Techstars. (2022, Dec. 30). *Techstars Investment Terms.* Techstars. Retrieved from: https://www.techstars.com/newsroom/investment-terms.

Texas Business Incubator Association. (2015). *Business Incubators Directory.* The State of Texas. https://gov.texas.gov/uploads/files/business/business_incubators_directory.pdf.

Tseng, A.A. and Raudensky, M. (2014). Performance evaluations of technology transfer offices of major us research universities. *Journal of Technology Management and Innovation*, 9(1): 93–102. https://doi.org/10.4067/S0718-27242014000100008.

Upsuite. (2021, May). *The 9 Types of Co-working Space Available Today.* Upsuite. https://www.upsuite.com/blog/types-of-coworking-space-available/.

Zasowski, N. (2020). *Disrupting the Venture Landscape: Why Startup Studios is where Investors find Capital Efficiency*. Global Studio Startup Network.

Part III

Measurement of Startup Performance

Chapter 12

Cultures of Evaluation

Leveraging Academia for Due Diligence in Angel Investments

Jerome A Katz

It is a truism that in academia, everything is graded – the students, the faculty, the schools, everything. In that sense, academic institutions often embody cultures of evaluation. How would such a culture translate into a real-world situation such as an angel investment? This chapter outlines the approach used by Saint Louis University (SLU) in their Billiken Angels Network (BAN). They created BAN using the same evaluation culture used in their entrepreneurship classes, with deep analysis of the submitted proposals, a committed team of experts evaluating the proposals, and a developmental attitude embedded in the evaluation process, so that evaluation is intended to help improve the recipient and their ideas. Over its ten-year lifespan, BAN achieved these goals and produced a track record of fewer negative results than was typical of angel investment, and a set of social outcomes still providing returns to the larger academic program. The details of BAN, its context, approach, and outcomes make up the broad outline of this chapter.

Developing a Culture of Evaluation: SLU's Entrepreneurship Program

The first entrepreneurship course at SLU was taught by Prof. Robert Brockhaus in 1974, and became a permanent class at SLU in 1978, making SLU one of the first 25 schools in the world to teach entrepreneurship as an ongoing course (Katz 1991). In 1987, Prof. Jerome Katz was recruited from Wharton, and Brockhaus and Katz

Brockhaus Endowed Chair of Entrepreneurship, Richard A. Chaifetz School of Business, Saint Louis University, 3674 Lindell Blvd., St. Louis MO 63146.

became the first director and associate director of the Institute for Entrepreneurial Studies in the College of Business at Saint Louis University. At that time the Entrepreneurship Program was also created to organize and run the courses in entrepreneurship, while cross-campus and community outreach activities were concentrated in the Center.

The SLU Entrepreneurship Program had developed a distinctive pedagogical style. In the 1970s, most classes in U.S. schools were focused on small business, and taught largely by entrepreneurs as adjunct instructors, usually without a textbook, but with a focus on personal "war stories." Brockhaus was a former entrepreneur who had pursued a BS in Mechanical Engineering, a master's in industrial administration, and a Ph.D. in Organizational Behavior and became one of the very first tenure-track faculty and published researchers teaching entrepreneurship in an AACSB accredited business school in the USA. He was determined to focus on small and medium sized businesses with particular attention to those students seeking to create high-growth ventures. He used research articles from other nascent entrepreneurship researchers as well as articles from business reviews in classes, eventually adopting textbooks as suitable options became available. While an entrepreneur in his own right, he taught with multiple speakers and panelists in nearly every class, encouraging them to tell their war stories specific to the night's topic with Bob playing the role of curator, inquisitor, and connector of entrepreneurs, students, best practices, and the emerging stream of entrepreneurship research.

The best practice focus initially surprised students and entrepreneurs both. But SLU in the 1980s housed the headquarters of the Missouri State Small Business Development Centers. That a small, private, Catholic, Jesuit university would run a state-Federal agency shows how little the University of Missouri System valued entrepreneurship or small business at the time. In most other states, SBDCs were part of the state university systems. By 1987, Brockhaus was the state director and Katz was made an associate director when he came to SLU. This positioned them to benefit from and contribute to the creation and promulgation of best practices in small business development at the national level, and test them out in their classes before practicing and aspiring entrepreneurs.

As a result, the hallmarks of professional best practices at that time, feasibility analyses, and business plans as developed for the SBDCs by Courtney Price, Richard Buskirk, and Mack Davis (1991) which is known today as Kauffman's FastTrac program. These models for feasibility analysis (FastTrac I) and business plans (FastTrac II) became a mainstay of SLU classes. For local entrepreneurs dealing with the SBDC or the SLU classes learned to organize and present these key documents. When the SBA began to require business plans as part of the loan package for SBA-guaranteed loans, the entrepreneurs in the SLU pipeline were already experts on the key documents, but more to the point, the evaluation of entrepreneurs' ideas, including student entrepreneurs, became legitimized nationally, firmly favoring cultures of evaluation where possible.

The SBDC and its parent the SBA were built using the highly successful model of the US Department of Agriculture's Agricultural Extension Program, which took and tested ideas developed in universities and passed-on the best practices to farmers (Katz 2003). The SBDC and related programs such as the Small Business Institute

(Solomon and Weaver 1983) were examples of government efforts to improve small business and foster entrepreneurship, and these values, best practices, and associated experiential exercises were built-into the SLU Entrepreneurship Program.

When Katz joined the program in 1987, he brought the analytic model of entrepreneurship being developed at Wharton to SLU but adopted Brockhaus' interactive town-and-gown model over the case method so much a part of the Wharton experience. Like Brockhaus, Katz had been an entrepreneur-turned-academic, initially taking roles in the extended family's businesses, and eventually creating a consulting firm which he later sold, Katz's additions to the process were to organize and professionalize the recruiting process (Katz 1995), and formalizing the culture of evaluation (Fetterman 2002, Murphy 1999). In this approach, students' ideas were constantly being created, viewed, and evaluated. The goal of the evaluation was to help the students make as many mistakes as possible inside the "ivory tower" where the errors cost little time or money and have time and support to fix these problems and solidify their entrepreneurial prospects as part of their academic career. Students would typically graduate not only with their business plan ready, but with an existing set of relationships with bankers, attorneys, potential suppliers, and customers, which included a mix of fellow SLU alumni and St. Louisans from other schools and backgrounds.

Making this work while honoring Brockhaus' successful town-gown model involved creating a pipeline for using outside experts, with most people being recruited initially as speakers (Fig. 1). Those speakers who did a good job of interacting with students and showed an appreciation of the idea of a discipline of entrepreneurship (i.e., they were willing to accept the idea that a strong business foundation could help make an entrepreneur better) were invited to become mentors and advisors to students with interests in entrepreneurship that aligned with the mentors. Mentors who achieved superior outcomes with students and got high marks from students were asked to become judges for the class projects (feasibility studies, industry analyses, and business plans). And when new co-instructors were needed, they were invariably recruited from the ranks of judges, since these people had proven themselves in the classroom and with the students and instructors and had learned and bought-into the model of developmental, best practice entrepreneurship, and the associated evaluation culture.

Over a typical year, more than 100 individuals would come through the pipeline. As noted in Fig. 1, some of these people did not progress, and occasionally

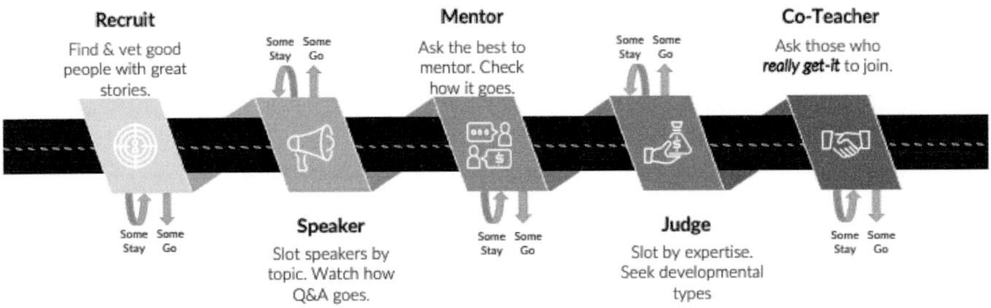

Figure 1. The SLU entrepreneur pipeline.

entrepreneurs were dropped for time conflicts, having trouble maintaining a developmental outlook when evaluating, or just burning out from the demands of the role. Some individuals would also "drop-back" in the pipeline, e.g., no longer judging, but still coming in as a speaker.

Underlying this model were tough standards wrapped in a positive, developmental attitude, consistent with Fetterman's approach. That said, grading for major projects was done not using the academic standards of the typical undergraduate class, but what was described as a "real-world standard." This was achieved by having multiple judges for each major project. Two of the judges were the co-teachers of the class and at least one other judge would be a pipeline member specific to the type of business. Sometimes if a pipeline judge worked on the first version of the plan a second pipeline judge would be added to the judging team for the final presentation. When grading, the newest judge would go first, then any other pipeline judge, the junior co-teacher, and finally the senior co-teacher to minimize Stockholm syndrome and deference to faculty. Simply put, the summary letter grade was based on the following standard in Table 1.

The advantage of this rubric was that entrepreneurs, lawyers, accountants, venture capitalists, and all manner of consultants could instantly grasp how to grade. The rubric above went through numerous revisions to best capture and position the typical ways business people thought about businesses and business ideas.

This model was supported by a multipage Likert scale scoring sheet and the marked-up version of the plan by each judge. Copies of the Likert-scale grading packet and the grading rubric are available for public download from http://tiny.cc/BAN-Scoring-Packet. In the early days, students were encouraged to take notes during feedback, but as the technology advanced, the presentations, Q&A, and feedback sessions were all recorded.

When discussing the grades, anything more than a 3-row difference (B- to B is two rows, B- to B+ is three) was seen as anomalous and was discussed. Most often, these became "forehead slapping" moments when someone forgot a key factor and the second most common driver was one judge having the greater insight or expert knowledge than the others. In either situation, the judges typically converged on grades very quickly. When there remained a difference, the norm was to take the lowest grade of the range at midterms, to reinforce the need to improve, and the highest grade of the range at the final presentation out of respect for the work the student did. From discussions among the teaching team, these adjustments were seen as ways to show the developmental element of the program.

Business plans were done at midterm with this approach, and despite their history in the program, most students underestimated the difficulty of delivering an "A" quality business plan. The last half of each semester is usually for those students one of the most intense periods of their academic lives. The co-teachers and even the judges had to become accustomed to calls and visits, later emails and text messages, at nearly all hours. Because the co-teachers treated this as the level of commitment and aspiration sought in the nascent entrepreneurs, the vast majority of the pipeline judges embraced the effort and would even brag to one another about their efforts. To make this work, the approach used was to explain "this is what the world will expect of you and we're here to help you get to where you need to be." Through this

Table 1. Business plan overall grade rubric.

Grade	Description of Grade
A	Plan is ready for investment. Business has potential to be highly profitable or do an excellent job of meeting entrepreneur's stated goals. "I'll invest right now!" or "I'll buy it right now!" or "I'll partner right now." "A sure winner!" **"I am ready to sign on the dotted line right now."**
A-	Almost everything is acceptable.
B+	Plan is usable **as is** or the business described in the plan will probably operate (1) as predicted **and** (2) profitably. Such as business would be described as "solid" rather than "stellar". "I would tell them to go ahead and do it." "The plan should work out.", "I would not personally invest in this one, but I'm sure others might consider it."
B	With minor revision, the plan would be fundable or would operate predictably/profitably. "I'd want to find out more, but if they can give the right answers, it looks promising to me." "If they iron some bugs out, I think they could make a go of it. **"I'm taking you out to a nice dinner. I have a few questions to ask you."**
B-	A little better than a C+, but not quite a B.
C+	A little better than a C, but not in the B range.
C	Plan offers the potential for a good business but would not be fundable or would appear to be predictable/ profitable given the existing plan. Major or substantive revision needed. "There are some major problems with this, but there is a possible moneymaker here." "The plan doesn't tell me enough, but I think there is some hope for the business." "This plan has a wholly inadequate *(put part of business plan here)* but is otherwise good." **"Coffee, no food. Free refills. Lots of questions."**
C-	A little better than a D, but not quite a C.
D	Plan's idea does not appear to be viable or likely to lead to profits, and the plan is substantially below average from a competitive standpoint. "This is not likely to work." "The plan is fatally flawed." "This plan has two or more major parts which are substantially sub par." **"Nothing to eat or drink. I'm detecting major flaws here."**
F	Plan and idea are wholly inadequate, not professional quality, and/or not likely to be competitive or profitable. "This wasn't ready." "Person needs to rethink going into business." **"You're embarrassing yourself and wasting my time. Get out of here!"**

kind of approach, SLU had created and sustained a positive culture of evaluation for more than 35 years.

In the end, the SLU Entrepreneurship program built a distinctive pedagogical and cultural infrastructure to promote entrepreneurship among students and the larger community. The key elements of this were, and remain to this day:

(1) Entrepreneurship education grounded in research and carefully curated experience. This culminated in the creation of a textbook embodying the processes, exercises, and stories of the program, *Entrepreneurial Small Business* (Katz and Green 2007). The seventh edition came out in 2023.

(2) A town-gown partnership in the classroom through a constant stream of speakers, mentors, and judges from the community, and the constant presence in the class of tenure-track entrepreneurship faculty and entrepreneurs from the community as co-teachers (Katz 1995).

(3) A highly practical experiential model, rooted in the milestone projects of entrepreneurship education, with some type of feasibility analysis in the early stages and a business plan or its equivalent at the final stage (Katz 2018).
(4) A culture rooted in constant evaluation, constant support, and constant collaboration (Fetterman 2002, Murphy 1999).

Brockhaus retired in 2005 and Katz took over as the head of the Entrepreneurship Program. Among the activities he set in motion was the publication of the text built on the lessons used in the Entrepreneurship Program (Katz and Green 2007), extending the co-teaching model to all entrepreneurship classes, and solving the enduring problem of funding student and alumni startups.

Given the academic program's success in generating startups, one of the points of concern was the difficulty the more ambitious ideas had getting early-stage funding in St. Louis. There was only one angel group in town, and connections to that group were sparse. While St. Louis was described as one of America's cities with an extraordinary percentage of people with unearned income from the legacies of the industrialists who made St. Louis a major manufacturing hub, the locals also described the situation as having "T-Rex Investors," i.e., people with "deep pockets and short arms." Finding potential angels was easy, but finding motivated potential angels was difficult. That said, notable numbers of SLU alumni fit the financial criteria of accredited investors and because of their prior involvement in the Entrepreneurship Program, there was reason to believe that these SLU alumni and friends of the Program might be interested in a university-based angel network, which led to the creation of BAN.

Leveraging SLU and its Culture to Create an Angel Network

In 2005, Katz felt that three elements would make a new university-based angel network possible. First, years of running the pipeline model resulted in many potential volunteers and supporters for an angel network. Second, a subset of those volunteers were already angel investors, but the vast majority could be considered "free radical" angels, floating around the entrepreneurial ecosystem, and hoping to come across a viable prospect in which to invest. They might think a steady supply of prospects might be worth their while. Third, the pipeline and the evaluation culture underpinning it could form the basis for a more developmentally oriented angel network, which would be in keeping with SLU's Jesuit culture and outlook of "men and women for others." Fourth, was that the entrepreneurial ecosystem in St. Louis was growing by leaps and bounds.

The 1980s and 1990s in St. Louis saw the rise of an innovation and entrepreneurial ecosystem heavily focused on biotechnology (Bayham et al. 2007). In fact, the area's name for its region in those times was the BioBelt. Monsanto (now Bayer), headquartered in St. Louis, and Barnes-Jewish Hospital, a part of Washington University, co-anchored biotech, and medical technologies and were central to the creation of local venture capital firms and facilities. They also lobbied local and state governments to fund the creation of innovation spaces in the St. Louis region, leading to the creation of several organizations and research parks in the area. The first angel group, the St. Louis Angel Network, was formed in the early 2000s but ceased operation after a few years. The St. Louis Arch Angels started in 2005 and continue to this day, with a strong orientation toward biotech and IT, with occasional forays into consumer products. With incubator and research park capacity growing and venture capital investments in the region also on the rise, the belief that a second angel group might be workable seemed reasonable. That it might be a university-backed group was a more speculative issue.

From 10 angel groups in 1996 (Cremades 2018), to 100 by 1999 and 255 by 2005 (ACA 2009), 11 angel groups had weak ties to local universities, while only 3 had formal university affiliations – Notre Dame's Irish Angels, Marquette's Golden Angels, and the University of Hawaii's Hawaii Angels (Hudson 2007). ACA helped connect SLU to Theresa Sedlak of the Irish Angels, which started as a part of Norte Dame's Gigot Center but was soon spun out as a separate organization, largely due to concerns about the university's liability. Connection to Tim Keane of Marquette's Golden Angels for early mentoring was facilitated by strong existing ties among Jesuit business schools.

Between ACA, the Irish Angels, and Golden Angels, Katz was able to put together a proposal for the Billiken Angels, and have it reviewed and vetted by Entrepreneurship Program mentors and ACA. Members of SLU's entrepreneurship pipeline, esp. those who were or had been members of SLU's Board of Trustees, were asked for their opinions on the idea and its prospects. The fact that two of the best-known university-based angel groups were also Catholic universities, helped sell the concept to internal decision-makers. Even so, working through the mechanics of the new angel group at the pace typical of academia meant the BAN would not start until 2008.

The Billiken Angels Network was organized from the start as a formal part of SLU, because of concerns that other spun-out versions could suffer from "piercing the veil" sorts of problems, where an independent BAN could be shown to have some sort of tie to the deeper-pocketed SLU, which would fail to protect SLU financially. So, it was easier and more directly defensible to run BAN as a part of the University. BAN would be based with Katz in the College of Business as its unpaid director, and it would use the secretarial services of the Institute for Entrepreneurship. It would be organized as a network, with angels holding onto their money and investing as they saw fit. There was an Executive Board, made up of the Director and 5 angels with long ties to the Entrepreneurship Program. Two angels had been long-term co-teachers. Another one was a long-time judge in the pipeline, and the last was one of the angel-co-teacher's partners, a long-time member of the Business School's Dean's Advisory Board. As the Fellows became more organized and central, three Lead Fellows (volunteers) were added to the Advisory Board.

The Process: Evaluating Startups for Presenting

To be considered for investment, opportunities would need "SLU DNA" having current or former SLU faculty, staff, students, or intellectual property. BAN would consider any industry, and firm at any stage, including growth capital for existing firms. BAN would not invest in non-profits. Investments would be equity-based, not loans, and where the fledgling angel groups on either coast talked about 30X targets in 5 years, based on the number of high-growth businesses being founded in the Midwest, and BAN's broad mandate, the target was described as 5X5, five times the amount invested in five years.

Investment amounts were initially stated to range from $25,000 to $250,000 per round, before co-investment. Investments in SLU current students' businesses could start at as little as $5000. Related to that co-investment, SLU created a $1 million sidecar fund, administered by the University Comptroller, who served as "SLU's angel." While this was initially structured as a two-for-one angel/SLU match, for every investment made by angels, this was later refined to permit SLU to selectively participate in co-investments or rounds of investment, and to scale their financial involvement as they saw fit. The sidecar fund made SLU rather distinctive. By 2016, with over 40 university-affiliated angel groups globally, only 4 schools had university co-investments – Baylor, New Mexico State, Akron, and SLU (Katz 2016).

Members would be accredited investors, but SLU DNA was not required. Members would pay $2500 a year in dues, done as two checks – one tax-deductible check for $1250 to support the academic Entrepreneurship Program and one non-deductible check for $1250 to underwrite the work of the BAN. This involved underwriting food and drink for BAN's 6 meetings every year, ACA membership, and a portion of associated legal fees. Members also pledged to help the Entrepreneurship Program out at least three times a year, schedule permitting. Because the Entrepreneurship Program did not have a dedicated endowment, the tax-deductible portion of the dues became the major source of financial support for the program.

Katz planned to leverage the pipeline to create a pool of volunteers for initial screening for BAN at the start of the investment process and due diligence later in the investment process. These volunteers were called the Billiken Angel Fellows, and like the angels, many had participated in the pipeline. For the Fellows, BAN was one of the most high-powered networking opportunities available to young and mid-career professionals in St. Louis. As one Fellow put it, "Where else does someone like me in a big firm get to rub elbows with two dozen millionaires, trade phone numbers, and get to help them out? I've gotten calls from angels asking my opinion or for referrals. It doesn't get any better than that!"

The Fellows fell into three groups: (1) Doctors of Medicine (MDs) and medical students, (2) Lawyers, (3) Accounting and Finance experts, (4) Engineers, and (5) "people with mad skills." Because of BAN's "any industry, any stage" approach, the range of ideas coming in was profound, and finding people who had expertise was an enduring challenge, and the source of the mad skills group. For example, this group included the lead buyer for a national clothing chain, a compounding pharmacist, a third-world agronomist, the technical analyst for the St. Louis Federal Reserve Bank, a digital marketing specialist (when this was leading-edge), and another dozen specialized consultants. The vast majority came through pipeline connections, and as additional types were needed, they came through referrals from SLU academics, BAN Fellows, and BAN members.

Despite the Fellow's quote above, the actual number of dues-paying angels was closer to three dozen than two. Over the 10 years, BAN was in operation 54 angels participated, although, for most of that period, there were 35–36 angels. Between 50% and 70% were SLU alumni, and 70% of the original group were judges and co-teachers from the pipeline. Recruitment of later angels is split roughly evenly among those recruited by the Dean and SLU's Development Office, Prof. Katz, other angels, and from the Fellows. Over the course of the decade, four of the Fellows became BAN members.

The meetings ran 3.5 hours on Wednesday nights at the Business School as shown in Table 2 below. There would typically be 3 presentations. Presentations ran 15 minutes, with questions and answers (angels first, then Fellows) running typically 30–45 minutes, followed by an immediate discussion by the angels present. If any of the angels expressed an interest in looking more closely at the opportunity, the attending Fellows would be polled for volunteers to work the due diligence. Tentatively one of the angels expressing an interest would serve as an "interim lead angel" polling other angels as to their questions and interests and working with an "interim lead Fellow" who will organize and communicate with the Fellows. If no angel present expressed interest, the videos of the session would be made available to the other members, and they would have until the 2nd Monday after the presentation (i.e., two full weekends, given the early finding that investing was for most angels a weekend activity) to express interest.

In 75% of presentations, one or more angels expressed a potential interest at the meeting so the beginnings of a due diligence team could be started by the meeting's end, and Katz would inform the entrepreneurs the following day. For opportunities that were going to move into due diligence that night, the angels, Fellows, and entrepreneur team would spend a few minutes together to map out the process. For the presenting companies where no angel present at the meeting expressed an interest, the entrepreneurs got an explaination on how the two-weekend process goes and that Katz would be in touch with them.

Table 2. Typical BAN meeting agenda.

5:30–6:00	Social (appetizers and bar will be available all evening)	
6:00	Welcome (5 minutes)	Katz
6:05–6:20	First Prospect Presentation (15 minutes)	First Prospect Team
6:20–6:50	Q&A on First Prospect (30 minutes)	First Prospect Team
6:50–7:00	Break (10 minutes)	
7:00–7:15	Second Prospect Presentation (15 minutes)	Second Prospect Team
7:15–7:45	Q&A on Second Prospect (30 minutes)	Second Prospect Team
7:45–8:00	Break (15 minutes)	
8:00–9:00	Discussion on Presentations (60 minutes)	Katz
9:00	Adjourn	

The Process: Opportunity Selection

During the 10 years of BAN's operation, 462 firms submitted business plans for consideration. The SLU DNA rule, which existed since BAN's start in 2008 was relaxed in 2012 so that to be considered the firm needs to be within a 2-hour drive of St. Louis, or if they have SLU DNA they could be located anywhere. In reality, there was an early investment in an SLU alum based in Los Angeles. The investing angels felt they would typically get out to California often enough to keep tabs on the business, but when it failed between visits, the angels concluded they want to be closer to their investments, although no rule changes were made.

As shown in Fig. 2, of 462 submissions from **All Firms**, 181 either were outside the region or lacked SLU DNA (before 2012). This determination from 2008–2012 was made solely by the Director, but in 2012 the position of "Senior Lead Fellow" was created to reflect the growing role of three Fellows who took on many of the details of running BAN's processes. Across the 2008–2018 period, 281 remaining **Qualifying Firms** went through an initial screening by the BAN Fellows.

The Qualifying Firms were then subjected to another screening process to identify the most promising prospects – identifying what BAN would call **Presenting Firms**. The process consisted of teams of 6–10 Fellows who took responsibility for supervising reviews of incoming business plans every month. Each month a different Fellow would track incoming business plans, and assign them to other Fellows based on expertise, fit to the business, and where the initial screeners flagged possible problem areas. These review leaders would draw on the 60-some Fellows, and when necessary, work with the Director to find Fellows with the needed expertise. There was a monthly meeting of the Fellows working on current screenings to decide which opportunities they would recommend presenting to BAN. In some cases, the screening team would ask entrepreneurs for more information, but in general, yes/no decisions were the major result of the meetings for most opportunities being reviewed.

The analytic model used by the Fellows followed the model for analyzing business plans used in the Entrepreneurship Program classes. This consisted of six categories:

- Plan provides a reasonable return to investors (2x–5x in 5 years)
- Product/Service/Process is valuable and needed

Figure 2. BAN's opportunity funnel.

- Plan provides a protected competitive advantage
- Product/service has a large market
- Marketing plan will achieve goals
- Company has people to execute the plan

In this application of SLU's culture of evaluation, a clear failure in any of the six areas would be enough to disqualify a firm from consideration, but a firm with a weakness in one or two areas might still be considered if (1) other components that are particularly strong or (2) the weakness might be addressable through funding or other forms of BAN support. Fellows used the Likert-scale questionnaires developed for the Business Plan classes, but with a simplified rubric: Present, Get More Data, Reject.

Given most plans were reviewed by 3 Fellows, they followed the model from the classes, where differing rubric outcomes are discussed and a collective decision arrived at. When a clear answer wasn't evident, the larger group of Fellows evaluating that month presented the findings and offered their analysis. Using this process and rubric over the 10 years, of 281 firms considered, 72 were identified as **Presenting Firms** to come before BAN, and 209 were rejected. Those 72 firms were 16% of **All Firms** submitted, and 26% of all **Qualifying Firms**. One concern about screening processes is if they are too restrictive, rejecting promising prospects. From an analysis done in 2016 and updated in 2022, 30 firms received outside funding after rejection by BAN. In 23 of 30 cases, these firms raised $100,000 or less and the companies eventually closed according to their records on Crunchbase.com. Three received funding from other angel groups. Nine firms received funding from accelerators. Four received accelerator funding and went on to receive follow-on funding from angels. Three obtained funding through crowdfunding, and four did not report the nature of their funding.

Described as the "ones that got away" 7 of the 209 declined firms received significant funding (i.e., $500,000 or more) to date. There were two major successes, with $25 million and $27.2 million raised to date, with the other five firms ranging from $900,000 to $4.8 million in angel investments (average for the 7 firms $9.3 million). In the particular math of angel investing, the key takeaway number is 2 out of 209 was the rate at which the screening missed key prospects.

Returning to Fig. 2, of the 72 **Presenting Firms** BAN decided to consider investment in 46 of them (representing 10% of **All Firms** that applied and 64% of **Presenting Firms**). Those **Due Diligence Done** firms would go through the full due diligence process outlined below. Of the 46 firms, 26 of the firms labelled as **Investment Offered** survived the due diligence process and received investment offers from BAN. This represented 6% of **All Firms** that applied and 57% of the firms that had **Due Diligence Done**. But not all investment offers go through. For BAN, 21 of the 26 went to completion, representing 5% of **All Firms** that applied, and 81% of **Investment Offered** firms.

Even with this rigorous screening process to select firms to present, experience suggested that some firms who survived the review did not live up to expectations. This would most often be because the presentation was lackluster, or the entrepreneur did not handle questions well or outside of the business plan they had not thought through the details of their businesses.

Invariably these turned out to be firms that were "on the cusp" in the initial screening evaluations. With 30 millionaires in a room and the "F" rubric of "Kid, you're embarrassing yourself and wasting my time. Get out of here!" would be very much in evidence when a company presentation did not live up to expectations. As a result, the standard for presenting probably did rise over the years precluding those "on the cusp" from presenting.

Traditionally, angel groups at the time tended to respond to applicants only when the firm was going to present. BAN provided all rejected entrepreneurs a phone call or meeting to explain how they were evaluated, what BAN found, and thought was missing, and where possible, pointing entrepreneurs to resources to help them in their next iterations. This approach was also consistent with the developmental aspect of the culture of evaluation as practiced by Fetterman, as well as Brockhaus and Katz.

The Process: Performing Due Diligence on Promising Prospects

The process for due diligence similarly followed the model taught in the Entrepreneurship Program classes. The Director would inform the BAN community via email of the results of that month's BAN meeting, and direct non-attendee angels to review the presentations and let the Director know if they had questions or any interest in any of the presenting firms. If angels expressed an interest in a firm on which due diligence was already being organized, they would get connected to the other involved angels. If there were no angels yet interested, the involved angel would get in touch with the entrepreneurs to get more details on the firm. If the angels liked what they heard the Director would inform the other angels of the interest to see if other angels might like to go in on the due diligence. The BAN Fellows would be polled for volunteers to work due diligence.

Under either model, the angels showing an interest in a particular **Due Diligence Done** firm from the BAN meeting (or via email or call after seeing the video of the meeting) would meet in person or on a conference call to discuss their thoughts and concerns. They would also organize the potential angel investor group for each particular opportunity. This would entail the identification of someone who would serve as the lead angel for the potential investment. When possible, a secondary would be identified to assure continuity given most of the angels were working full-time.

A similar process would occur with the Fellows around those same opportunities. From these meetings, a Lead Fellow for each due diligence effort would be identified, along with a secondary lead where possible.

Once angels were clear on their interests, the lead angel or angels would meet with the BAN Fellows to discuss the specifics of the due diligence process, the questions or concerns the angels had about the particular business, and the Fellows' initial thoughts on the firm. Where needs for specialized expertise or resources are needed, the due diligence team would poll their members and Fellows, and if that didn't produce a satisfactory resolution, they would as the Director for the needed resources or expertise. Between being able to draw on the resources of a research university with a full suite of academic departments including a Medical School and the Rolodex built from nearly 20 years of operation of the Entrepreneurship Center and Program, even esoteric needs were consistently able to be met. Part of what made this work was the civic-mindedness of St. Louisans. Compared to Katz's experience in entrepreneurial communities in Boston and Philadelphia, St. Louisans were much more willing to talk and to help one another out.

The Criticality of Rigorous Due Diligence Processes

In one research report, Wiltbank and Boecker (2007) showed that angel groups who conducted high levels of due diligence (more than 20 hours) had significantly better financial returns. Given that the teams of BAN angels and particularly Fellows were willing to put time into due diligence, BAN set it as a norm a particularly thorough process. Wiltbank and Boecker noted that deals that involved more than 40 hours of due diligence reported a 7.1x multiple (vs. 5.9x for over 20 hours and 1.1x for exists with less than 20 hours of due diligence).

Early due diligence efforts resulted in processes over 90 days in length. This became problematic for entrepreneurs, to the point that the Director stipulated a target of 45 days from the day of presentation to complete due diligence. This quickly brought the time spent to under 60 days but achieving the 45 targets remained a hit-or-miss thing until BAN closed. Still, most of the due diligence efforts involved 60 hours or more.

The specific process followed by BAN for Due Diligence followed a consistent sequence, shown in Fig. 3. The number in each box corresponds to the sequence below.

The specifics of the due diligence process were informed by several works, notably Camp (2002) and Green and Carroll (2000) a book developed by SLU accounting Ph.D. Richard Green and edited by Prof. Katz provided rudiments of a model for developmental due diligence, in this case, applied to investigations of

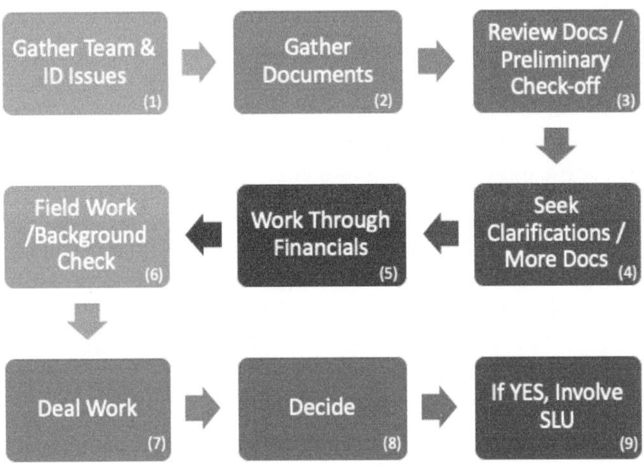

Figure 3. The BAN due diligence process.

existing small and medium businesses. Another key early influence was the Angel Research Institute's due diligence model from their "Power of Angel Investing" seminar, which BAN participated in 2008, and again in 2014. BAN Fellow Christoph Bausch and Prof. Katz developed the initial matrix for due diligence, building off the above documents, along with due diligence templates from Go4funding.com, Kauffman's angel guidebook (Preston 2004), and Docstoc.com. The model can be accessed at http://tiny.cc/BAN-DD-Framework. Looking at Fig. 3, the specifics for each step were as follows:

(1) **Build Evaluator Team:** Gather a Team of angels and Fellows and Identify the Issues: That was described above. For steps 2 through 7, the lead angel would be in contact with the Lead Fellow or other members of the team, asking questions, connecting Fellows to resources, and maintaining a relationship with the entrepreneur in anticipation of a future deal.

(2) **Gather Documents:** The list built from the above sources (i.e., Camp 2002, Green and Carroll 2000) listed a total of 112 types of documents in 16 categories – general corporate materials, financial information, debt financing, real property, intellectual property, litigation, HRM, compliance with laws, products or services, customer information, insurance, professionals, other agreements, tax matters, acquisitions & divestitures, and public relations. This can be found in http://tiny.cc/BAN-DD-Framework in worksheet 2 (See Documents and Materials at the end of this chapter). This step usually was done by Fellows who were generalists or lawyers.

(3) **Review Documents/Preliminary Check-off:** For most of the firms, esp. startups, it was typical that the company might only have 5–10 of the possible types of documents. The Fellows would pour through the documents and seek additional inputs as needed. Most teams had at least one finance-focused member, one legal focused member, and one-technically focused member. BAN quickly came to a go/no-go decision when a firm lacked crucial documents.

(4) **Seek Clarification or More Documents:** Part of this could be answered through Q&A with the entrepreneurs or their attorneys or accountant, but invariably there would be some documents that simply did not exist and those formed one of the enduring risks of these investments. The Fellows informed the angels of these situations, and the angels made the final determination of how to handle it (continue to consider investing, ask the entrepreneur to generate the document, get other clarification, or end due diligence without making an offer).

(5) **Work Through Financials:** If the documentation was adequate or forthcoming, a team of finance and accounting Fellows would begin work, doing a cursory review of the financials, and beginning deep dives to assess the soundness of the financials and the supporting documentation for the assumptions and the numbers themselves. The goal here was to come to an evaluation of the believability of the financial projections and based on that the soundness of the valuation assigned to the business and the offering.

(6) **Field Work and Background Check:** Fellows (often accompanied by one or more of the angels) would visit the company's location to check out the physical

plant or office space, the employees, and the culture of the firm. Part of this was almost always looking at additional data, which also let the team see how well organized and backed-up the startup was. The cultural issues the team looked into in field visits, interviews, and background checks related to how people got along with the leadership team and one another, their comfort with the idea of new investors coming in, and their reaction to new ideas. Parallel to this effort is another element of fieldwork checking with others in the industry (including competitors), customers, and the history of the entrepreneurs and key employees. Given the local focus of BAN, much of this was done in person, but phone calls were a constant during the process.

(7) **Deal Work:** With an understanding of the company in its documentation, financials, and in-person, the Fellows and angels would then perform a fresh analysis of the deal being offered, with a particular eye on whether the deal is supported by the company's proposed plan, financials, and human resources. At this stage extensive comparative analysis is done using investment databases (Gust, Crunchbase, etc.) and angel group connections to get a clearer sense of what investments in competing firms in the industry have looked like, how they have done, and what valuations had been supported. All of which goes back to the question of whether the proposed deal is investible, and if not, what would it take to make it so?

(8) **Decide:** Typically, this is a meeting with the angels and all the Fellows involved in the due diligence effort. The Fellows typically create a summary report, ranging from 6 to over 20 pages detailing their findings. Table 3 shows an outline of a typical report. In some cases, the report was sent to the full group (angels and Fellows interested in the prospective investment) ahead of the meeting, in other cases, the report was given out at the meeting. Either way, the angels and Fellows would discuss the company, the Fellow's findings, and the next steps. At this step all of the angels are brought into the process (if they hadn't been active earlier) and the next steps would be decided. Some of these offers were occasionally made directly after the decision meeting, but most often there would be discussions between the lead angel and the entrepreneur about specific concerns that remained after the due diligence effort. As the entrepreneur gave answers or worked on the issues raised, the prospects for investing grew. As shown in Fig. 2, of 46 firms that went into due diligence, BAN made investment offers to 26, while 20 firms fell out of consideration. When BAN's angels were ready to make an investment offer, they would contact SLU.

The due diligence reports were often supplemented with a dedicated data room for BAN-only access through Gust.com, which would contain the company's documents and those work products developed by the Fellows and angels. The Fellows present their assessments, and the angels and Fellows do what often turned into a spirited question and answer session. The goal here was to decide whether to invest, and if investing, what would be offered with what kind of deal terms. If one or more of the angels do decide to go ahead, they would inform the rest of the BAN angels of the pending investment to see if other angels would like to join the deal. The angels would be given the due diligence report and

Table 3. Topical outline of a BAN due diligence report.

1. Company Timeline (.5–1 page)
2. Product/Service Overview (1 page)
3. Company Leadership (.5–1 page)
4. Target Market (.5–1 page)
5. The Deal (.5–1 page)
6. Selected Financial Projections (1–2 pages)
7. Risk and Opportunities (.5–1 page)
8. Marketing Strategy
9. Exit Strategy
10. Due Diligence Approach (i.e., Notes) and Recommendation (1 page)

access to the data room. Once the number and investments of the angels are known, they go to the next step.

(9) Involving SLU: The lead angel and the Director would connect with SLU's Comptroller to check into the possibility and specifics of SLU's involvement through the sidecar fund. With SLU's involvement known, the lead angel begins negotiating with the entrepreneur.

Due Diligence in a Culture of Evaluation

The culture of the evaluation was evident in two ways during the due diligence processes. First, among the BAN angels and Fellows, there was an openness and even an ambition to learn all that the group could. The due diligence process places a premium on insight, and the variety of points of view in BAN gatherings was exceptionally broad. This meant that a host of extremely smart and motivated people were looking at the same company and trying to add to the collective understanding – doing a better job of evaluating. Often participants surprised one another with unexpected insights. These "aha!" moments became the stuff of BAN legend, as the group learned one another's capabilities and ways of solving problems.

This resulted in a remarkable number of impromptu "class sessions" where the Director, one of the angels, Fellows, or visiting experts would give a 10-minute explanation of some specialized topic that appeared in the process of due diligence. Examples included a Ph.D. microbiologist explaining recombinant DNA technologies, the head buyer for a national women's wear company describing how fast fashion revolutionized that industry, or a private equity fund principal explaining the tricks of the trade for figuring tranche sizes for angel investments. Because BAN was embedded in a university, and students were often attending as guests to learn about high-growth entrepreneurship first-hand, there was a strong norm supporting an educational approach to the angel investing process, and it was a strong additional social incentive for members of the BAN community. What would also happen was that the lessons would get repeated in SLU's entrepreneurship classes as BAN attendees took the latest lessons back to the classroom.

This social incentive idea turned out to be a crucial one for the success of the angel group. For the angels, while being around other angels was a positive experience, it was one they encountered all the time since they travelled in the same St. Louis circles. What was distinctive and exciting for the angels was having the Fellows around. There was easily a 20+ year age difference between the two groups, but this was part of the attraction. The angels saw themselves as mentors for the Fellows, but also discovered the Fellows were doing a lot of reverse mentoring. Every angel became tremendously more capable online because of their contact with and goading from the Fellows to up their online capabilities. Among the Fellows, there was a clear interest in showing the others the latest techniques and findings relevant to startups and their evaluation, and these constant streams of new ideas were a source of energy and inspiration for all of BAN.

If there was a contrary element in this effort, it was a natural consequence of being in a roomful of smart people. It was the desire not to make a mistake. Interestingly, this concern was more evident among angels. Basically, as very successful multi-millionaires and highly respected in their industries, they all came to BAN with their reputations ahead of them. But the nature of BAN's supremely broad mandate – any industry, any stage – meant that even the most expert of them would often find themselves out of their element. More than one angel remarked to the Director privately that they held off saying things because they weren't sure if they might steer the others down the wrong path. Fellows made analogous comments, worrying about looking foolish in front of potential mentors, clients, and employers.

The Director worked with the Executive Board to re-emphasize the educational role and mission of BAN. Angels and Fellows were reminded that the collective goal was the come up with a successful model for a "kinder, gentler" approach to angel investing, consistent with the developmental model inherent in the Entrepreneurship Program's culture of evaluation, the educational role SLU played in the St. Louis community, and the University's Jesuit social mission of developing people and communities although doing so in ways that made a profit for the angels and the University was part of that larger mission too. This gave many of the meetings a feeling more like a large, occasionally unruly, classroom and resulted in angels and Fellows making comments about new things they learned at each meeting. It also became a staple to create a to-do list for the Director to research topics or techniques and report back to BAN at a future meeting or email.

The upshot of this was that the social element of BAN membership became its strongest benefit. The angels and Fellows loved being together and talking about startups, commerce, and a host of life lessons. In the breaks between presentations at BAN meetings, the Director would frequently have to use his "outside voice" to order everyone to their seats so the next company could present. One entrepreneur observing this said, "I figure if you like my business even a tenth as much as you people clearly like one another, I should have a lock on money from BAN."

The second way the culture of the evaluation was evident in BAN's due diligence process was in the way BAN worked with companies. Every one of the 281 **Qualifying Firms** who met BAN criteria for consideration received feedback on their submissions. For 209 of these firms, the feedback would come from the Director or one of the Fellows involved in the initial screening. Of the 72 **Presenting Firms,** the 26 in which no angel expressed any investment interest received feedback from the Director. Of the 46 **Due Diligence Done** companies, BAN came to a decision to end due diligence before completion on 20 firms. In each of these cases, BAN's Director, Fellows, or angels would meet with the entrepreneurs and share their findings and their reasoning. The model used was directly modeled on the process used in the due diligence classes with students. The goal was for the entrepreneur to come away with a better understanding of the angel investment process in general, BAN's approach, and how their firm might improve for future rounds of funding or just improve their operations based on BAN's findings. The kind of expertise and data-gathering BAN's experts could do was often far beyond the knowledge or resources of the startups, and many entrepreneurs, even when disappointed, went away with a more detailed understanding of their business because of the process.

The Long-Term Results

For the 46 firms where due diligence was finished, BAN decided to make offers to 26 of the firms. BAN was able to close deals with 21 firms, and over its lifetime the group made 31 total investments in those 21 firms. The total invested by BAN was $3.6 million, of which $1 million was SLU's co-investment. From 2015 to 2018, the HALO Report ranked the Billiken Angels as one of the top 5 angel groups in the Great Plains region in the number of deals made.

At the time of BAN's moving to harvest mode in 2018, of the 21 firms invested, 5 had closed. By 2022, 3 more had closed. The remaining 13 firms had 2 acquisitions (10%) which resulted in returns in the 2–3x range, and the 11 continuing firms (52% of the portfolio) had collectively raised $289 million in additional funding by May 2022 and are still in business. The average investment amount for these portfolio firms compares favorably with the total raised by the firms rejected by BAN but received significant funding elsewhere –$65.2 million across 7 firms.

This means that BAN to date has a failure rate of 38%, which is better than the 69% reported in the most-cited study of angel returns, Wiltbank's research for the Angel Research institute (Wiltbank and Boeker 2007, Wiltbank and Brooks 2017), as shown in Fig. 4. These results are also better than the findings from 21000+ venture capital investments tracked by Correlation Ventures (Skillicorn 2018). The results suggest that the due diligence model BAN developed and used provided superior results to those typical for angel networks.

The other side of this analysis would reflect that while 13 out of 21 (62%) is a reasonable survival rate, in terms of harvests and returns the BAN portfolio, and potentially the associated due diligence approach, delivered suboptimal financial outcomes. The classic model for angel (and venture capital) investment is based on returns across a 5-to-10-year period. At the end of the 10 years of BAN operation, 16 firms were operating, but none had produced a harvest event. This means that the angels' funds remained tied-up in these investments, and while these privately-held firms are growing in valuation, or more specifically the valuation of the private stock of the firms, the only way to achieve liquidity is through sales back to the company or other stockholders, or where permitted by law and investment documents sale on a secondary market for private company stock (Likos 2021, Likos and Lake 2020) such as EquityZen, Forge or NASDAQ's Private Market.

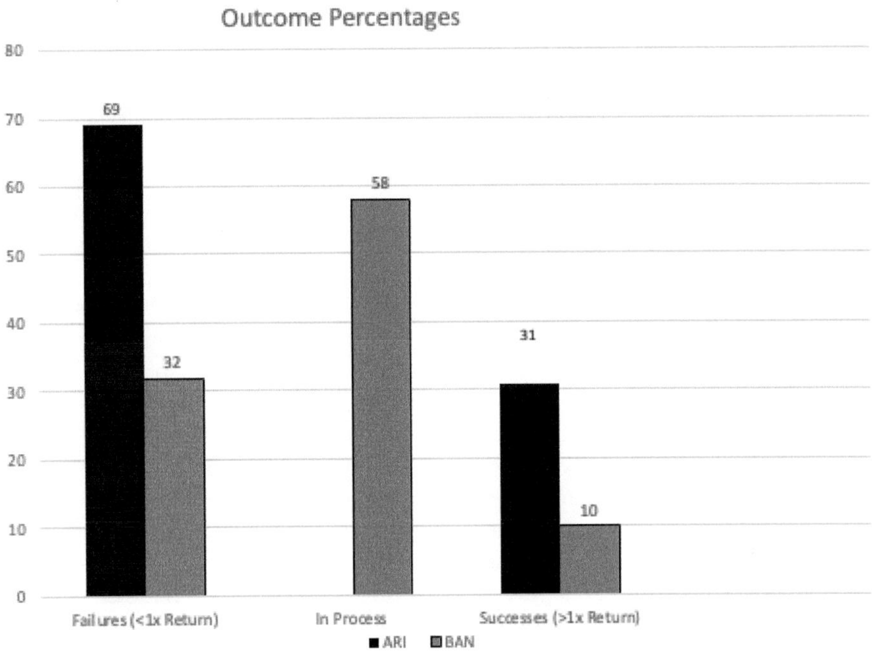

Figure 4. Returns on angel Investment from the ARI studies and BAN.

Source: Adapted from Wiltbank, R.E. and Brooks, W.T. Tracking angel returns: 2016 report with 2017 update. Angel Resource Institute, Fig. 1, https://angelresourceinstitute.org/reports/tracking-angel-returns-2017-update.pdf.

Part of the outlook of BAN which has made the longer paydays acceptable for angels came from an aspect of BAN's founding culture. The three founding members of BAN's Executive Board came from the private equity industry. While private equity firms on the east and west coast were shifting their focus in the early 2000 toward a faster-paced, earlier-stage-investment model like venture capitalists, private equity in the Midwest retained its historic roots as "patient money" (United States House of Representatives, 2005). Those traditional private equity firms believed in long time horizons and creating enduring value.

Final Reflections

The due diligence approach used by BAN was relatively labor intensive, but also seemed to produce superior results, if slower ones, than traditional due diligence efforts. In reflecting on the distinctive features BAN faced and crafted, the two clearest factors came from BAN's academic setting and its possibly unusual social structures.

BAN being a university-based angel network meant that it emerged from a culture of evaluation with a strong developmental element. This meant that evaluation as a process was valued and taught in the classroom by the Director and many of the original angels, so bringing that approach to BAN was a natural extension of a known and successful approach. "Successful" is backed by the fact that the Entrepreneurship

Program at SLU had an unbroken string of top-40 national rankings from *USNews*, Princeton Review, and *Entrepreneur* Magazine from 1994 onward.

That academic basis also determined a number of characteristics. Given entrepreneurship classes were co-taught by standing tenure-track faculty and local entrepreneurs and experts, this partnership carried over into BAN. With the university sidecar fund, the portion of dues going to the Entrepreneurship Program and the ongoing presence of students in meetings and eligible for $5000 launch investments, there was a clear side mission to help Saint Louis University and its entrepreneurship efforts, which introduced a virtuous cycle in the angel investment process.

In the Angel Resource Institute training program for new angel groups, there is little time spent on the social structures and elements of angel groups, but for BAN these played an important role in the operation of the group and its particular approach to the due diligence process. As noted above, there was a considerable spillover of processes, models, and collaboration from the classroom to the Billiken Angels. This was particularly true regarding mutual education and maintaining a development focus for entrepreneurs and students. The far-ranging gathering of experts to help make the "any industry, any stage" promise workable also paralleled a similar attitude in classes. Building on the mentoring inherent in classes, BAN's Fellows also promoted a strong reverse mentoring effort tied to the Internet, biotech, and other advanced technologies which deepened the mutual reliance and respect of the angels and Fellows. This in turn resulted in the development of strong cross-age friendships and a tradition of mutual referrals.

The other social structure underlying BAN's approach came from its investment culture, which was probably a mix of Midwestern conservatism and the traditional "patient money" approach of the founding members from private equity. This approach made choices such as requiring business plans, and data rooms, having a lengthy, complete, and detailed due diligence process, having written reports, and having multiple perspectives involved in the due diligence process obvious and quickly agreed to by all involved.

That said, those same cultures and structures did result in longer times to make decisions, and skepticism about companies promoting "J-curve" financial projections promising extraordinary returns within only 3–5 years. Looking at BAN's investment record, and its exceptionally long payout period, with its record of companies still growing and raising increasing amounts of funding suggest that while BAN's story is not fully written yet, there is some hope of positive outcomes in the future.

Acknowledgements

As the chapter suggests, the Billiken Angels was a true collaboration between academia and the larger community. These collaborators included the Executive Team of Laura Burkemper (angel), Mary Elizabeth Coleman (Fellow), Jerome Katz (professor) and Jeanne Rhodes (administrative assistant). The BAN Executive Board included those three as well as angel members, Joe Garea, David Kocs, Steve Rull, Rick Thornberry, and SLU comptroller members Gary Whitworth and later Jim Fugel. Joe Pimmel and Andrea Fan led the Fellows screening and due diligence efforts at different points and were supported by the rest of the BAN Fellows Leadership

Committee which included Mike Cisar, James Dorna, and Andrew Fehlman. Additionally, three Fellows worked with Professor Katz to help develop aspects of the due diligence process. Christoph Bausch did the heavy lifting of creating the Due Diligence Framework spreadsheet (https://tiny.cc/BAN-DD-Framework) while Brandon Beal worked on implementing screening process improvements, and Josh Martsolf did a variety of duties as BAN's MBA intern. The Business School deans who backed the creation and support of the program included Ellen Harshman, Leroy Grossman, Scott Safranski, and Mark Higgins.

The Fellows who put in the thousands of hours of work outlined in this chapter included Orren Adams, Nick Akers, Matthew Althage, Melanie Auberry, J.J. Bollozos, Brandon Beal, Jake Bender, Dan Broderick, Chris Carman, Chris Carril, Robert Caruso, Melinda Chu, Dave Costenaro, Thomas Davidson, Shannon Dickson, A.J. Eads, Joseph Fischer, Daniel Glasscock, Andrew Hall, Justin Hall, Jim Hammelman, Aaron Hand, Amy Harkins, Suzy Hartmann, Tim Hassett, Dominic Imgrund, Rodger Jungk, J.P. Keating, Paul Klug, Robert Kramer, Ron Kramer, Narayan Krishnaswami, Veeraya Kullavanijaya, Angie Lawing, Charles Ma, Joyce McCauley, Tim Marsden, George Michaels, Andrew Miller, John Murphy, Manas Parikh, Tim Petron, Brian Posnanski, Sean Ridley, Jim Rittenbaum, Marie Sahrmann, Joe Schuh, Eric Schrenker, Neal Shah, Mitch Seman, Brian Slawin, Clyde Smith, Michael Stagoski, Jason Stone, James Strong, Joseph Stroot, Nanthiya Sujijantarat, Brian Turner, Jason Urban, Nicholas Paul Vaughn, Samuel Weigley, Daniel Willingham, and Crystal Winkeler.

The Billiken Angels, whose interest, passion and money underpinned this entire enterprise also deserve acknowledgement. One of their most enduring requests of me was to not disclose their names. Those on the Executive Board accepted the role as the public face of the BAN, and to this day the rest of the 54 angels remain anonymous but they retain my deepest thanks.

References

Angel Capital Association. (2009). Important things for entrepreneurs to know about angel investors. Overland Park, KS: Angel Capital Association, URL: https://slideplayer.com/slide/5831296/, slide 8.

Bayham, E.L., Katz, J.A., Calcaterra, R. and Zahner, J. (2007). The St Louis BioBelt – Centre for plant and life sciences: A triumph of converging individual efforts. pp. 265–395. *In*: F. Therin (ed.). *Handbook of Research on Techno-Entrepreneurship*. Cheltenham UK: Edward Elgar.

Camp, J.J. (2002). *Venture Capital Due Diligence: A Guide to making Smart Investment Choices and Increasing your Portfolio Returns*, Vol. 102. John Wiley & Sons.

Cremades, A. (2018). "How Angel Investors and Angel Groups Work." Forbes.com, September 25, 2018. https://www.forbes.com/sites/alejandrocremades/2018/09/25/how-angel-investors-and-angel-groups-work/?sh=53d8c7ca76dc.

Fetterman, D.M. (2002). 2001 Invited address: Empowerment evaluation: Building communities of practice and a culture of learning. *American Journal of Community Psychology*, 30(1): 89–102.

Green, R.P. and Carroll, J.J. (2000). *Investigating Entrepreneurial Opportunities: A Practical Guide for Due Diligence*. Sage Publications.

Hudson, M. (2007). Personal communication, May 30.

Katz, J.A. (1991). The institution and infrastructure of entrepreneurship. *Entrepreneurship Theory and Practice*, 15(3): 85–102.

Katz, J.A. (1995). Managing practitioners in the entrepreneurship class. *Simulation & Gaming*, 26(3): 361–375.

Katz, J.A. (2003). The chronology and intellectual trajectory of American entrepreneurship education: 1876–1999. *Journal of Business Venturing*, 18(2): 283–300.

Katz, J.A. (2016). The path forward. Presentation to the Chaifetz School Dean's Advisory Board.

Katz, J.A. (2018). The business plan: Reports of its death have been greatly exaggerated. In *Annals of Entrepreneurship Education and Pedagogy–2018*. Edward Elgar Publishing.

Katz, J.A. and Green II, R.P. (2007). *Entrepreneurial Small Business*. New York, NY: McGraw-Hill Education.

Likos, P. (2021). How to invest in startups. *U.S. News.* Retrieved from: https://money.usnews.com/investing/investing-101/articles/how-to-invest-in-startups.

Likos, P. and Lake, R. (2020). How to Become an Angel Investor. *U.S. News.* Retrieved from: https://money.usnews.com/investing/investing-101/articles/how-to-become-an-angel-investor.

Murphy, D.F. (1999). Developing a culture of evaluation. *The Journal. TESOL France*, 6: 5–13. Retrieved from: https://www.tesol-france.org/uploaded_files/files/dermot-murphy-2003.pdf.

Preston, S.L. (2004). *Angel Investment Groups, Networks, And Funds: A Guidebook to Developing the Right Angel Organization for Your Community.* Kansas City, MO: Kauffman Foundation. Retrieved from: https://angelresourceinstitute.org/research/report.php?report=109&name=Kauffman%20Guidebook%20to%20Develop%20the%20Right%20Angel%20Organization%20for%20Your%20Community.

Price, C.H., Buskirk, R.H. and Davis, R.M. (1991). *Program for Writing Winning Business Plans*. Denver, CO: Premier Entrepreneur Programs, Inc.

Skillicorn, N. (2018). 65% of Venture Capital-Backed Deals Fail to Return Investment, and Only 4% Make Substantial Returns. *Idea to Value.* Retrieved from: https://www.ideatovalue.com/inno/nickskillicorn/2018/10/65-of-venture-capital-backed-deals-fail-to-return-investment-and-only-4-make-substantial-returns/.

Solomon, G.T. and Weaver, K.M. (1983). Small business institute economic impact evaluation. *American Journal of Small Business*, 8(1): 41–51. https://doi.org/10.1177/104225878300800110.

United States House of Representatives. (2005). Private Equity for Small Firms: The Importance of The Participating Securities Program, Pub. L. No. Serial No. 109–10, § Committee On Small Business. Retrieved from: https://www.govinfo.gov/content/pkg/CHRG-109hhrg21285/html/CHRG-109hhrg21285.htm.

Wiltbank, R.E. and Boeker, W. (2007). *Returns to Angel Investors in Groups.* Angel Capital Association. Retrieved from: https://www.angelcapitalassociation.org/data/Documents/Resources/AngelGroupResarch/1d%20-%20Resources%20-%20Research/6%20RSCH_-_ACEF_-_Returns_to_Angel_Investor_in_Groups.pdf.

Wiltbank, R.E. and Brooks, W.T. (2017). Tracking angel returns: 2016 report with 2017 update. Angel Resource Institute. Retrieved from: https://angelresourceinstitute.org/reports/tracking-angel-returns-2017-update.pdf.

NOTE

The BAN documents listed can be downloaded at these locations:

BAN_Due_Diligence_Framework: http://tiny.cc/BAN-DD-Framework [https://www.dropbox.com/s/yioye0n9p82667w/BAN_Due_Diligence_Framework.xlsx?dl=0].

BP scoring packet: http://tiny.cc/BAN-Scoring-Packet

https://www.dropbox.com/s/x0y5ay4t8e0qkj7/BAN%20Scoring%20%20Packet.docx?dl=0

Chapter 13

More Than Money

Considering Nonfinancial Measures of Organizational Performance in Startups

*Rosalyn G Sandoval** and *Holly D Holladay-Sandidge*

Understanding the key drivers of firm performance is a primary aim of organizational research, with much work dedicated to this topic (Barney 2001, Dess and Robinson Jr 1984, Hamann et al. 2013, Lubatkin and Shrieves 1986, March and Sutton 1997, Miller et al. 2013, Richard et al. 2009). But what, precisely, is "firm performance"? For decades, scholars have critiqued the ambiguity of the concept of firm performance, highlighting the critical limitations (i.e., concept redundancy, theoretical conflation, misuse of measures, mis-specification of hypotheses, and low statistical conclusion validity; MacKenzie 2003) associated with both inconsistent definitions and measurement (Barney 2001, Chakravarthy 1986, Eccles 1991, Fryxell and Barton 1990, Hult et al. 2008, Kanter and Brinkerhoff 1981, Keats 1988, Miller et al. 2013, Morgan and Strong 2003, Pennings and Goodman 1977, Richard et al. 2013, Rowe and Morrow Jr 1999, Shenhav et al. 1994, Venkatraman and Ramanujam 1986). The diversity of definitions for firm performance that arise in the organizational literature—ranging from profit maximization and/or present value (Jensen and Meckling 1976) to "high returns over longer periods of time" (Wernerfelt 1984, p. 172) or, even more broadly, "fulfillment of the economic goals of the firm" (Venkatraman and Ramanujam 1986, p. 803), among others (Miller et al. 2013)—complicates the ability to understand the driving forces behind firm survival. This is especially true for new ventures. While vastly different across important dimensions (e.g., time), existing definitions rely heavily on the central assumption that organizational performance is inextricably linked to financial performance,

[1] California State University, Monterey, 931 Holovits Court, Marina, CA 93933, 209-642-206.
[2] University of North Carolina at Charlotte, 2000 Echo Forest Drive, Apt 207, 803-473-0999.
Email: hhollada@uncc.edu
* Corresponding author: rossandoval@csumb.edu

leading researchers to suggest that accounting profitability, economic value creation, and shareholder value creation (i.e., how much money the firm brings in) are key drivers of firm performance (Rothaermel 2017). However, these metrics are often unable to explain the survival of startup organizations, which may operate with limited and highly inconsistent financial capital, ultimately calling into question the relationship between financial and firm performance.

Historically, researchers and practitioners have equally contributed to the assumption that financial performance equates to firm performance (Hitt 1988). Chief among the earliest concerns surrounding the use of financial measures to assess organizational performance was an overreliance on *short-term* financial measures (Rappaport 1978). Research demonstrates that this overreliance can mask critical issues and drive practitioners to make inaccurate decisions, subsequently hindering the organization's ability to survive long-term (Hitt 1988, Reilly and Fuhr Jr 1983, Ross and Goodfellow 1980). Some scholars contend that researchers' continued proclivity for making broad claims about firm performance based on singular (often short-term) financial measures is rooted in a quest for legitimacy within the realm of practice (March and Sutton 1997, Miller et al. 2013). In other words, these scholars suggest that in seeking to prove the relevance of research to upper-level managers, business analysts, and others outside of academia—individuals who have a vested interest in thinking about firm performance broadly—performance researchers have adopted their practice of broadly defining and measuring this construct, to the detriment of increasing knowledge and refining science (Miller et al. 2013).

While critics have voiced concerns since the late 1970s (e.g., Rappaport 1978), the understanding of firm performance has progressed considerably over the past two decades. Notably, practitioners, in particular, have helped to move the needle forward. Recognizing that environmental, social, and governance (ESG) factors impact organizational functioning, as well as financial performance (Eccles et al. 2015), some companies have begun to engage in *integrated reporting*, "a process founded on integrated thinking that results in a periodic integrated report by an organization about value creation over time and related communications regarding aspects of value creation," (The International Integrated Reporting Council [IIRC]; Serafeim 2015, p. 34). Essentially, integrated reporting combines "nonfinancial performance" (e.g., ESG; Eccles et al. 2015, p. 8) or a sustainability report (Churet et al. 2014) with financial measures in order to present "a single 'narrative'… intended mainly for investors in which top management provides its views on how sustainability issues and initiatives are expected to contribute to the long-term growth strategy of the business" (Churet et al. 2014, p. 56; see Hart et al. this volume). In addition to acknowledging the importance of nonfinancial performance, the International <IR> Framework also highlights the need to consider financial value creation for the short-, medium-, and long-term, standing in stark contrast to the sole focus on short-term revenue generation that has dominated reporting in the past (Churet et al. 2014).

Despite this progress among some practitioners, considering nonfinancial performance has yet to become mainstream in research or practice. Richard et al. (2009) provide some insight into a potential explanation for the reticence of academics to adopt a broader conceptualization of performance; they contend that the construct of organizational performance is limited to "three specific areas of firm outcomes: (a) financial performance (profits, return on assets, return on investment, etc.); (b) product market performance (sales, market share, etc.); and (c) shareholder return (total shareholder return, economic value added, etc.)" (Richard et al. 2009, p. 722), whereas *organizational effectiveness* is the broader construct, "captur[ing] organizational performance plus...internal performance outcomes...and other external measures that relate to considerations that are broader than those simply associated with economic valuation (either by shareholders, managers, or customers), such as corporate social responsibility" (Richard et al. 2009, p. 722). Thus, it can be deduced that perhaps the lack of scholarly work considering nonfinancial performance measures may be rooted in the belief that the inclusion of these factors should be reserved for studies on organizational effectiveness. While management scholars have adopted other measures from the organizational effectiveness domain as their dependent performance variable (e.g., organizational innovation and organizational efficiency), financial measures still serve as the predominant operationalization of performance (Hamann et al. 2013, Katsikeas et al. 2016, Richard et al. 2009).[5]

Of these financial measures, research remains primarily fixated on *return on equity* (ROE), *return on assets* (ROA), and total shareholder return, which only reflects *shareholder* value at a single point in time, rather than the value *generated by firms* (Steigenberger 2014). The popularity of these measures reflects the tendency to prioritize revenue/value generation for shareholders over other organizational stakeholders (Katsikeas et al. 2016, Richard et al. 2009). As Steigenberger (2014) observes, employing such measures is fundamentally at odds with a resource-based view (RBV) of the firm, which argues that competitive advantage relies upon "'[a firm's ability to] deliver greater benefits to their customers for a given cost' (Peteraf and Barney 2003, p. 311)" (p. 47). This perspective underscores more than the misalignment of measurement to the theoretical construct in question: It positions customers as high legitimacy stakeholders, along with managers and shareholders (Mitchell et al. 1997, Richard et al. 2009), though their interests often receive only tertiary consideration as performance measures. This view thus begs the question: who are the organizational stakeholders who matter (or should matter) to firms when evaluating performance?

Churet et al. (2014) argue that "*all* stakeholders (not just investors) need to gain a deeper understanding of the interconnectedness between business results and the changing dynamics that characterize today's business environment" (Churet et al. 2014, p. 56; emphasis in the original). Again, who "counts" as an organizational stakeholder was questioned. Recently, the greater business community has provided some insight into whom they believe to possess such interest in their endeavors: in August 2019, Business Roundtable—a professional association of CEOs from

[5] Richard et al. (2009) include a comprehensive listing of financial performance measures (broken down into accounting measures, financial market measures, and mixed accounting/financial market measures). The definitions of said measures can be found in the management literature. The authors encourage readers to seek out and review this article if interested in financial performance measures.

American companies—issued an updated version of their *Statement on the Purpose of a Corporation*, naming employees, customers, suppliers, local communities, and shareholders as critical stakeholders for whom firms have a responsibility to generate value (Business Roundtable Association [BRA] 2019). This list largely mimics that outlined by the <IR> Framework, with a few key omissions: "business partners" and government representatives such as "legislators, regulators, and policy-makers" (Eccles et al. 2015, p. 4).

Since its release, the updated statement has garnered much skepticism (Gelles and Yaffe-Bellany 2019, Rappaport 2019, Winston 2019). Signed by 181 CEOs of major corporations (e.g., Apple, Amazon, and General Motors; Gelles and Yaffe-Bellany 2019), this treatise denounces the association's previous commitment to prioritizing shareholder profit maximization. Regardless of motive (some critics posit that the statement is "empty rhetoric"; Winston 2019), this declaration provides new insights into how researchers and practitioners alike should conceptualize and measure organizational performance. Now, rather than resorting to the practice of cherry-picking a financial measure that is likely to yield statistically significant results, academics and professionals can establish a framework to guide future evaluations of new venture performance rooted in practical relevance.

Next, the types of value creation new ventures might offer various stakeholders and how such utility can be used as a measure of performance need to be determined. New ventures are an ideal sample to rely upon when conceptualizing nonfinancial performance measures since ventures at their earliest stages of development often do not have any revenue or sales indicators to measure financial performance (De Mol et al. 2020). Therefore, other measures—such as pitch or business plan quality (Der Foo et al. 2005), early funding amount (Rosenbusch et al. 2011), and failure, survival, and growth (Cooper et al. 1994)—are often used as proxies for traditional financial measures (i.e., ROE, ROA, shareholder return, number of users/user growth, financial projections based on market size) when assessing performance. Furthermore, although startups and new ventures represent an ideal group for conceptualizing nonfinancial performance measures, all organizations should consider nonfinancial performance measures at each stage in their life cycle (i.e., temporal constraints and benefits). Table 1 outlines the commitments offered in the Business Roundtable's 2019 statement of purpose and examples of existing or proposed corresponding performance measures in the extant literature.[6] Though these examples have been used or proposed as performance measures in existing organizations, little work has discussed how they may be used to measure new venture performance. This chapter also notes that some of these examples have been discussed more extensively in literature from domains other than management, strategy, or entrepreneurship; for example, Katsikeas et al. (2016) illustrate that much of the scholarly work evaluating performance via customer-related constructs lies in marketing research. However, even this field has experienced a shift to more shareholder-centric measures in recent years.

[6] This is not a comprehensive list.

Table 1. Stakeholder considerations and example measures.

Firm Commitment to Stakeholders (Business Roundtable Association, 2019, BRA)		**Example Constructs from Existing Literature**	**Relevant Citation(s)**
Employees	"Investing in our employees. This starts with compensating them fairly and providing important benefits. It also includes supporting them through training and education that help develop new skills for a rapidly changing world. We foster diversity and inclusion, dignity, and respect."	Employee Satisfaction Pay/Compensation Satisfaction Affective Well-Being at Work General Well-Being Psychological Well-Being Thriving at Work Work Well-Being Employee Well-Being	Russell and Daniels (2018) Warr (1990) Dagenais-Desmarais and Savoie (2012) Porath et al. (2012) Parker and Hyett (2011) Zheng et al. (2015) Worker Well-Being Questionnaire (NIOSH)
Customers	"Delivering value to our customers. We will further the tradition of American companies leading the way in meeting or exceeding customer expectations."	Customer Mindset Constructs Customer Satisfaction Perceived Quality Perceived Value Customer/Consumer Behavior Customer Acquisition/ Retention Word of Mouth Consumer Well-Being	Katsikeas et al. (2016) Lee et al. (2002)
Suppliers	"Dealing fairly and ethically with our suppliers. We are dedicated to serving as good partners to the other companies, large and small, that help us meet our missions."	Business-supplier relationships	See Tangpong et al. (2015) for a full review
Communities	"Supporting the communities in which we work. We respect the people in our communities and protect the environment by embracing sustainable practices across our businesses."	Community Well-Being. Community Flourishing Community Quality of Life	VanderWeele (2019) Shultz et al. (2017) Sirgy and Cornwell (2001)
Shareholders	"Generating long-term value for shareholders, who provide the capital that allows companies to invest, grow and innovate. We are committed to transparency and effective engagement with shareholders."	Shareholder Returns Stock Returns	Katsikeas et al. (2016)

The multi-level structure of the stakeholder list outlined in the Business Roundtable's statement adds to the complexity of conceptualizing performance, particularly for new ventures. Employees, customers, and shareholders are individuals for whom organizations should be generating value. How they do so for these stakeholders may differ broadly from how they may generate value for suppliers and communities, which are group-level, country-level, and global-level entities (See Hart et al. this volume). Notably, "suppliers" are also firm-level entities (e.g., business-to-business relationships). At the same time, communities are still larger collectives in which organizations operate. Figure 1 illustrates this multi-level structure and informs some of the proposed strategies for founders that follow. The next section places conceptualizing nonfinancial performance in the context of new venture creation before discussing value generation and associated performance measurement for each of the stakeholder groups named in Table 1, excluding only shareholders. Given the amount of scholarly work dedicated to this group, this chapter focuses on those less often considered in the literature, that are particularly relevant for new ventures.

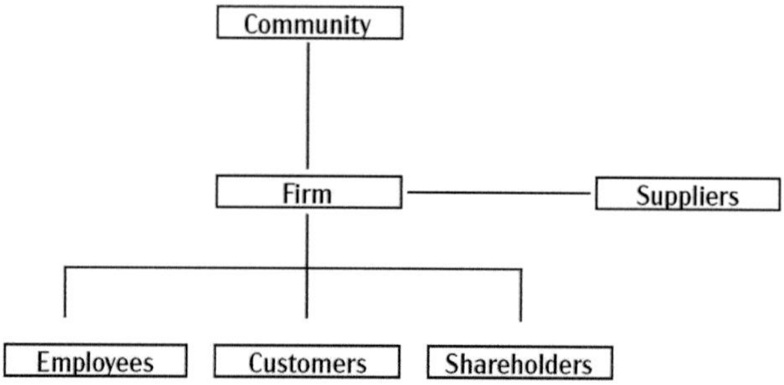

Figure 1. Multilevel structure of stakeholder-firm relationships.

Life Cycles of Organizations and New Ventures

The nonfinancial aspects of performance discussed thus far have implications for entrepreneurs during the venture creation cycle. Specifically, this chapter argues that nonfinancial performance measures should be considered and interwoven into founders' earliest visions for a new venture. Throughout the various stages of an organization's life cycle (e.g., the startup stage, the emerging growth stage, the mature stage, and the decline/transition stage), firms will prioritize different stakeholders, depending upon stakeholder potential to provide critical resources relevant to that stage (Jawahar and McLaughlin 2001). This chapter focuses on the startup and emerging growth stages and the stakeholders most relevant in these phases of the life cycle.

In the startup stage of new ventures, entrepreneurs will engage in several key activities revolving around their stakeholders, including—but not limited to—hiring employees, and suppliers, defining market opportunities and customers, and developing their products or services within local communities (Carter et al. 1996, Gartner et al. 2004, Hertel et al. 2021). During even the earliest of these organizing activities, entrepreneurs must consider how they will deliver value to stakeholders and measure their organizational performance. These early decisions around essential stakeholder relationships can significantly impact stakeholder well-being and the new venture's survival. Furthermore, defining primary stakeholder groups (e.g., employees, customers, suppliers, community) can provide meaningful indicators for the nonfinancial performance of the new venture. By focusing on generating value that supports the well-being of their various stakeholder groups, founders and organizations can invest in these parties that are so vital to their survival in ways that will be mutually beneficial long-term.

Stakeholder Well-being as a Measure of New Venture Performance

Employees

During the startup phase, new ventures often focus on a few critical needs that can influence performance and survival: startup funds, customer acceptance, and employee relations (Dodge et al. 1994, Dodge and Robbins 1992). Due to the time and financial commitment invested in hiring them (i.e., recruiting, finding the right employee, onboarding, and training), employees are a vital consideration for new ventures. Thus, the hiring process is a significant organizing activity for new ventures. Entrepreneurs often make their first employee hires beyond the founding team in the emerging growth stage—after a commercially viable product (or service) has been created—typically occurring within 18–24 months of their existence (Delmar and Shane 2004).

Furthermore, employee and team human capital are crucial for a new venture's survival and growth (Demir et al. 2017, Gimeno et al. 1997, Gjerløv-Juel and Guenther 2019, Unger et al. 2011, Weber and Zulehner 2010), making the assembly of new venture teams and employees one of a venture's more crucial resources. New ventures that hire three employees or fewer in the first year tend to remain stronger teams than ventures that start with more employees (Cooper et al. 1989). Finding employees and building teams equally passionate about the business as the founder should be a central aim. Research indicates that the quality of business ideas and long-term firm-level performance may suffer when team members differ in passion for their work (De Mol et al. 2020); meanwhile, having team members on the same page (i.e., sharing positive emotions, shared passion, shared decision-making strategies, shared cognition) regarding the venture can lead to many positive outcomes such as greater idea generation, trust, and promotion of more enduring social resources among the group members that can positively influence the venture overall (Carton and Tewfik 2016, Early and Mosakowski 2000, Cardon et al. 2017, Santos and Cardon 2019, De Jong and Dirks 2012, Ensley and Pearce 2001)—as such, finding employees and building teams equally passionate about the business as the founder is essential.

In contrast, employees and teams differing in their passions for the work can diminish the quality of business ideas and the firm's long-term funding (De Mol et al. 2020). At the same time, a collective identity can yield better interactions among individuals and help people work more effectively toward their goals (Earley and Mosakowski 2000, Wry et al. 2011). Prior work demonstrates that employees and more connected teams with a shared identity are less likely to leave the venture (Gjerløv-Juel and Guenther 2019). When employee turnover is low, early employment growth can lead to higher survival in the long run (Gjerløv-Juel and Guenther 2019). However, despite their importance to the firm's fate, employee well-being and related constructs are rarely used as the outcome measure of performance in new ventures (Shepherd et al. 2019).

Conceptualizing Employee Well-Being Performance Measures

Employee well-being is critical to new venture survival (Bates and Holton III 1995, Bosma et al. 2004, Colombo and Grilli 2005, Cooper et al. 1994, Delmar and Shane 2004, Leung and Fong this volume). Of all stakeholder groups mentioned heretofore, employees are perhaps the most studied in terms of their well-being; though often overlooked as a measure of performance, employee well-being is a well-established, complex, multi-faceted, and multidimensional construct comprised of economic, psychological (eudemonic and hedonic), physical, and social well-being components (Grant et al. 2007, Van De Voorde et al. 2012). Research indicates that numerous organizational characteristics impact employee well-being, including organizational support (Eisenberger and Stinglhamber 2011, Panaccio and Vandenberghe 2009), social environment (Freier and Hughes this volume), work-family balance (Bagger and Li 2014, Hammer et al. 2009, Piszczek 2020), human resource systems and job control (Jensen et al. 2013), and leadership behaviors and supervisor support (Babalola et al. 2021, Lorinkova and Perry 2017, van Dierendonck 2011). These established organizational features play a major role in employees' quality of life and well-being, which, in turn, impact the productivity of the individual and the organization.

As the Business Roundtable's 2019 statement indicates, organizations are responsible for investing in their employees. New ventures face the challenge of building a future workplace for employees—a consideration that often gets lost when the venture is striving to get off the ground. However, employee investment ought to be considered a critical component for an organization's survival. To deliver better value to employees, organizations should implement strategies to support employee well-being along each of the subdimensions of the construct. Interestingly, the Business Roundtable's statement provides examples of investing in employees that align with the economic, psychological, and social subdimensions, specifically addressing three ways organizations can contribute to employees' well-being. According to the statement, organizations can help to foster employees' economic well-being (i.e., the positive status of financial self-sufficiency and economic outlook; Judge et al. 2010, McKee-Ryan et al. 2005) by "compensating them fairly and providing important benefits" (BRA 2019). Secondly, organizations can encourage employees' *eudaimonic well-being*—the component of psychological well-being that encapsulates happiness found in the expression of virtue, the experience of purpose, and the realization of personal potential (Ryan and Deci 2001,

Ryff 1989, 2019, Ryff and Keyes 1995)—by "supporting them through training and education that help develop new skills for a rapidly changing world" (BRA 2019). Finally, by fostering "diversity and inclusion, dignity and respect" (BRA 2019), firms can promote employees' psychological and social well-being (i.e., the quality of an employee's relationships with other employees and the workplace community; Grant et al. 2007, Keyes 1998). An inclusive and diverse work environment helps to strengthen employees' ability to make strategic decisions under reduced constraints (Finkelstein and Hambrick 1990), improves workgroup functioning (Ely and Thomas 2001, Polzer et al. 2002), increases productivity, innovation, and creativity (Swann Jr et al. 2003, Watson et al. 1993), creates less groupthink (Wanous and Youtz 1986), enhances community relations (Montgomery and McGlynn 2009), helps with litigation prevention (Hirsh and Cha 2018), and reduces attrition (Obenauer 2019). Organizations can further facilitate employee social well-being through such means as taking steps to ensure supervisors are supporting subordinates (Babalola et al. 2021), through effective leadership strategies (Lorinkova and Perry 2017, van Dierendonck 2011), and creating a supportive work environment (Eisenberger and Stinglhamber 2011, Kurtessis et al. 2017).

Additionally, to fully satisfy employee psychological well-being, organizations should also strive to also invest in employee *hedonic well-being* (i.e., happiness found in the experience of pleasure versus displeasure; Diener et al. 1999, Fisher 2014; Ryan and Deci 2001) of employees by increasing *affective* well-being (i.e., feelings of pleasure and activation; Wright 2014), and job satisfaction (i.e., employees positive attitude toward work; Locke 1976) while lowering emotional exhaustion (i.e., feelings of being depleted of physical and emotional resources; Maslach et al. 2001) and work-family conflict (i.e., "role pressures from work and family are mutually incompatible such that participation in one role is made more difficult by virtue of participation in the other role" (Greenhaus and Allen 2011, pp. 165–166). Finally, organizations can also contribute to employees' physical well-being—their physiological health and subjective bodily experiences (Grant et al. 2007, Judge et al. 2010)—by ensuring they have safe physical working conditions, assisting with accommodations for disabilities, providing opportunities for physical exercise during and outside of working hours, and by adjusting work schedules to help ensure (to the best of their ability) that employees can rest and rejuvenate through sleep.

Practices That Founders Can Implement to Help with Employee Well-being

In line with recommendations offered by Hill and Stewart (2000), this chapter offers several ways in which new ventures can find, motivate, and retain high-quality employees while also considering their well-being. These efforts can (and should) be incorporated into the earliest organizing activities of the venture; thus, establishing a framework for supporting employees *begins with the founder*. Founders can outline efforts to support employees on each well-being subdimension in their earliest drafts of their business plans and build them into initial operating budgets and forecasts. For example, founders can plan to support employee economic well-being by implementing motivation-enhancing HR practices, such as incentive and reward practices, when they establish the venture or as it grows and becomes financially successful. Such practices have been useful in helping new ventures compensate for other HR-related disadvantages and play a key role in attracting and retaining talented employees (Rauch and Hatak 2016). Early planning for such initiatives can allow founders to factor in these ideas when first seeking funding for their ventures, which may help them secure sufficient financial investments to put these practices into play sooner than anticipated.

Organizations can invest in employee psychological well-being by providing training and other learning opportunities to help employees realize their potential. Training and learning opportunities can also apply to leaders/founders who engage in leadership and people management training to better guide their staff. While founders should include these plans in their initial budgets and business plans, they can also set a precedent for offering employees informal learning opportunities that do not require direct financial investment. Research suggests that when employees are exposed to new roles and opportunities to expand their skill sets, they enjoy and find their work meaningful (Batra and Pollitt 2014), contributing to their overall psychological well-being. The venture also benefits directly from having high-quality employees who can perform multiple roles during the various stages of organizational growth (Heneman et al. 2000). Founders can create multiple opportunities for employee psychological well-being, which should not require additional financial capital, by writing them into their business plans and job descriptions or related hiring materials. For example, founders can consider their employees' work-family needs and write flexible policies and procedures into their core documents. They should also consider free or low-cost opportunities to bring in guest speakers who might give presentations or workshops to employees. In this way, founders can support employee well-being by creating a culture of informal learning and individual growth that also benefits the venture.

Additionally, founders can promote social and physical well-being by establishing an organizational culture that promotes employee flexibility and health and provides a supportive and caring work environment. New ventures have a great deal of flexibility, without the constraints of hierarchy and bureaucracy that often plague established organizations. Thus, the ability to give employees some level of freedom and autonomy, which have been linked to greater creativity and innovation (Amabile 1996), is a strength of new ventures. Often new ventures are limited in their financial resources, so they need to combat this weakness with other attractive attributes, such as greater emphasis on employee health benefits and generous leave/flexible work policies. By providing these and similar incentives, founders can demonstrate their commitment to fostering an organizational culture of care and support—one that is genuinely interested in employees' well-being—and can strongly influence potential employees' perceptions of a new venture (Moser et al. 2017).

Communicating a clear organizational vision can also contribute to employee well-being. As with creating a supportive, caring culture, founders should have an enticing vision (Moser et al. 2017) and communicate it clearly to employees. Research suggests that involving employees in the vision and strategic decisions for the venture can create a greater organizational identity and strengthen teams (Batra, 2017). Establishing a clear, enticing vision often requires little-to-no direct financial investment, and founders can engage employees in the process of refining the vision internally almost immediately after the venture's creation (Moser et al. 2017). Lastly, founders can empower employees to make crucial decisions and be part of organizational decision-making, leading to greater idea sharing and better handling of internal and external issues (Abbott 2003, Batra 2017).

Customers

Research on the well-being of remaining stakeholders of interest in this chapter is far more limited than that on employee well-being. Therefore, the remaining discussion on these groups is abbreviated, and much of it is conceptual. Researchers are encouraged to consider engaging in future empirical work in these areas.

Customers are often at the heart of founders' initial ideas for their firms. Founders create new ventures based on an opportunity, market gap, and expertise (Shane and Venkataraman 2000), believing that they have a product or service that they can sell to potential customers. Since new ventures rely on customers for survival, this stakeholder group is arguably the most critical consideration of a new venture. Without customers, the firm will be unable to generate revenue, and the venture will fail. Thus, entrepreneurs must consider how to generate value and contribute to the well-being of their customers early in their product or service development, taking care to identify their target demographic correctly and their customers' wants/needs as the firm's sustainability relies on their ability to do so.

Perhaps of all stakeholder groups, customers have the most potential to influence new venture survival since they represent the source of continuing cash flow (Jawahar and McLaughlin 2001). New ventures need to proactively address customer well-being as part of responsible new venture creation, whether the business model relies on repeat customers or depends upon satisfied customers spreading positive information about the organization via word of mouth to generate new customers. A failure to do so will lead to a decline in other performance metrics and the potential failure of the new venture (Gimeno et al. 1997). Beyond affecting financial outcomes and new venture survival, customer well-being can also influence important entrepreneurial activities such as innovation, opportunity recognition, legitimacy, and reputation (Rowley et al. 2007).

Conceptualizing Customer Well-Being Performance Measures

Like other well-being constructs, recent work on customer well-being is built upon frameworks that emphasize the influence of eudaimonic and hedonic well-being (Anderson et al. 2013) on an individual's quality of life. Individuals seek to meet their well-being needs through significant activities in various life domains (Pancer 2009). In their role as customers, individuals strive to fulfill the eudaimonic and hedonic components of well-being by satisfying a set of needs relating to their health and happiness, while also meeting societal needs for sustainability and social responsibility (Sirgy et al. 2007). Measurement often captures individual self-reports of *life satisfaction* (i.e., "evaluation of one's life according to subjectively determined standards" (Schimmack et al. 2002, p. 582)), thus suggesting that customer well-being is the satisfaction derived from consuming goods and services that address consumer needs (i.e., eudaimonic and hedonic well-being; Day 1987, Lee et al. 2002, Mick 2008, Sirgy et al. 2006). Though customer well-being may be operationalized in many ways, existing work often measures it using customer satisfaction assessments (Katsiekas et al. 2015), which rarely probe these underlying factors driving well-being.

Interestingly, some research argues that customer well-being is "a state in which consumers' experiences with goods and services—experiences related to the acquisition, preparations, consumption, ownership, maintenance, and disposal of specific categories of goods and services in the context of their local environment—are *judged to be beneficial to both consumers and society*" (Sirgy and Lee 2006, p. 43; emphasis added). This definition fails to account for product and service offerings that have both positive and negative effects on consumers. For example, recent attention to the positive and negative effects of technology and social media on consumers complicates this understanding of consumer well-being. Notably, a key missing element to understanding consumer well-being is time. Products and services that may offer immediate benefit to customer well-being may also be harmful long-term. These observations suggest that organizations should consider both the short- and long-term benefits and costs to customer well-being when developing and marketing their products.

Practices That Founders Can implement to Help with Customer Well-Being

New ventures need to consider customers' current/unmet needs to be able to address their customers' well-being (Slater and Narver 1995). To address customer well-being, founders can implement several practices. First, customer well-being is closely tied to employee well-being. When new ventures have quality employees that are taken care of and satisfied in their organization, they can respond better to customer and market needs (Reed 2000), which significantly impacts customer satisfaction (Rogg et al. 2001). Additionally, there is a positive relationship between employee training and their expertise in providing superior customer service (Chandler and Hanks 1994). Because consumption is a crucial part of customer satisfaction (Lee et al. 2002), the *process* of purchasing is essential to consider. Second, new ventures need to consider their customer service strategy (Edelman et al. 2005). Customers are more likely to feel their needs are met and their concerns heard when new ventures have a quality product or service, pursue a customer loyalty strategy (Carter et al. 1994), and have "passionate" responsiveness to customers (Hills and Narayana 1989). Getting customer buy-in is also a way to foster customer engagement with products and organizations. More work on crowd-funding ventures and products should explore how this engagement fosters customer well-being. Third, similarly to new ventures providing informal learning activities for their employees' well-being, new ventures can engage in a learning orientation (i.e., a consistent commitment to learning, a shared vision, and open-mindedness; Sinkula et al. 1997) to enhance customer well-being. This orientation helps establish sound information processing and capabilities needed to understand customer needs (Boulding et al. 2005). Fourth, new ventures can adopt a total market orientation (Narver et al. 2004), consisting of responsive and proactive market orientations. Responsive market orientation is "a business's attempt to understand and satisfy customers' expressed needs." In contrast, proactive market orientation is "the attempt to understand and satisfy customers' latent needs" (Narver et al. 2004, p. 336). In entrepreneurship and new ventures, this is often satisfied through innovation that provides a more efficient or effective means and/or ends (Casson 1982, Shane and Venkataraman 2000). To stay true to their roots, founders can identify novel and meaningful ways to satisfy customers' needs, leading to positive new venture outcomes such as enhanced creativity and more effective innovations (Im and Workman Jr 2004). Lastly, founders can rely heavily on integrating marketing activities to support the venture's acquisition and dissemination of knowledge about customer needs and marketing mix activities to disseminate knowledge to customers regarding the venture's product or service (Webb et al. 2010, Yli-Renko et al. 2001). These activities can help founders understand and communicate with customers, address their customers' needs and well-being, and help founders from focusing too much on obsolete or a waning set of customers (Baker and Sinkula 1999, Christensen and Bower 1996).

Suppliers

While customers may represent the primary source of cash flow for organizations, suppliers, also directly impact the ability to generate revenue and fulfill the firm's goals. Interest in the relationship between organizations and their suppliers dates back to the 1980s (Tangpong et al. 2015). Prior to this time, firms (specifically those based in North America) largely disregarded the critical role that suppliers play in providing key tools and services central to their aims and relied upon cost-minimizing strategies (e.g., competitive bidding and the weighted factor approach) for selecting suppliers (Stuart 1993). Following research suggests that price is not always a central issue for organizations when selecting suppliers (Dobler et al. 1990, Stuart 1993) and that even when it is of primary concern, competitive bidding selection procedures may only benefit firms in the short run while driving up their costs long term (Hahn et al. 1986, Stuart 1993)—in turn, potentially harming chances of survival—scholars began to investigate the limitations of this approach. This line of inquiry has since prompted a large body of work on *supplier partnerships* and *strategic alliances* (Stuart 1993), much of which is closely linked to procurement and supply chain management scholarship.

Scholars have historically contended that business-supplier relationships (BSRs) fall into one of two camps (Tangpong et al. 2008): those in which relational exchanges characterize the relationship (e.g., cooperation, trust, and commitment: Tangpong et al. 2015) and those in which it is characterized by power-dependence exchanges (e.g., exchanges that are rooted in the power-dependence dynamic between parties and are largely transactional: Tangpong et al. 2015). The nature of exchanges can directly impact BSRs, as a willingness of one party to exploit power and control or to prevent opportunism of the other can impact the performance and survival (Provan and Skinner 1989, Tangpong et al. 2015, Wang and Wei 2007, Wathne and Heide 2000). However, in their review of the BSR literature, Tangpong et al. (2015) suggest that the binary distinction between relational and power-dependence exchange-based BSRs is somewhat misguided, proposing instead that all BSRs are built upon exchanges characterized by some degree of dependence (e.g., supplier dependence—the degree to which a supplier firm depends upon a buyer firm—or buyer dependence—the degree to which a buyer depends upon the supplier) and relationalism.[7] They then identify eight types of BSRs, which inform the subsequent section on conceptualizing supplier well-being. Tangpong et al.'s (2015) table is replicated below, summarizing their proposed BSR types and Influence on performance for both buyers and suppliers below in Table 2.

Conceptualizing Supplier Well-Being Performance Measures

Tangpong et al.'s (2015) work serves as a roadmap for conceptualizing supplier well-being. First, it should be emphasized that this chapter has shifted perspectives from how firms (at the organizational level) interact with individuals (i.e., employees and customers) to how they interact with other firms. This shift is an essential consideration because, in some situations, suppliers may have more power to affect firms' performance than individual entities due to their size/the level at which they operate. In dealing with individuals, the loss of a single employee, customer, or shareholder likely will not dramatically impact an organization's opportunities for survival, excepting extreme cases (e.g., loss of a co-founder or highly influential individual in the organization; loss of a majority shareholder or one with influence to block future investments from others, etc.). In contrast, the loss of a significant supplier can stop an organization from producing its goods/services and from being able to fulfill its goals for an indefinite period. Thus, these relationships are crucial to the functioning of the organization.

Despite scholarly attention to the nature of these relationships, little work has considered the concept of supplier well-being. Tangpong et al. (2015) illustrate that several of the relationships proposed in Table 2 may be beneficial for either party at a given time. To address supplier well-being—in other words, the supplier's ability to achieve its organizational goals—buyers should seek relationships that help foster supplier success. Ideally, such relationships would simultaneously foster buyer success; however, buyers should consider the potential benefits of putting their goals second to those of suppliers in some cases. For example, in supplier-led collaborations, buyers put suppliers first initially, operating on their timeline to fulfill their goals. However, buyers also reap benefits in the long run. This area of research needs further exploration, particularly in the strategic management and entrepreneurship literature.

[7] Tangpong et al. (2015) define relationalism as "the degree to which buyer and supplier firms promote behaviors that maintain or improve their relationship (Noordewier et al. 1990, Smith 1998) reflect[ing] long-term cooperative relationships as opposed to short-term discrete transactions or adversarial market relationships (e.g., Boyle et al. 1992, Kaufmann and Dant 1992)" (p. 160).

Table 2. Summary of BSR types and performance-influencing mechanisms (replicated from Tangpong et al. 2015, p. 166).

BSR Type	Relationalism	Supplier Dependence	Buyer Dependence	Buyer's Performance-Influencing Mechanisms	Supplier's Performance-Influencing Mechanisms
Market/ discrete relationship	Low	Low	Low	Using market mechanisms and spot contracts to govern exchange relationships, resulting in low coordination costs Only able to make adjustments within the limits of contracts, thus susceptible to high transaction costs under the condition of high uncertainty	Using market mechanisms and spot contracts to govern exchange relationships, resulting in low coordination costs Only able to make adjustments within the limits of contracts, thus susceptible to high transaction costs under the condition of high uncertainty
Captive-buyer/ supplier-dominant relationship	Low	Low	High	Leveraging the supplier's expertise/ capabilities and securing access to critical components to prevent supply disruption Sub-optimizing the overall performance due to ongoing power exploitation for self-interests by the supplier	Reaping financial gains through the exercise of bargaining power at the buyer's expense Developing a poor reputation and negative track record in relationship management practices, thus potentially discouraging prospective buyers in the future

Table 2 contd. ...

...Table 2 contd.

BSR Type	Relationalism	Supplier Dependence	Buyer Dependence	Buyer's Performance-Influencing Mechanisms	Supplier's Performance-Influencing Mechanisms
Captive-supplier/buyer-dominant relationship	Low	High	Low	Reaping financial gains through the exercise of bargaining power at the supplier's expense Developing a poor reputation and negative track record in relationship management practices, thus potentially discouraging prospective suppliers in the future	Dedicating efforts to maintain the ongoing relationship with the buyer, which accounts for a significant portion of outputs and sales Sub-optimizing the overall performance due to ongoing power exploitation for self-interests by the buyer
Strategic/bilateral partnership	High	High	High	Sharing risks and rewards with the supplier and adjusting the internal assets and operations to fit the supplier's customized parts and components Promoting bilateral communication and cooperation to achieve incremental and continuous improvement in the overall performance of both parties	Sharing risks and rewards with the buyer and adjusting the customized parts and components to fit the buyer's internal assets and operations Promoting bilateral communication and cooperation to achieve incremental and continuous improvement in the overall performance of both parties

Table 2 contd. ...

...Table 2 contd.

BSR Type	Relationalism	Supplier Dependence	Buyer Dependence	Buyer's Performance-Influencing Mechanisms	Supplier's Performance-Influencing Mechanisms
Supplier-led collaboration	High	Low	High	Leveraging the supplier's expertise/ capabilities and securing access to critical components to prevent supply disruption Strengthening its competitive position through supplier-enabled innovations	Leveraging the buyer as a vehicle to bring its new products/ technology to the market expeditiously Optimizing its product/ technology performance through joint design/ development efforts with the buyer
Buyer-led collaboration	High	High	Low	Being specialized in certain key supply chain activities and delegating other activities to its supplier network Controlling and coordinating a broad range of supply chain activities to attain incremental and continuous improvement in overall supply chain performance	Having access to the buyer's demand information thus makes its production planning more effective Learning from the buyer as well as other suppliers in the network to incrementally and continuously improve its overall capabilities

Table 2 contd. ...

...Table 2 contd.

BSR Type	Relationalism	Supplier Dependence	Buyer Dependence	Buyer's Performance-Influencing Mechanisms	Supplier's Performance-Influencing Mechanisms
Competitive/ win-lose partnership	Low	High	High	Sub-optimizing the overall performance due to episodic self-interest seeking maneuvers by the supplier Breeding periodic unproductive conflicts in the BSR, as each party attempts to exert control over the other and pursue its own agenda in a zero-sum game	Sub-optimizing the overall performance due to episodic self-interest seeking maneuvers by the buyer Breeding periodic unproductive conflicts in the BSR, as each party attempts to exert control over the other and pursue its own agenda in a zero-sum game
Free will/ voluntary collaboration	High	Low	Low	Cross-fertilizing or synergizing its unique strengths with those of the supplier to attain a breakthrough improvement in overall performance Promoting bilateral and open communication, idea generation, knowledge sharing, and team orientation in the BSR through the equal status, autonomy, and shared leadership between both parties	Cross-fertilizing or synergizing its unique strengths with those of the buyer to attain a breakthrough improvement in overall performance Promoting bilateral and open communication, idea generation, knowledge sharing, and team orientation in the BSR through the equal status, autonomy, and shared leadership between both parties

Practices That Founders Can Implement to Help with Supplier Well-Being

Considerations of supplier well-being are virtually non-existent in the entrepreneurship literature. As a result, the primary suggestions for founders remain broad. Researchers are encouraged to engage with scholarship in procurement and supply chain management to help contribute to these ideas moving forward. This chapter advocates for founders to consider the types of relationships they want to engage in with their suppliers. Tangpong et al.'s (2015) typology provides context for founders. Business schools and entrepreneurship programs should encourage students to consider the various types of relationships they might build with suppliers rather than focusing on the immediate financial gratification of competitive, cost-cutting strategies. Thinking about the long-term benefits of potential relationships with suppliers early can help build a foundation for future success, even if doing so may require patience, more upfront capital (which founders can build into new venture budgets/funding rounds), and other early sacrifices.

Communication between parties is perhaps the most critical component of buyer-supplier relationships. Founders of buying firms should make every effort to engage their suppliers in outlining the goals for both parties in the relationship as early as possible. Knowing whether both parties may be interested in (potentially) keeping the door open for future collaboration can significantly impact the initial relationship and equip both parties to navigate current interactions better. Furthermore, founders should be open to evolving relationships with their suppliers. Research suggests that as new product development has increasingly become more of a collaboration between buyers and suppliers, BSRs have become a critical resource for "buyer firms' product innovativeness and supplier firms' innovative capacity" (Tangpong et al. 2015: p. 167). To capitalize on such opportunities and create value for their supplier partners, founders must be open to growing and evolving with their suppliers and should approach early collaborations with this mindset.

Communities

As discussed previously, entrepreneurs will focus on specific stakeholders dependent upon their needs at the time, as access to stakeholder resources is critical for creating a new venture (Burns et al. 2016). Although not always immediately recognized for their value to new ventures, local or regional communities can provide all types of critical resources (i.e., social, cultural, financial, human, and physical) necessary for new venture creation (Welter et al. 2018). For example, social networks are vital for organizational success. The local community is a network of potential opportunities/resources that entrepreneurs can utilize (Barraket et al. 2019), as connections with the local community can add creativity, consistency, and connectivity to a new venture (Branzei et al. 2018).

Nevertheless, first, entrepreneurs must go through the process of embedding themselves within the local community by establishing those social relationships that enable the entrepreneur to become part of the local structure (Jack and Anderson 2002). In this way, the relationship between the entrepreneur and the local community works two ways, providing knowledge, credibility, contacts, and resources (Jack and Anderson 2002). Local communities can also be influential in shaping entrepreneurial opportunities (Murphy et al. 2020) and the entrepreneurial process (Bacq et al. 2020), providing access to financial resources via crowdfunding (Josefy et al. 2017) and microfinancing (Khavul et al. 2013). Other community-level environments can influence entrepreneurs and the new venture creation process, including industry- or sector-based communities and national and transnational communities (Jennings et al. 2013).

Conceptualizing Community Well-Being Performance Measures

Although new ventures may not influence every aspect of community well-being, and some ventures are specifically created to address social issues (e.g., social entrepreneurship), *all* ventures should consider their role in fostering community well-being. Community well-being is a multidimensional construct that encompasses six major areas of influence in the community: a strong mission, healthy practices, proficient leadership, good relationships, flourishing individuals, and a satisfied community (VanderWeele 2019), in a networked system centered on "understanding of community and fulfilling needs and desires of its members" (Sung and Phillips 2018, p. 64). Some key ways new ventures can affect indicators of these six community well-being aspects include hiring local employees and contributing to their health and well-being; taking part in community events and volunteering within the community; generating business growth and contributing to local spending, which fosters community wealth; engaging in environmentally-conscious practices such as sustainable energy use, biodiversity, and waste management; creating an environment accepting of diverse cultures and backgrounds; and participating in/ encouraging civil engagement (Bacq et al. 2020, Baktir and Watson 2020, Davern et al. 2011, Dluhy and Swartz 2006, Dubb 2016, Hertel et al. 2021, Holden 2009, Ramos and Jones 2005).

Practices That Founders Can Implement to Help with Community Well-Being

As with the other suggestions, entrepreneurs are encouraged to consider how they may directly impact community well-being during the earliest stages of venture formation. A strong organizational mission—which founders begin to form (perhaps even unconsciously) in their initial ideas for the venture—that encourages the adoption of improved business practices lays a foundation for striving to make the world a better place (Baktir and Watson 2020). By continuously modeling efforts to learn and improve, entrepreneurs and new ventures can encourage and challenge their local communities to work toward bettering the world around them, which VanderWeele (2019) references as a critical goal for these entities. Similarly, founders can involve the local community in establishing a clear shared vision and convey that vision to the broader community (Hertel et al. 2021). When entrepreneurs give a sense of identification with the venture to local community supporters, they are more successful in gaining local resources, and the community feels more represented and connected with the venture (Hertel et al. 2021).

As individuals in positions of power and influence, entrepreneurs can help foster well-being at the collective community level and for individuals within their communities. Understanding of "community" is ever-evolving and with advances in technology, entrepreneurs' reach and influence can extend further each day, providing them with opportunities to promote healthy practices that allow relationships to develop and strengthen, build on community sustainability and goal attainment, and guide appropriate handling of conflicts and disputes (VanderWeele 2019). Additionally, entrepreneurs should provide proficient leadership within their communities (VanderWeele 2019), as their standing within their communities will affect access to community resources (Jack and Anderson 2002, Vestrum and Rasmussen 2013). Furthermore, entrepreneurs can inspire others to consider their communities (be they physical/local communities, race/ethnicity-centric communities, international communities, or groups of which they are a part) and drive positive social change by embedding a commitment to community well-being in their visions for their ventures (Bailey and Lumpkin 2021). Founders should strive to establish good relationships within their communities—rooted in trust and respect (Baktir and Watson 2020, VanderWeele 2019)—early in the venture's formation. Some ways they might do so include presenting the business plan and team to their communities while still in the development stages, offering various involvement opportunities to community members, and emphasizing a shared community-based ownership structure (Hertel et al. 2021). Entrepreneurs and their employees can also participate in community events and volunteer in an effort to support shared causes (Bailey and Lumpkin 2021). By hiring employees from within their communities and providing them with a place to flourish, founders can contribute to the individual (see employee well-being above) and community well-being. By considering how they can better their communities and factor opportunities to foster community well-being into their business plans, founders can help drive satisfied communities and better employee-customer relations (Baktir and Watson 2020, VanderWeele 2019). A summary of key suggestions for each area of stakeholder well-being are presented in Table 3.

Reconsidering New Venture Creation: Co-creative Entrepreneurship

This chapter now offers a brief description of an alternative perspective on the new venture creation process and commentary on how service design tools can be used to enhance the well-being of various stakeholders and the social impact of new ventures.

Drawing upon the recognition that organizations should strive to generate value for all stakeholders, this chapter now questions how stakeholders can simultaneously contribute to new ventures in their development stages. Entrepreneurship—or new venture creation—is generally viewed as a process involving one or a few select individuals in the earliest stages as the initial idea is refined and takes shape. However, an alternative perspective argues that new venture creation is a social action that involves engaging stakeholders collaboratively to create value (Buchanan and Vanberg 1991, Grönroos 2006). This perspective is gaining traction: In his most recent book, *The Third Wave*, Steve Case notes that the new wave of entrepreneurship is about organizations partnering together (i.e., co-creating with their stakeholders) (Case 2017).

Table 3. Final takeaways regarding stakeholder well-being.

Practices that founders can implement to help with...	
Employee Well-being	Founders can… • Outline efforts to support employees on each well-being subdimension in their earliest drafts of their business plans and build them into initial operating budgets and forecasts. • Provide training and other learning opportunities to help employees realize their potential. Offer employees informal learning opportunities. • Write employees into business plans and job descriptions or related hiring materials. • Establish an organizational culture that promotes employee flexibility and health and provides a supportive and caring work environment. • Communicate a clear organizational vision. • Empower employees to make crucial decisions and be part of organizational decision-making, leading to greater idea sharing and better handling of internal and external issues.
Customer Well-being	Founders can… • Have quality employees that are taken care of and satisfied in their organization. When they are employees can respond better to customer needs. • Provide a learning orientation (i.e., a consistent commitment to learning, a shared vision, and open-mindedness) to enhance customer well-being. This orientation helps establish sound information processing and capabilities needed to understand customer needs. • Adopt a total market orientation, consisting of responsive and proactive market orientations. • Rely heavily on integrating marketing activities to support the venture's acquisition and dissemination of knowledge about customer needs and marketing mix activities to disseminate knowledge to customers regarding the venture's product or service.
Supplier Well-being	Founders can… • Consciously consider the types of relationships they want to engage in with their suppliers. • Communication effectively as a critical component of the buyer-supplier relationship. • Have a collaborative relationship with their suppliers.
Community Well-being	Founders can… • Model efforts to learn and improve and encourage and challenge their local communities to work toward bettering the world around them. • Promote healthy practices within the community that allows relationships to develop and strengthen, build on community sustainability and goal attainment, and guide appropriate handling of conflicts and disputes. • Inspire others to consider their communities and drive positive social change by embedding a commitment to community well-being in their visions for their ventures. • Participate in community events with their employees and volunteer within the local community.

As entrepreneurs go through the various stages of new venture creation and consider the well-being of their employees, customers, suppliers, and local communities, they should think of their ventures as co-created with their stakeholders. Karami and Read (2021, p. 13) contend that "co-creation represents a powerful engine available to resource, case, and reputation-strapped entrepreneurs. It is high time we focus on how to harness that engine". Co-creation that is focused on collaboration, facilitating interaction, and formalizing commitments will help to advance the entrepreneurial process, provide, and enable effective use of resources, form new institutional arrangements, will benefit all involved in the co-creation, help resolve uncertainty, and help with human capital and resource constraints (Karami and Read 2021). Table 4 offers several practical co-creation solutions using a hypothetical home energy AI startup that can be adapted for other startups.

Table 4. Co-creative action examples from an AI startup.

Issue	**Co-creative Action**
Who to hire?	Why hire? Create an open innovation challenge on reducing home energy use with AI.
Who to work with?	Use Facebook and LinkedIn to share the problem statement with a wide range of possible collaborators.
Where to start?	Articulate the problem in a way that invites potential collaborators to contribute to creating a solution.
What to prioritize?	Enabling interactions between collaborators and encouraging commitments that advance the process.
What do I need?	Let's see what my stakeholders commit. Different paths will be prioritized according to commitments that include resources.
What is valuable?	I don't know upfront. But I do know commitments are what is valuable, and associated resources come from them.
What institutions[8] do I need?	I don't need any upfront. However, my stakeholders may be existing institutions, may identify institutions that can help, or may co-create new ones.
How do I interact with institutions?	Don't take institutions as given. Engage existing institutions in the process so they can shape our solution and potentially shape themselves as well.
Offering Opportunity	Let my stakeholders determine one or many solutions. Let my stakeholders suggest applications for our solutions. Some stakeholders may become customers, but I don't know where the opportunity is upfront.
Value	Let my stakeholders tell the other stakeholders and me where the value is and how it relates differently to each stakeholder.
Uncertainty	Iterations of stakeholder commitments and the resources stakeholders provide combine to control uncertainty.
Constraints	Leverage prior success to secure additional cash, or more likely, engage existing and new committed stakeholders who benefit from adding resources.

Note. Adapted from Karami and Read (2021).

[8] Institutions are systems of established and prevalent social rules that structure social interactions (Hodgson 2006).

Tools and Methods to Enhance Social Impact

As key contributors to global economies, entrepreneurs can make a tremendous social impact on multiple levels of well-being. This chapter has already discussed in detail how entrepreneurs can influence the well-being of their employees, customers, suppliers, and local communities. Additionally, this chapter discusses a few broad methods and tools that entrepreneurs can use to create and sustain collective well-being and make a positive social impact. Specifically, this chapter discusses seven tools and methods often used in service design: design probes, service blueprints, appreciative inquiry, contextual interviews, actor maps, sustainable business models, and service prototyping (Alkire et al. 2020).

Design Probes

According to Mattelmäki (2006), "design probes are an approach of user-centered design for understanding human phenomena and exploring design opportunities." Design probes are an exploratory tool used to understand better the problems and potential solutions faced in a given situation. In the case of new venture formation, design probes can be illuminating. As stated, entrepreneurs found organizations to meet an observed need within an industry or existing market. Design probes can flip this process on its head: Instead of entrepreneurs assuming they know what problem they want to solve, they can explore new opportunities through user participation, which can provide context and enhance their perceptions to ultimately enrich their business offering for the marketplace (Mattelmäki 2006).

Service Blueprints

Service blueprints are diagrams used to "map" the customer journey, providing a visual representation of the relationship between different service components (Gibbons 2017) of the organization. Service blueprints can be especially useful in pinpointing dependencies between employee-facing and customer-facing processes and can help optimize the well-being of both stakeholder groups. When applying service design approaches to organizational management, venture leaders strive to improve employees' experiences directly while simultaneously indirectly enhancing customers' experiences by planning and organizing the venture's resources in specific ways, often utilizing a service blueprint to achieve these aims (Gibbons 2017).

Appreciative Inquiry

The main goal of appreciative inquiry as a service design tool is to help individuals move toward a shared vision for the future by engaging others in strategic innovation (Ludema and Fry 2008). For appreciative inquiry to be effective, it must include all stakeholder groups involved in the venture; organizational leaders must encourage participation from all those involved in creating the new venture in seeking ideas, even tapping unlikely sources (Ludema and Fry 2008). Appreciative inquiry requires organizational leaders to ask "good" questions (i.e., "how" questions versus "why" questions, probes to elicit details, questions with a logical flow, non-leading questions, and open-ended questions) in an effort to learn from other stakeholders in developing and executing that shared vision. Such "good" questions are built upon an appreciation of stakeholders: recognizing their strengths and using said strengths to build the foundation of the new venture.

Contextual Interviews

Contextual interviews are another valuable tool for assessing the collective well-being of stakeholders, serving to provide information on gaps in products or services, potential shortcomings, and optimal design of the organization's products and/or services. Contextual interviews are designed to probe the context within which stakeholders (i.e., customers, communities, suppliers, and employees) will use or interact with the organization's product(s) or service(s). In contextual interviews, entrepreneurs and organizational stakeholders build a partnership based on a mutual interpretation of the product and/or service in question and focus on improving the opportunity for all involved (Salazar 2020).

Actor Maps

Entrepreneurs can also employ *actor maps*—visual depictions of key individuals or organizations that make up the ecosystem surrounding their ventures—to better understand how the ecosystem affects its players (i.e., the venture and stakeholders) and how their actions influence the system. Actor mapping visually demonstrates the relationships among actors, illustrating connections between them and their relationships to the new venture (Forman and Discenza 2012). Actor mapping is distinct from stakeholder analysis—the "process of systematically gathering and analyzing qualitative information to determine whose interests should be taken into account when developing and/or implementing a policy or program" (Forman and Discenza 2012)—which essentially involves prioritizing a list of key individuals or groups to target as part of an action plan. In contrast, the visual depiction represented in the actor map can help entrepreneurs identify the context, social connections, structural and system patterns, and perspectives of those who are or should be involved in a new venture.

Sustainable Business Models

Aiming to develop sustainable business models from the outset can also help entrepreneurs identify important stakeholders for their ventures. Sustainable business models aim to generate value for all stakeholder groups without sacrificing "the natural, economic, and social capital" (Breuer and Lüdeke-Freund 2014, p. 3) upon which the organization relies. As discussed in previous points, considering all potential actors as early as possible is critical for entrepreneurs and new ventures: Doing so is key to reconfiguring the understanding of "value" across stakeholder groups and thus how organizations go about generating said value (Breuer and Lüdeke-Freund 2014).

Service Prototyping

Lastly, service prototyping is another useful method to inclusively incorporate stakeholders into the process of creating a new venture. Service prototyping models the service (could also apply to products) a new venture offers to stakeholders *before* offering it to customers, allowing stakeholders to provide feedback on their experiences and encourage their involvement in the service innovation (Kuure et al. 2014). Together, these seven methods are a good starting point for incorporating stakeholders into the new venture creation process, ultimately incorporating their needs and well-being into the venture's product, service, and organization.

Future Research

While the most common level of analysis in measuring performance in entrepreneurship research is at the firm level, researchers are encouraged to use other performance measures to capture value, growth, and entrepreneurial contributions. Specifically, this chapter discussed four other areas of nonfinancial performance that researchers can examine in their studies as dependent variables: employee well-being, customer well-being, supplier well-being, and community well-being.

A recent review of all dependent variables previously explored in the entrepreneurship literature implies that no single study from this domain has focused on employee performance (Shepherd et al. 2019). While employee well-being and performance research are well-studied topics in disciplines such as organizational behavior, human resources, industrial-organizational psychology, and management, entrepreneurship scholars have yet to thoroughly examine the role of employees in new ventures, leaving a considerable gap in the knowledge base. Entrepreneurship is a relevant and growing field (Shane and Venkataraman 2000) and startups are vastly different from established organizations, which should lend interest in studying other performance measures. Shepherd et al. (2019) contend that employee performance is a critical variable for future entrepreneurship work, stating:

> *Particularly in growing and knowledge-intensive firms, it appears that employees are extremely important stakeholders, and their performance is a critical proximal outcome. The extent to which entrepreneurs or entrepreneurial firms can manage employees to retain high-performing employees, compensate desired employee behaviors, and otherwise facilitate employee well-being is likely critical to firm performance. For example, how do employee compensation and well-being relate to the infusion of venture capital funding, or does rapid growth affect employee retention? (Shepherd et al. 2019, p. 176).*

Researchers should focus on the specific processes and practices (or lack thereof) that new ventures engage in to create environments and organizations focusing on employee well-being. With the volatility and uncertainty characterizing new venture creation, what encourages employees to stay with the venture, how they contribute to its growth (beyond just growth in employee numbers), and how founders incorporate their employees in their decision-making are essential questions to investigate. Researchers should also investigate how these processes and practices emerge throughout new venture creation cycles.

Customer and supplier well-being are two other prime areas for entrepreneurship researchers since much of this research remains siloed in other disciplines. Several years ago, Webb et al. (2011) called entrepreneurship researchers to consider marketing activities and customers in their research on the entrepreneurship process. Since then, the intersection of customer, supplier, and entrepreneurship research has grown but is still limited. Although much work is focused on the subjective "opportunity" (Korsgaard et al. 2016), readers are encouraged to consider how entrepreneurs are addressing customers' wants, needs, and, more importantly, their *unmet needs* that may still be unidentified and rooted in customers' subconscious (Bylund and Packard 2021). Suppliers are vital in assisting entrepreneurs in obtaining the necessary tools and materials for addressing their customers' needs. In some cases, they can be customers, as well. Furthermore, suppliers' willingness to adapt to an entrepreneur's needs is based on mutual commitment and trust in their working relationships (Hasaballah et al. 2019). In this way, the value created by the new venture can be measured as an increase in subjective satisfaction and/or customer or supplier well-being (Bylund and Packard 2021). Researchers should explore how these relationships progress to create trust, mutual understanding, and commitment.

Additionally, it is exciting to see how community well-being is becoming a more studied topic in entrepreneurship (Bacq et al. 2020, Baktir and Watson 2020, Barraket et al. 2019, Hertel et al. 2021, Murphy et al. 2020). While researchers can investigate how entrepreneurial action influences individual-level well-being by studying employee, customer, and, in many cases, supplier well-being, those interested in how entrepreneurial action influences community well-being must take a more collective approach. Communities are group-level entities, thus requiring collective measures of well-being. Researchers might explore multiple avenues of influence on a community's well-being, including the vision and mission of the new venture, the involvement of the venture's founder, team, and employees in the community, and how entrepreneurs include the community in their business planning. Lastly, researchers are encouraged to explore how these relationships change based on community characteristics such as potential threats and hazards in the area, racial/ethnic composition, and overall community wealth. In this way, researchers can consider the power, stability, and commitment between the venture and the community.

Finally, researchers are encouraged to take a more holistic approach to studying entrepreneurship by attempting to incorporate multiple outcome levels and aspects of well-being. Just as founders were encouraged to consider stakeholders in the co-creation of their venture, researchers should take a similar approach, factoring in all relevant stakeholders, including employees, customers, suppliers, and the community, as crucial parties in venture co-creation (Karami and Read 2021).

Conclusion

Recent interest in the role of business in American society has sparked questions regarding the nonfinancial value that organizations provide for various stakeholders. While measures of organizational performance have historically focused on revenue generation, and particularly on creating financial gains for shareholders, the Business Roundtable's 2019 "Statement on the Purpose of a Corporation" advocates that organizations (and their representatives) have responsibilities to generate value for all stakeholders—including customers, employees, suppliers, and communities—in addition to those with a direct financial investment (BRA 2019). Identifying this responsibility to all stakeholders has since generated numerous questions regarding *who* is considered an organizational stakeholder, *what value* organizations can provide for these various stakeholders, and subsequently, *how this value can be measured.* This chapter has unpacked some of the complexity of these questions by exploring the statement that first outlined the position of commitment to nonfinancial value creation across stakeholder groups. It discussed how alternative conceptualizations of organizational value and performance might impact entrepreneurs at various stages in the venture creation cycle, emphasizing how nonfinancial performance measures should be considered and interwoven into the earliest visions of a new venture. Finally, this chapter outlined future directions for research and suggested ways in which scientists and organizations can work together to build a robust body of work advancing theory and practice in this area.

References

Abbott, J. (2003). Does employee satisfaction matter? A study to determine whether low employee morale affects customer satisfaction and profits in the business-to-business sector. *Journal of Communication Management*, 7(4): 333–339.

Alkire, L., Mooney, C., Gur, F.A., Kabadayi, S., Renko, M. and Vink, J. (2020). Transformative service research, service design, and social entrepreneurship: An interdisciplinary framework advancing wellbeing and social impact. *Journal of Service Management*, 31(1): 24–50.

Amabile, T.M. (1996). *Creativity and Innovation in Organizations*, Vol. 5. Boston: Harvard Business School.

Anderson, L., Ostrom, A.L., Corus, C., Fisk, R.P., Gallan, A.S., Giraldo, M., Mende, M., Mulder, M. Rayburn, S.W., Rosenbaum, M.S., Shirahada, K. and Williams, J.D. (2013). Transformative service research: An agenda for the future. *Journal of Business Research*, 66: 1203–1210.

Babalola, M.T., Mawritz, M.B., Greenbaum, R.L., Ren, S. and Garba, O.A. (2021). Whatever it takes: How and when supervisor bottom-line mentality motivates employee contributions in the workplace. *Journal of Management*, 47(5): 1134–1154.

Bacq, S.C., Hertel, C.J. and Lumpkin, G.T. (2020). Societal impact at the nexus of entrepreneurship and community—Taking stock and looking ahead. In *Academy of Management Annual Meeting Proceedings*, Vol. 20, No. CONF. Academy of Management.

Bagger, J. and Li, A. (2014). How does supervisory family support influence employees' attitudes and behaviors? A social exchange perspective. *Journal of Management*, 40(4): 1123–1150.

Bailey, R.C. and Lumpkin, G.T. (2021). Enacting positive social change: A civic wealth creation stakeholder engagement framework. *Entrepreneurship Theory and Practice*, 10422587211049745.

Baker, W.E. and Sinkula, J.M. (1999). The synergistic effect of market orientation and learning orientation on organizational performance. *Journal of the Academy of Marketing Science*, 27(4): 411–427.

Baktir, Z. and Watson, F. (2020). Trust-driven entrepreneurship for community well-being of refugees and their local hosts. *Journal of Macromarketing*, 41(2): 251–266.

Barney, J.B. (2001). Resource-based theories of competitive advantage: A ten-year retrospective on the resource-based view. *Journal of Management*, 27(6): 643–650.

Barraket, J., Eversole, R., Luke, B. and Barth, S. (2019). Resourcefulness of locally-oriented social enterprises: Implications for rural community development. *Journal of Rural Studies*, 70: 188–197.

Bates, R.A. and Holton III, E.F. (1995). Computerized performance monitoring: A review of human resource issues. *Human Resource Management Review*, 5(4): 267–288.

Batra, S. (2017). How do new ventures attract and retain talented employees? The case of Shaadisaga. *Human Resource Management International Digest*, 25(2): 1–3.

Batra, S. and Pollitt, D. (2014). Building a culture of creativity at Moving Pixels Company: Informality, openness and respect support hard-working employees. *Human Resource Management International Digest*.

Bosma, N., Van Praag, M., Thurik, R. and De Wit, G. (2004). The value of human and social capital investments for the business performance of startups. *Small Business Economics*, 23(3): 227–236.

Boulding, W., Staelin, R., Ehret, M. and Johnston, W.J. (2005). A customer relationship management roadmap: What is known, potential pitfalls, and where to go. *Journal of Marketing*, 69(4): 155–166.

Branzei, O., Parker, S.C., Moroz, P.W. and Gamble, E. (2018). Going pro-social: Extending the individual-venture nexus to the collective level. *In Vol.* 33: 551–565. Elsevier.

Breuer, H. and Lüdeke-Freund, F. (2014). Normative innovation for sustainable business models in value networks. *The Proceedings of XXV ISPIM Conference-Innovation for Sustainable Economy and Society*, 1–17.

Buchanan, J.M. and Vanberg, V.J. (1991). The market as a creative process. *Economics & Philosophy*, 7(2): 167–186.

Burns, B.L., Barney, J.B., Angus, R.W. and Herrick, H.N. (2016). Enrolling stakeholders under conditions of risk and uncertainty. *Strategic Entrepreneurship Journal*, 10(1): 97–106.

Business Roundtable Association. (2019). Business Roundtable "Statement on the Purpose of a Corporation" Proposes New Paradigm.

Bylund, P.L. and Packard, M.D. (2021). Subjective value in entrepreneurship. *Small Business Economics*, 1–18.

Cardon, M.S., Post, C. and Forster, W.R. (2017). Team entrepreneurial passion: Its emergence and influence in new venture teams. *Academy of Management Review*, 42(2): 283–305.

Carter, N., Gartner, W. and Reynolds, P.D. (1996). Exploring startup event sequences. *Journal of Business Venturing*, 11(3): 151–166.

Carter, N.M., Stearns, T.M., Reynolds, P.D. and Miller, B.A. (1994). New venture strategies: Theory development with an empirical base. *Strategic Management Journal*, 15(1): 21–41.

Carton, A.M. and Tewfik, B.A. (2016). Perspective—A new look at conflict management in work groups. *Organization Science*, 27(5): 1125–1141.

Case, S. (2017). *The Third Wave: An Entrepreneur's Vision of the Future*. Simon and Schuster.

Casson, M. (1982). *The Entrepreneur: An Economic Theory*. Rowman & Littlefield.

Chakravarthy, B.S. (1986). Measuring strategic performance. *Strategic Management Journal*, 7(5): 437–458.

Chandler, G.N. and Hanks, S.H. (1994). Founder competence, the environment, and venture performance. *Entrepreneurship Theory and Practice*, 18(3): 77–89.

Christensen, C.M. and Bower, J.L. (1996). Customer power, strategic investment, and the failure of leading firms. *Strategic Management Journal*, 17(3): 197–218.

Churet, C., Robecco, S.A.M. and Eccles, R.G. (2014). Integrated reporting, quality of management, and financial performance. *Journal of Applied Corporate Finance*, 26(1): 55–67.

Colombo, M.G. and Grilli, L. (2005). Founders' human capital and the growth of new technology-based firms: A competence-based view. *Research Policy*, 34(6): 795–816.

Cooper, A.C., Gimeno-Gascon, F.J. and Woo, C.Y. (1994). Initial human and financial capital as predictors of new venture performance. *Journal of Business Venturing*, 9(5): 371–395.

Cooper, A.C., Woo, C.Y. and Dunkelberg, W.C. (1989). Entrepreneurship and the initial size of firms. *Journal of Business Venturing*, 4(5): 317–332.

Dagenais-Desmarais, V. and Savoie, A. (2012). What is psychological well-being, really? A grassroots approach from the organizational sciences. *Journal of Happiness Studies*, 13(4): 659–684.

Davern, M.T., West, S., Bodenham, S. and Wiseman, J. (2011). Community indicators in action: Using indicators as a tool for planning and evaluating the health and wellbeing of a community. In *Community Quality-of-life Indicators: Best Cases*, 319–338. Springer.

Day, R.L. (1987). Relationships between life satisfaction and consumer satisfaction. *Marketing and Quality-of-life Interface*, 289–311.

De Jong, B.A. and Dirks, K.T. (2012). Beyond shared perceptions of trust and monitoring in teams: Implications of asymmetry and dissensus. *Journal of Applied Psychology*, 97(2): 391.

De Mol, E., Cardon, M.S., de Jong, B., Khapova, S.N. and Elfring, T. (2020). Entrepreneurial passion diversity in new venture teams: An empirical examination of short-and long-term performance implications. *Journal of Business Venturing*, 35(4): 105965.

Delmar, F. and Shane, S. (2004). Legitimating first: Organizing activities and the survival of new ventures. *Journal of Business Venturing*, 19(3): 385–410.

Demir, R., Wennberg, K. and McKelvie, A. (2017). The strategic management of high-growth firms: A review and theoretical conceptualization. *Long Range Planning*, 50(4): 431–456.

Der Foo, M., Wong, P.K. and Ong, A. (2005). Do others think you have a viable business idea? Team diversity and judges' evaluation of ideas in a business plan competition. *Journal of Business Venturing*, 20(3): 385–402.

Dess, G.G. and Robinson Jr, R.B. (1984). Measuring organizational performance in the absence of objective measures: The case of the privately-held firm and conglomerate business unit. *Strategic Management Journal*, 5(3): 265–273.

Diener, E., Suh, E.M., Lucas, R.E. and Smith, H.L. (1999). Subjective well-being: Three decades of progress. *Psychological Bulletin*, 125(2): 276.

Dluhy, M. and Swartz, N. (2006). Connecting knowledge and policy: The promise of community indicators in the United States. *Social Indicators Research*, 79(1): 1–23.

Dobler, D.W., Burt, D.N. and Lee, L. (1990). *Purchasing and Materials Management*, 5th ed. McGraw Hill: New York.

Dodge, H.R. and Robbins, J.E. (1992). An empirical investigation of the organizational life cycle. *Journal of Small Business Management*, 30(1): 27.

Dodge, H.R., Fullerton, S. and Robbins, J.E. (1994). Stage of the organizational life cycle and competition as mediators of problem perception for small businesses. *Strategic Management Journal*, 15(2): 121–134.

Dubb, S. (2016). Community wealth building forms: What they are and how to use them at the local level. *Academy of Management Perspectives*, 30(2): 141–152.

Earley, C.P. and Mosakowski, E. (2000). Creating hybrid team cultures: An empirical test of transnational team functioning. *Academy of Management Journal*, 43(1): 26–49.

Eccles, R.G. (1991). The performance measurement manifesto. *Harvard Business Review*, 69(1): 131–137.

Eccles, R.G., Krzus, M.P. and Ribot, S. (2015). Meaning and momentum in the integrated reporting movement. *Journal of Applied Corporate Finance*, 27(2): 8–17.

Edelman, L.F., Brush, C.G. and Manolova, T. (2005). Co-alignment in the resource–performance relationship: Strategy as mediator. *Journal of Business Venturing*, 20(3): 359–383.

Eisenberger, R. and Stinglhamber, F. (2011). *Perceived Organizational Support: Fostering Enthusiastic and Productive Employees*. American Psychological Association.

Ely, R.J. and Thomas, D.A. (2001). Cultural diversity at work: The effects of diversity perspectives on work group processes and outcomes. *Administrative Science Quarterly*, 46(2): 229–273.

Ensley, M.D. and Pearce, C.L. (2001). Shared cognition in top management teams: Implications for new venture performance. *Journal of Organizational Behavior: The International Journal of Industrial, Occupational and Organizational Psychology and Behavior*, *22*(2): 145–160.

Finkelstein, S. and Hambrick, D.C. (1990). Top-management-team tenure and organizational outcomes: The moderating role of managerial discretion. *Administrative Science Quarterly*, 484–503.

Fisher, C.D. (2014). Conceptualizing and measuring wellbeing at work.

Forman, J.B. and Discenza, R. (2012). Got stake (Holder) management in your project.

Fryxell, G.E. and Barton, S.L. (1990). Temporal and contextual change in the measurement structure of financial performance: Implications for strategy research. *Journal of Management*, 16(3): 553–569.

Gartner, W.B., Gartner, W.C., Shaver, K.G., Carter, N.M. and Reynolds, P.D. (2004). *Handbook of Entrepreneurial Dynamics: The Process of Business Creation*. Sage.

Gelles, D. and Yaffe-Bellany, D. (2019). Shareholder value is no longer everything, top CEOs say. *The New York Times*.

Gibbons, S. (2017). Service Blueprints: Definition. *https://www.nngroup. com/articles/service-blueprints-definition/last access, 13*, 2019.

Gimeno, J., Folta, T.B., Cooper, A.C. and Woo, C.Y. (1997). Survival of the fittest? Entrepreneurial human capital and the persistence of underperforming firms. *Administrative Science Quarterly*, 750–783.

Gjerløv-Juel, P. and Guenther, C. (2019). Early employment expansion and long-run survival. *Journal of Business Venturing*, 34(1): 80–102.

Grant, A.M., Christianson, M.K. and Price, R.H. (2007). Happiness, health, or relationships? Managerial practices and employee well-being tradeoffs. *Academy of Management Perspectives*, 21(3): 51–63.

Greenhaus, J.H. and Allen, T.D. (2011). Work-family balance: A review and extension of the literature.

Grönroos, C. (2006). On defining marketing: finding a new roadmap for marketing. *Marketing Theory*, 6(4): 395–417.

Hahn, C.K., Kim, K.H. and Kim, J.S. (1986). Costs of competition: Implications for purchasing strategy. *Journal of Purchasing and Materials Management*, 22(3): 2–7.

Hamann, P.M., Schiemann, F., Bellora, L. and Guenther, T.W. (2013). Exploring the dimensions of organizational performance: A construct validity study. *Organizational Research Methods*, 16(1): 67–87.

Hammer, L.B., Kossek, E.E., Yragui, N.L., Bodner, T.E. and Hanson, G.C. (2009). Development and validation of a multidimensional measure of Family Supportive Supervisor Behaviors (FSSB). *Journal of Management*, 35(4): 837–856.

Hasaballah, A.H.A., Genc, O.F., Mohamad, O.B. and Ahmed, Z.U. (2019). Exploring the interface of relationship marketing and export performance: A conceptual perspective. *Journal of Research in Marketing and Entrepreneurship*, 21(2): 126–148.

Heneman, R.L., Tansky, J.W. and Camp, S.M. (2000). Human resource management practices in small and medium-sized enterprises: Unanswered questions and future research perspectives. *Entrepreneurship Theory and Practice*, 25(1): 11–26.

Hertel, C., Binder, J. and Fauchart, E. (2021). Getting more from many—a framework of community resourcefulness in new venture creation. *Journal of Business Venturing*, 36(3).

Hill, R. and Stewart, J. (2000). Human resource development in small organizations. *Journal of European Industrial Training*.

Hills, G.E. and Narayana, C. (1989). Profile characteristics, success factors and marketing in highly successful firms. *Frontiers of Entrepreneurship Research*, 69–80.

Hirsh, E. and Cha, Y. (2018). For law and markets: Employment discrimination lawsuits, market performance, and managerial diversity. *American Journal of Sociology*, 123(4): 1117–1160.

Hitt, M.A. (1988). The measuring of organizational effectiveness: Multiple domains and constituencies. *Management International Review*, 28–40.

Holden, M. (2009). Community interests and indicator system success. *Social Indicators Research*, 92(3): 429–448.

Hult, G.T.M., Ketchen, D.J., Griffith, D.A., Chabowski, B.R., Hamman, M.K., Dykes, B.J., Pollitte, W.A. and Cavusgil, S.T. (2008). An assessment of the measurement of performance in international business research. *Journal of International Business Studies*, 39(6): 1064–1080.

Im, S. and Workman Jr, J.P. (2004). Market orientation, creativity, and new product performance in high-technology firms. *Journal of Marketing*, 68(2): 114–132.

Jack, S.L. and Anderson, A.R. (2002). The effects of embeddedness on the entrepreneurial process. *Journal of Business Venturing*, 17(5): 467–487.

Jawahar, I. and McLaughlin, G.L. (2001). Toward a descriptive stakeholder theory: An organizational life cycle approach. *Academy of Management Review*, 26(3): 397–414.

Jennings, P.D., Greenwood, R., Lounsbury, M.D. and Suddaby, R. (2013). Institutions, entrepreneurs, and communities: A special issue on entrepreneurship. *Journal of Business Venturing*, 28(1): 1–9.

Jensen, J.M., Patel, P.C. and Messersmith, J.G. (2013). High-performance work systems and job control: consequences for anxiety, role overload, and turnover intentions. *Journal of Management*, 39(6): 1699–1724.

Jensen, M.C. and Meckling, W.H. (1976). Theory of the firm: Managerial behavior, agency costs and ownership structure. *Journal of Financial Economics*, 3(4): 305–360.

Josefy, M., Dean, T.J., Albert, L.S. and Fitza, M.A. (2017). The role of community in crowdfunding success: Evidence on cultural attributes in funding campaigns to "save the local theater." *Entrepreneurship Theory and Practice*, 41(2): 161–182.

Judge, T.A., Ilies, R. and Dimotakis, N. (2010). Are health and happiness the product of wisdom? The relationship of general mental ability to educational and occupational attainment, health, and well-being. *Journal of Applied Psychology*, 95(3): 454.

Kanter, R.M. and Brinkerhoff, D. (1981). Organizational performance: Recent developments in measurement. *Annual Review of Sociology*, 7(1): 321–349.

Karami, M. and Read, S. (2021). Co-creative entrepreneurship. *Journal of Business Venturing*, 36(4).

Katsikeas, C.S., Morgan, N.A., Leonidou, L.C. and Hult, G.T.M. (2016). Assessing performance outcomes in marketing. *Journal of Marketing*, 80(2): 1–20.

Keats, B.W. (1988). The vertical construct validity of business economic performance measures. *The Journal of Applied Behavioral Science*, 24(2): 151–160.

Keyes, C.L.M. (1998). Social well-being. *Social Psychology Quarterly*, 121–140.

Khavul, S., Chavez, H. and Bruton, G.D. (2013). When institutional change outruns the change agent: The contested terrain of entrepreneurial microfinance for those in poverty. *Journal of Business Venturing*, 28(1): 30–50.

Korsgaard, S., Berglund, H., Thrane, C. and Blenker, P. (2016). A tale of two Kirzners: Time, uncertainty, and the "nature" of opportunities. *Entrepreneurship Theory and Practice*, 40(4): 867–889.

Kurtessis, J.N., Eisenberger, R., Ford, M.T., Buffardi, L.C., Stewart, K.A. and Adis, C.S. (2017). Perceived organizational support: A meta-analytic evaluation of organizational support theory. *Journal of Management*, 43(6): 1854–1884.

Kuure, E., Miettinen, S. and Alhonsuo, M. (2014). Change through Service Design–Service Prototyping as a Tool for Learning and Transformation.

Lee, D.-J., Sirgy, M.J., Larsen, V. and Wright, N.D. (2002). Developing a subjective measure of consumer well-being. *Journal of Macromarketing*, 22(2): 158–169.

Locke, E.A. (1976). The nature and causes of job satisfaction. *Handbook of Industrial and Organizational Psychology*.

Lorinkova, N.M. and Perry, S.J. (2017). When is empowerment effective? The role of leader-leader exchange in empowering leadership, cynicism, and time theft. *Journal of Management*, 43(5): 1631–1654.

Lubatkin, M. and Shrieves, R.E. (1986). Towards reconciliation of market performance measures to strategic management research. *Academy of Management Review*, 11(3): 497–512.

Ludema, J.D. and Fry, R.E. (2008). The practice of appreciative inquiry. *The Sage Handbook of Action Research: Participative Inquiry and Practice*, 280–296.

MacKenzie, S.B. (2003). The dangers of poor construct conceptualization. *Journal of the Academy of Marketing Science*, 31(3): 323–326.

March, J.G. and Sutton, R.I. (1997). Crossroads—organizational performance as a dependent variable. *Organization Science*, 8(6): 698–706.

Maslach, C., Schaufeli, W.B. and Leiter, M.P. (2001). Job burnout. *Annual Review of Psychology*, 52(1): 397–422.

Mattelmäki, T. (2006). *Design Probes*. Aalto University.

McKee-Ryan, F., Song, Z., Wanberg, C.R. and Kinicki, A.J. (2005). Psychological and physical well-being during unemployment:A meta-analytic study. *Journal of Applied Psychology*, 90(1): 53.

Mick, D.G. (2008). *The Moment and Place for a Special Issue*. In Vol. 35, 377–379. The University of Chicago Press.

Miller, C.C., Washburn, N.T. and Glick, W.H. (2013). Perspective—The myth of firm performance. *Organization Science*, 24(3): 948–964.

Mitchell, R.K., Agle, B.R. and Wood, D.J. (1997). Toward a theory of stakeholder identification and salience: Defining the principle of who and what really counts. *Academy of Management Review*, 22(4): 853–886.

Montgomery, A. and McGlynn, C. (2009). New peace, new teachers: Student teachers' perspectives of diversity and community relations in Northern Ireland. *Teaching and Teacher Education*, 25(3): 391–399.

Morgan, R.E. and Strong, C.A. (2003). Business performance and dimensions of strategic orientation. *Journal of Business Research*, 56(3): 163–176.

Moser, K.J., Tumasjan, A. and Welpe, I.M. (2017). Small but attractive: Dimensions of new venture employer attractiveness and the moderating role of applicants' entrepreneurial behaviors. *Journal of Business Venturing*, 32(5): 588–610.

Murphy, M., Danis, W.M. and Mack, J. (2020). From principles to action: Community-based entrepreneurship in the Toquaht Nation. *Journal of Business Venturing*, 35(6): 106051.

Narver, J.C., Slater, S.F. and MacLachlan, D.L. (2004). Responsive and proactive market orientation and new-product success. *Journal of Product Innovation Management*, 21(5): 334–347.

Obenauer, W.G. (2019). Are all voluntary attritions created equally? Understanding the need to incorporate employee diversity into attrition modeling. *Industrial and Organizational Psychology*, 12(3): 302–305.

Panaccio, A. and Vandenberghe, C. (2009). Perceived organizational support, organizational commitment and psychological well-being: A longitudinal study. *Journal of Vocational Behavior*, 75(2): 224–236.

Pancer, E. (2009). What is Consumer Well Being? An Historical Analysis. Proceedings of the Conference on Historical Analysis and Research in Marketing,

Parker, G.B. and Hyett, M.P. (2011). Measurement of well-being in the workplace: The development of the work well-being questionnaire. *The Journal of Nervous and Mental Disease*, 199(6): 394–397.

Pennings, J.M. and Goodman, P.S. (1977). *New Perspectives on Organizational Effectiveness*. Jossey-Bass.

Peteraf, M.A. and Barney, J.B. (2003). Unraveling the resource-based tangle. *Managerial and Decision Economics*, 24(4): 309–323.

Piszczek, M.M. (2020). Reciprocal relationships between workplace childcare initiatives and collective turnover rates of men and women. *Journal of Management*, 46(3): 470–494.

Polzer, J.T., Milton, L.P. and Swarm Jr, W.B. (2002). Capitalizing on diversity: Interpersonal congruence in small work groups. *Administrative Science Quarterly*, 47(2): 296–324.

Porath, C., Spreitzer, G., Gibson, C. and Garnett, F.G. (2012). Thriving at work: Toward its measurement, construct validation, and theoretical refinement. *Journal of Organizational Behavior*, 33(2): 250–275.

Provan, K.G. and Skinner, S.J. (1989). Interorganizational dependence and control as predictors of opportunism in dealer-supplier relations. *Academy of Management Journal*, 32(1): 202–212.

Ramos, O.T. and Jones, K. (2005). Comprehensive community indicators systems. *National Civic Review*, 94(2): 74–78.

Rappaport, A. (1978). Executive incentives vs corporate growth. *Harvard Business Review*, 56(4): 81–88.

Rappaport, A. (2019). How CEOs can forge a new kind of shareholder value. https://www.bloombergquint.com/opinion/business-roundtable-statement-of-purpose-fails-on-specifics.

Rauch, A. and Hatak, I. (2016). A meta-analysis of different HR-enhancing practices and performance of small and medium sized firms. *Journal of Business Venturing*, 31(5): 485–504.

Reed, T.S. (2000). Entrepreneurship in the new competitive landscape. *Entrepreneurship as Strategy: Competing on the Entrepreneurial Edge*, 23.

Reilly, B.J. and Fuhr Jr, J.P. (1983). Productivity: An economic and management analysis with a direction towards a new synthesis. *Academy of Management Review*, 8(1): 108–117.

Richard, O.C., Kirby, S.L. and Chadwick, K. (2013). The impact of racial and gender diversity in management on financial performance: How participative strategy making features can unleash a diversity advantage. *The International Journal of Human Resource Management*, 24(13): 2571–2582.

Richard, P. J., Devinney, T.M., Yip, G.S. and Johnson, G. (2009). Measuring organizational performance: Towards methodological best practice. *Journal of Management*, 35(3): 718–804.

Rogg, K.L., Schmidt, D.B., Shull, C. and Schmitt, N. (2001). Human resource practices, organizational climate, and customer satisfaction. *Journal of Management*, 27(4): 431–449.

Rosenbusch, N., Brinckmann, J. and Bausch, A. (2011). Is innovation always beneficial? A meta-analysis of the relationship between innovation and performance in SMEs. *Journal of Business Venturing*, 26(4): 441–457.

Ross, G.H. and Goodfellow, J.L. (1980). A fitness approach to corporate survival. *Journal of Business Quarterly*, 45: 19–25.

Rothaermel, F. (2017). *Strategic Management* (3 ed.). McGraw Hill Education.

Rowe, W.G. and Morrow Jr, J.L. (1999). A note on the dimensionality of the firm financial performance construct using accounting, market, and subjective measures. *Canadian Journal of Administrative Sciences/Revue Canadienne des Sciences de l'Administration*, 16(1): 58–71.

Rowley, J., Kupiec-Teahan, B. and Leeming, E. (2007). Customer community and co-creation: A case study. *Marketing Intelligence & Planning*.

Russell, E. and Daniels, K. (2018). Measuring affective well-being at work using short-form scales: Implications for affective structures and participant instructions. *Human Relations*, 71(11): 1478–1507.

Ryan, R.M. and Deci, E.L. (2001). On happiness and human potentials: A review of research on hedonic and eudaimonic well-being. *Annual Review of Psychology*, 52(1): 141–166.

Ryff, C.D. (1989). Happiness is everything, or is it? Explorations on the meaning of psychological well-being. *Journal of Personality and Social Psychology*, 57(6): 1069.

Ryff, C.D. (2019). Entrepreneurship and eudaimonic well-being: Five venues for new science. *J Bus Ventur*, 34(4): 646–663.

Ryff, C.D. and Keyes, C.L.M. (1995). The structure of psychological well-being revisited. *Journal of Personality and Social Psychology*, 69(4): 719.

Salazar, K. (2020). Contextual inquiry: Inspire design by observing and interviewing users in their context. *Nielsen Norman Group*, 6.

Santos, S.C. and Cardon, M.S. (2019). What's love got to do with it? Team entrepreneurial passion and performance in new venture teams. *Entrepreneurship Theory and Practice*, 43(3): 475–504.

Schimmack, U., Radhakrishnan, P., Oishi, S., Dzokoto, V. and Ahadi, S. (2002). Culture, personality, and subjective well-being: Integrating process models of life satisfaction. *Journal of Personality and Social Psychology*, 82(4): 582.

Serafeim, G. (2015). Integrated reporting and investor clientele. *Journal of Applied Corporate Finance*, 27(2): 34–51.

Shane, S. and Venkataraman, S. (2000). The promise of entrepreneurship as a field of research. *Academy of Management Journal*, 25(1): 217–226.

Shenhav, Y., Alon, S. and Shrum, W. (1994). 'Goodness' concepts in the study of organizations: a longitudinal survey of four leading journals. *Organization Studies*, 15(5): 753–776.

Shepherd, D.A., Wennberg, K., Suddaby, R. and Wiklund, J. (2019). What are we explaining? A review and agenda on initiating, engaging, performing, and contextualizing entrepreneurship. *Journal of Management*, 45(1): 159–196.

Shultz, C.J., Rahtz, D.R. and Sirgy, M.J. (2017). Distinguishing flourishing from distressed communities: Vulnerability, resilience and a systemic framework to facilitate well-being. In *Handbook of Community Well-being Research*, 403–421. Springer.

Sinkula, J.M., Baker, W.E. and Noordewier, T. (1997). A framework for market-based organizational learning: Linking values, knowledge, and behavior. *Journal of the Academy of Marketing Science*, 25(4): 305–318.

Sirgy, M.J. and Cornwell, T. (2001). Further validation of the Sirgy et al.'s measure of community quality of life. *Social Indicators Research*, 56(2): 125–143.

Sirgy, M.J. and Lee, D.-J. (2006). Macro measures of consumer well-being (CWB): A critical analysis and a research agenda. *Journal of Macromarketing*, 26(1): 27–44.

Sirgy, M.J., Lee, D.-J. and Kressmann, F. (2006). A need-based measure of consumer well being (CWB) in relation to personal transportation: Nomological validation. *Social Indicators Research*, 79(2): 337–367.

Sirgy, M.J., Lee, D.-J. and Rahtz, D. (2007). Research on consumer well-being (CWB): Overview of the field and introduction to the special issue. *Journal of Macromarketing*, 27(4): 341–349.

Slater, S.F. and Narver, J.C. (1995). Market orientation and the learning organization. *Journal of Marketing*, 59(3): 63–74.

Steigenberger, N. (2014). Only a matter of chance? How firm performance measurement impacts study results. *European Management Journal*, 32(1): 46–65.

Stuart, F.I. (1993). Supplier partnerships: influencing factors and strategic benefits. *The Journal of Supply Chain Management*, 29(4): 22–29.

Sung, H. and Phillips, R.G. (2018). Indicators and community well-being: Exploring a relational framework. *International Journal of Community Well-Being*, 1(1): 63–79.

Swann Jr, W.B., Kwan, V.S., Polzer, J.T. and Milton, L.P. (2003). Fostering group identification and creativity in diverse groups: The role of individuation and self-verification. *Personality and Social Psychology Bulletin*, 29(11): 1396–1406.

Tangpong, C., Michalisin, M.D. and Melcher, A.J. (2008). Toward a typology of buyer-supplier relationships: a study of the computer industry. *Decision Sciences*, 39(3): 571–593.

Tangpong, C., Michalisin, M.D., Traub, R.D. and Melcher, A.J. (2015). A review of buyer-supplier relationship typologies: progress, problems, and future directions. *Journal of Business and Industrial Marketing*, 30(2): 153–170.

Unger, J.M., Rauch, A., Frese, M. and Rosenbusch, N. (2011). Human capital and entrepreneurial success: A meta-analytical review. *Journal of Business Venturing*, 26(3): 341–358.

Van De Voorde, K., Paauwe, J. and Van Veldhoven, M. (2012). Employee well-being and the HRM–organizational performance relationship: a review of quantitative studies. *International Journal of Management Reviews*, 14(4): 391–407.

van Dierendonck, D. (2011). Servant leadership: A review and synthesis. *Journal of Management*, 37(4): 1228–1261.

VanderWeele, T.J. (2019). Measures of community well-being: A template. *International Journal of Community Well-Being*, 2(3): 253–275.

Venkatraman, N. and Ramanujam, V. (1986). Measurement of business performance in strategy research: A comparison of approaches. *Academy of Management Review*, 11(4): 801–814.

Vestrum, I. and Rasmussen, E. (2013). How community ventures mobilise resources: Developing resource dependence and embeddedness. *International Journal of Entrepreneurial Behavior & Research.*

Wang, E.T.G. and Wei, H. (2007). Interorganizational governance value creation: Coordinating for information visibility and flexibility in supply chains. *Decision Sciences*, 38(4): 647–674.

Wanous, J.P. and Youtz, M.A. (1986). Solution diversity and the quality of groups decisions. *Academy of Management Journal*, 29(1): 149–159.

Warr, P. (1990). The measurement of well-being and other aspects of mental health. *Journal of Occupational Psychology*, 63(3): 193–210.

Wathne, K.H. and Heide, J.B. (2000). Opportunism in interfirm relationships: Forms, outcomes, and solutions. *Journal of Marketing*, 64(4): 36–51.

Watson, W.E., Kumar, K. and Michaelsen, L.K. (1993). Cultural diversity's impact on interaction process and performance: Comparing homogeneous and diverse task groups. *Academy of Management Journal*, 36(3): 590–602.

Webb, J.W., Ireland, R.D., Hitt, M.A., Kistruck, G.M. and Tihanyi, L. (2010). Where is the opportunity without the customer? An integration of marketing activities, the entrepreneurship process, and institutional theory. *Journal of the Academy of Marketing Science*, 39(4): 537–554.

Webb, J.W., Ireland, R.D., Hitt, M.A., Kistruck, G.M. and Tihanyi, L. (2011). Where is the opportunity without the customer? An integration of marketing activities, the entrepreneurship process, and institutional theory. *Journal of the Academy of Marketing Science*, 39(4): 537–554.

Weber, A. and Zulehner, C. (2010). Female Hires and the Success of Startup Firms. *American Economic Review*, 100(2).

Welter, F., Xheneti, M. and Smallbone, D. (2018). Entrepreneurial resourcefulness in unstable institutional contexts: T he example of E uropean U nion borderlands. *Strategic Entrepreneurship Journal*, 12(1): 23–53.

Wernerfelt, B. (1984). A resource-based view of the firm. *Strategic Management Journal*, 5(2): 171–180.

Winston, A. (2019). Is the Business Roundtable Statement just empty rhetoric. *Harvard Business Review*, 30.

Wright, T.A. (2014). Putting your best "face" forward: The role of emotion-based well-being in organizational research. *Journal of Organizational Behavior*, 35(8): 1153–1168.

Wry, T., Lounsbury, M. and Glynn, M.A. (2011). Legitimating nascent collective identities: Coordinating cultural entrepreneurship. *Organization Science*, 22(2): 449–463.

Yli-Renko, H., Autio, E. and Sapienza, H.J. (2001). Social capital, knowledge acquisition, and knowledge exploitation in young technology-based firms. *Strategic Management Journal*, 22(6-7): 587–613.

Zheng, X., Zhu, W., Zhao, H. and Zhang, C. (2015). Employee well-being in organizations: Theoretical model, scale development, and cross-cultural validation. *Journal of Organizational Behavior*, 36(5): 621–644.

Chapter 14

Incentivizing Investors to Make Impactful Investments

Introducing a Model for Impact-Linked Carry

Jessica Hart,[1,*] *Karthik Varada,*[2] *Tom Schmittzehe*[3] and *Tomas Rosales*[4]

Impact investing is well on its journey to becoming mainstream. For those new to the topic, impact investing introduces the concept of the double bottom line: investments are expected to produce both financial and impact returns. For example, a venture that connects local artisans to global markets may produce both financial returns as well as impact returns (i.e., the creation of good jobs for the artisans). Many consider impact investing an important solution to the problem of investment decisions being made without regard for their broader societal impact. In the traditional investing model, decisions are based solely on financial metrics and this can sometimes result in harm to the environment or communities. Impact investing attempts to consider both the financial and non-financial. In fact, many impact investors seek business models whereby an increase in profits corresponds to an equal increase in impact delivered. A good example of this would be a medical diagnostic company: for each medical device sold there is an increase in revenues and the number of individuals receiving health interventions.

[1] Boston Consulting Group, 15 Coppice Walk, London, N20882, UK, 44 7772-994-629.
[2] University of Tokyo, 7-3-1, Hongo, Bunkyo-ku, Tokyo 113-8654.
[3] Moonshot Ventures.
[4] Pilipilway, off Rhapta Road, Westlands, Nairobi, 00200, Kenya.
Emails: karthik.varada123@gmail.com; rosa.tomas@gmail.com
* Corresponding author: jess.warren.hart@gmail.com

At the turn of the century, the idea of impact investing was very much on the fringes of the investing world, considered an extension of philanthropy where investors were expected to accept lower returns on their capital in exchange for investing in worthy causes. Today, impact targets are being applied across the capital spectrum. At the more concessionary end of the spectrum, there are venture philanthropists like Acumen who provide longer-term capital and typically lower commercial rates of return. On the more commercial side of the spectrum are traditional private equity (PE) firms like TPG and Bain Capital establishing successful impact funds that are targeting and achieving market rate returns.

There is a missing link in the way impact outcomes are incentivized in today's model. Whilst seemingly minor and technical, the incentive and reward systems of investors drive behavior and therefore shape the types of ventures that receive investment. The current system of incentives for investors belongs to the old world of investing, which only sought financial and not impact returns.

An incentive system that rewards investors for both commercial and impact outcomes is needed. Without such a system impact investors are instructed to invest for returns and impact but are only rewarded for financial outcomes. When faced with a choice between impact and commercial returns it will be very hard for the investor to choose impact. Why would an individual who knows the rules of a game choose to play a move which will make them lose? Such an impact-focused incentive system would also send a powerful message to the market; investors who want to use their capital to drive impact would be assured that the guardians of their capital share both their impact and financial goals.

This chapter lays out the current practice and state of the impact investing industry to illustrate why the impact incentive system is important. It reflects the best understanding of the market today that was developed through 20+ stakeholder interviews with leading impact funds in Europe, Asia, and the United States. This chapter proposes a model for linking incentives to impact which builds on examples the authors have seen in the market. It is intended to be used as an open-source blueprint for researchers studying VC investment decisions or organizational performance, as well as venture capital (VC) and PE fund managers that are looking to introduce impact-linked incentives. Key design choices are highlighted and debated using examples from today's practitioners. The second half of the chapter offers a case study of how impact-linked carry is adopted by the Indonesia Women Empowerment Fund (IWEF), a gender-lens investment fund investing in technology-driven solutions that address systemic barriers to women's economic opportunities and livelihoods.

Impact on the Rise

The last decade has seen a massive increase in capital flowing into environmental, social, and governance (ESG) and impact investing. The Forum for Sustainable and Responsible Investment (US SIF) reported $17.1 trillion ESG assets under management in the U.S. in 2020, a 43% increase from $12 trillion in 2018 (US SIF 2020). Similarly, the Global Impact Investors Network (GIIN) estimated impact assets under management to be $715 billion worldwide in 2020 (Hand et al. 2020). The largest PE players in the world are all lining up to launch impact vehicles: TPG started with their $2 Billion Rise Fund in 2016, followed by Bain Capital launching a $390 Million Double Impact Fund in 2017, and both now raising second funds (Kreutzer 2020). In 2020 KKR and Apollo Global Management added their names to the impact list and in 2021 Brookfield and TPG launched flagship climate funds (Mitchenall and Stutts 2021).

In the last decade, the climate crisis has risen to the top of the social and political agenda. More recently, the COVID-19 pandemic has highlighted other social issues at home and overseas. Demands for change come from multiple stakeholders; asset owners and the capital providers typically referred to as limited partners (LPs), customers, and top talent. A 2020 survey of LPs worldwide by the fund placement agent Rede Partners found that more than 75% of LPs surveyed were likely to be more focused on ESG or impact than before the global pandemic. Increasingly affluent consumers are seeking products and services which also have a positive impact. Forbes notes a significant rise in B corporations (a type of certification for businesses to demonstrate their environmental, social, and governance credentials) tripling in numbers in the last five years to 3500 companies now in 60 countries (Kohan 2021). An impact focus can also be a competitive advantage when attracting talent: a study by CONE found that 75% of millennials would take a pay cut to work for a socially responsible company (Butler 2019).

Impact Washing and Green Washing

Whilst the growth of impact investing as an asset class is certainly welcome, it is tempered by concerns about impact washing and green washing. Green washing is when a company markets itself as sustainable but in reality, is not (Furlow 2010). For example, an oil and gas extraction business which uses a small rainforest protection project implies that the whole business is not harming the environment. Impact measurement and impact management have become increasingly important to counter the practice of green washing and a whole sub-industry has developed to meet this need. In the past, the impact sector was criticized for incoherent and initiative-specific standards for impact, but there has been progress on this front. Industry leaders include the Impact Management Project (IMP)[9] — a coalition of impact stakeholders aiming to define and measure impact. Industry-specific impact metrics can be found through GIIN's IRIS+,[10] MSCI,[11] and others produce ESG scores, whereas the Impact Weighted Accounts Initiative (IWAI)[12] at Harvard Business School aims to price environmental and social externalities.

---

[9] Impact Management Project: https://impactmanagementproject.com/.

[10] IRIS+ by the GIIN: https://iris.thegiin.org/.

[11] MSCI ESG ratings: https://www.msci.com/our-solutions/esg-investing/esg-ratings.

[12] Impact Weighted Accounts: https://www.hbs.edu/impact-weighted-accounts/Pages/default.aspx.

Advances in impact measurement allow impact investors to be much more effective. Whereas previously impact was justified ex-post with anecdotes, it is now possible to set ex-ante impact key performance indicators (KPIs) and actively manage for impact during the investment period. In light of these improvements in impact measurement (and hopefully more coming up in the future), this chapter argues that investors can now meaningfully tie their incentives to impact performance. Doing so is a powerful way for investors to signal both confidence in their investing acumen and commitment to collinearity. The term collinearity is used here more loosely than in statistics but describes business models where an increase in profitability corresponds to an increasing impact achieved.

Financial Incentives for Impact

The idea of linking incentives to ESG outputs has already taken hold in the credit markets through sustainability-linked debit instruments. Sustainability-linked debt rewards borrowing firms for hitting clear and specific sustainability goals with a lower cost of debt (interest rate). The debt provider is essentially 'paying' for environmental outcomes or penalizing corporate borrowers for not attaining them. Large corporations like Enel have raised large sums in this way, most recently in a £500 million (~ 600 million USD) offering which was six times oversubscribed.

At the same time, public corporate CEOs are increasingly finding their compensation linked to ESG metrics; a 2020 report by Willis Towers Watson found that 51% of S&P 500 companies use ESG metrics in their executive incentive plans (Newbury and Delves 2020). The rationale for impact-linked debt can be translated directly to the equity markets. Furthermore, it should be relatively easy for impact investors to tie their incentives to the achievement of impact goals if they are truly investing according to their mission (i.e., with intentionality) and have rigorous impact measurement and management in place. Funds that do may also find it to be a differentiator and source of competitive advantage.

Current Compensation Practices in VC and PE

The current incentives in VC/PE provide an effective mechanism for LPs (i.e., capital providers) to ensure that the goals of the investment professionals are closely aligned with their own goals (i.e., the greatest return on investment). The typical roles within a fund are outlined in Fig. 1. Compensation for fund managers and general partners (GPs) at PE firms has an annual component and a longer-term component. Annually, GPs receive a fixed base salary as well as a variable performance-linked bonus. Over a longer period, GPs accrue carried interest. Carried interest, informally known as "carry", is the share of the fund's profit that is earned by a GP over the fund's lifetime. More junior investment professionals are typically compensated with a fixed salary and annual bonus. Although increasingly in the U.S. there are examples of carried interest being shared beyond GPs with more junior investors.

The annual compensation (salaries and bonuses) is funded through an annual management fee that is charged to the LPs. The management fee is traditionally

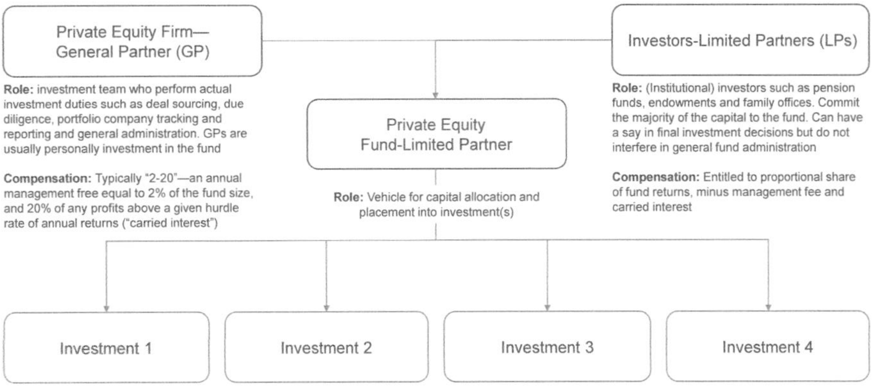

Figure 1. Roles within a private equity fund.

2% of committed capital. Both salary and annual bonus can vary widely and are often subject to negotiation; despite large amounts of assets under management, the headcount at most PE/VC funds is small and as a result, their people and performance processes are not as comprehensive as large corporate firms. Annual compensation is often negotiated on an individual basis, and bonuses are based on semi-subjective performance evaluations.

Carry is usually between 10%–20% of the increase in value of the fund. Twenty percent is the most common, hence the management fee and carry model is often known as "2–20". That is, 2% of the committed capital fee and a 20% increase in fund value. LPs and GPs agree on a "hurdle rate" which is the return a fund must produce before a carry is paid out. Depending on the risk and return target for a specific fund, the hurdle rate is typically between 6%–10% annual return. At the end of the fund's lifetime, 20% of the value of the fund above the hurdle rate goes into a carry pool (i.e., the combined overall carry from the fund's investments) for distribution amongst the fund's GP. Distribution of the carry pool to individual GPs (and the broader team) varies across funds, but typically reflects seniority and tenure at the fund, with senior partners receiving a larger share. Through this mechanism, the senior leadership at a fund is incentivized over the long-term to seek and secure high commercial rates of return. See below for an illustration.

There are too many technicalities to the carry system to detail here but three points are pertinent to the discussion of impact carry. Firstly, in funds with a longer lifespan (say 7–13 years) carry provides a truly long-term incentive since it is technically not paid out until the fund has reached its end of life (in practice most GPs receive some carry if they exit successful investments earlier). Secondly, most European funds aggregate carry on a fund level and carry is only distributed after the full fund has been deployed and the annualized hurdle met. This provides a shared incentive for everyone who is part of the carry pool to support the full portfolio of investments. In contrast, most U.S.-based funds have "deal-by-deal-carry", which can lead to "claw backs" (i.e., capital being returned by GPs) after the full fund has been disbursed if some investments are less successful than earlier ones. Finally, the financial rewards for individual GPs via carry can be enormous, as can be seen in the example outlined in Table 1. Imagine a fund that invests \$100 million. In the terms agreed up-front the hurdle rate is 10% and the carried interest for GPs will be 20% of profits above the hurdle. If the fund delivers a 3x return the value of the fund at closing will be \$300 million. After the hurdle rate, the carry pool is \$38 million. Divided by a typical team of 4–6 people, it provides a strong, \$6 million incentive to each member of the investment team.

Table 1. Example payout from a successful fund.

Terms of the Fund		**Amount**	**Units**
	Invested capital	100	*\$M*
	Hurdle rate	10%	
	Carried interest	20%	
Outcomes			
	Capital after exit	300	*\$M*
	ROI	300%	
	Returns above hurdle	190	*\$M*
	Carry pool	38	*\$M*
	1/6th share	6.3	*\$M*

Impact-Linked Carry Model

The Current Research on Market Progress on ILC and Barriers to Adoption

The carry model outlined in Table 1 can be modified to be more suitable for impact-focused investors. This was the subject of the authors' qualitative research study where subject matter expert interviews were conducted with some of the top global VC and PE funds. Through this, the authors were able to assess and document the current state of the market on Impact Linked Carry and understand perceived barriers to adoption.

Qualitative Research Methods

The authors conducted over 25 practitioner interviews with (i) leading impact investment funds globally that either have or have considered implementing impact incentives, (ii) interest organizations such as the Global Impact Investment Network (GIIN), the Global Steering Group for Impact Investing (GSG), the Impact Management Project (IMP), and the Predistribution Initiative, and (iii) leading academics in impact investments, finance, accounting, and incentives at the Harvard Business School.

Research Findings

The findings from the authors' qualitative research indicated that a small but increasing number of players across the world link their carried interest to their impact goals (see Table 2). Vox Capital and GAWA Capital have built models that they have used for years and are in the process of refining them into new vehicles. Through the research several newer funds that are linking carry to impact were identified: The Drawdown Fund, Ocean 14 Capital, Norrsken VC, and several other impact funds have all implemented some form of impact carry. Notably, the European Investment Fund (EIF), a European Union Agency encourages all the impact mandate GPs in which it invests to implement impact-linked carry using their Gamma Model (Grabenwarter 2013).

Findings also indicated that in a parallel market, Development Finance Institutions (DFIs) such as the CDC Group and the IFC have been incentivizing mission-aligned investments for many years. Their models differ significantly because their public sources of capital prohibit them from rewarding investment professionals with carried interest. In light of those challenges, they have, however, developed detailed frameworks for measuring impact potential and outcomes and rewarding their teams accordingly, which private sector firms could learn from.

Table 2. Example models in the market.

Fund Manager	**Type and Theme**	**Impact Measurement**	**Impact-Based Incentive Structure**
The Drawdown Fund	Climate focused market rate growth equity	Proprietary methodology based on the Drawdown Project	Ex-ante CO_2 emission reduction targets. Measured on exit
GAWA Capital	Mix debt- and equity provider to MFIs. Focused on building opportunities in low-income communities, used debt & equity	KPIs based on readiness to delivery social impact, verified by third party auditor	Half of carry is linked to impact performance, measured by both ex-ante binary goals and quantitative outcomes
The European Investment Fund (EIF)	EU agency LP. Provides financing to SMEs in Europe via banks and funds	Helps set bespoke KPIs for each investment of their GPs	Suggests all GPs it invests in to implement impact-linked carry
Norrsken VC (EIF is an LP)	Market rate impact VC fund investing across environment, education and health	Bespoke KPIs set for each investment and independently verified, LP committee approves	100% of carry linked to achievement of impact goals, 60% impact goal threshold to receive any impact carry (starting at 50% and then sliding scale)
Ocean 14 Capital (EIF is an LP)	Impact growth fund. Invests in venture and growth companies with sustainable solutions for oceans	Sets KPIs that are linked to fund-level impact goals, e.g., end over-fishing, LP committee approves	30% of carried interest tied to achievement of 1–5 impact KPIs per investment
Vox Capital	Impact VC. Invests in innovative solutions for low-income Brazilians	Determines a GIIRS score and sets an ex-ante target for improvement	Half of carry tied to achievement of GIIRS target scores

Note: This is not an exhaustive list of models.

When asked why they had implemented impact carry, the overwhelming response from managers who had implemented impact carry when asked why was that it was the logical step for a fund with a clear mission that was already managing impact well. Managers from Vox Capital expressed that "impact and financial returns can and should go hand in hand. It was natural for us to link impact to incentives and very much in line with our strategy." Additionally, some managers believed their funds had a strong signaling effect to investors and potential portfolio companies. Interestingly, they believed that it did not significantly increase their ability to attract talent (Norrsken VC: "everyone was here for impact anyway") but was helpful for general alignment in the team.

Biggest Barriers to Implementation

The authors asked those who have not implemented impact-linked carry what they saw as the biggest hurdles and they asked those that had implemented impact-linked carry what they were most concerned about before implementing. Findings largely indicated that there are three primary barriers to implementing impact-linked carry: (1) sound measurement of impact, (2) administrative burdens, and (3) developing an adequate incentive program to motivate the right behaviors.

Sound Measurement of Impact. The first and most common answer was an issue that the industry is generally struggling with—a general lack of robust management and measurement of impact. Most firms have clear conceptual understandings of what type of impact they are trying to create but lack quantitative means to measure and manage that impact. While all managers interviewed had identified or built an impact framework for themselves, few set ex-ante targets when underwriting investments, and only some of them actively managed and monitored relevant KPIs. Furthermore, the frameworks used varied from industry standards such as GIIRS or IRIS+ to fully in-house developed measurements and management tools.

Even in a scenario where general KPIs and targets could be agreed upon for a particular investment up front, there was a real concern that those KPIs could change in scope or scale during the investment period. "If I tie my carry to a specific impact KPI such as low-income students reached, and then realize that program completion rate is a better way to understand outcomes, how do you account for that?" An environmentally focused fund manager also mentioned the potentially skewing effects of mergers and acquisitions (M&A) or technology breakthroughs: "If I grow a company through M&A, and all of a sudden my CO_2 reduction is twice the level it used to be – should I then be rewarded for business growth or penalized for the increased GHG emissions?"

Complexity of Implementation

The second issue any managers alluded to was added complexity. The calculation and allocation of carried interest is already an administrative burden on many firms and adding an impact layer would make this an even more complex calculation. In impact funds that are part of a broader platform, the question of how to connect impact carry into the overall carry model across funds was mentioned multiple times.

Additionally, many firms struggled to synthesize impact across investments, making fund aggregation of impact carry a challenge. As one manager put it: "how do I aggregate and compare impact KPIs for a dyslexia software company with the impact of a CO_2 emissions measurement tool? And how do I decide which one is more impactful?"

Market rate return funds also struggled with how a commercially driven LP base would react to carry not tied to financials. Many worried that an LP base used to thinking of impact and financial returns as mutually exclusive or at least not fully collinear, would react negatively to an impact carry model. When asked, none of those firms had actually discussed the option with their LP base, however.

Distorted Incentives. Finally, some managers admitted that they feared that impact carry and bonuses could distort incentives, creating perverse second order effects. Impact carry by itself could drive behavior designed to "game the system" through impact washing or optimizing for specific KPIs only. As a representative of an interest organization put it "incentives are a powerful tool – you need to be very careful with how you use them".

Approach for Impact Linked Carry for a Single Investment

This chapter proposes a model which ties some of the carry pool to impact metrics. The model is based on the best practices of practitioners. Early and leading examples include GAWA Capital, Vox Capital, and funds supported by the European Investment Fund which includes Ocean 14 Capital. At the investment level, it is a straightforward addition to an existing carry structure. The model is best explained through a hypothetical example (see Fig. 2) in which a specific investment has given financial returns yielding $100 into a carried interest pool.

As illustrated in Fig. 2, the starting point is the same: the financial performance of an investment contributes to the total available carry pool, here an illustrative $100. The total carry pool is then divided and allocated into a financial bucket and an impact bucket. In this example, the carry is split 50/50 between financial and impact success. Fund managers can choose the allocation which suits their fund. Impact measurement is a large and complex subject, and it is assumed that the impact of a fund's investment can be objectively measured; that target can be set ex-ante (i.e., based on forecasts); and that target achievement can be measured on a scale between 0 and 100% on an annual basis. 'Financial only' or 'non-impact' carry goes directly into the final carry pool. The remaining portion, 'impact carry', is subject to impact performance.

For example, let's examine a single investment, where the impact KPI is metric tons of CO_2 equivalents reduced, and the target set ex-ante is 15 metric tons by the time of exit. If at the exit, the company has achieved 12 metric tons of reduced CO_2, that would translate into an impact score of 0.8. With that impact score at exit, the investment team will earn 80% of the impact carry. The 'unearned impact carry'

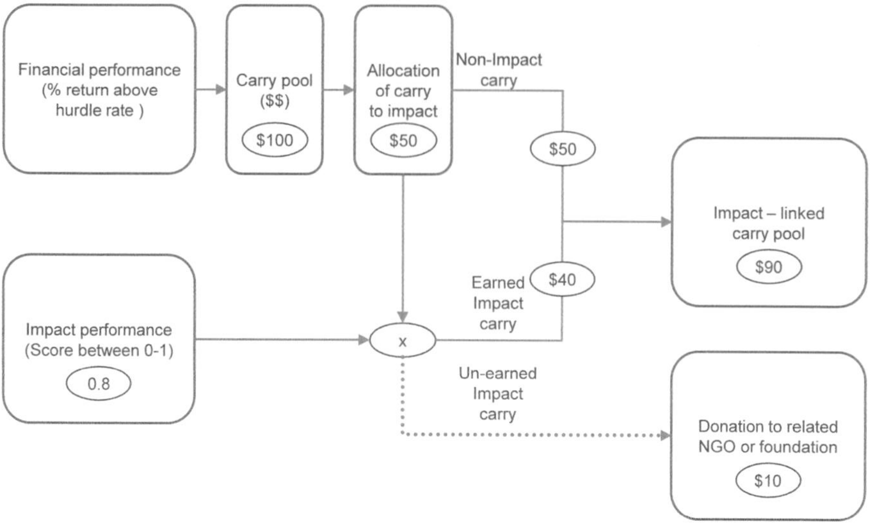

Figure 2. Impact-linked carry model for single investment.

is the 20% of the carry pool which the investment team does not receive because they did not hit the impact goals. This would be donated to a related non-profit organization in recognition of the missed impact.

Assumptions

The model makes a few important, sizable assumptions about the fund and its operations. The first assumption is that the fund already has a process for calculating and attributing the financial carried interest to the GPs. The second assumption is that the fund has a robust impact management and measurement system, and a structure in place to set and track targets. This chapter does not recommend a particular structure or framework for impact management, but instead has developed a checklist for a robust impact-measurement system that will plug directly into the proposed model:

- Impact measurement KPIs should be research-based and material for the investment in question.
- Measurement should be based on a recognized standard or framework.
- Targets for impact achievement are set ex-ante at underwriting and monitored at least annually.
- Impact KPI targets should be easy to understand and range between 1–3.
- Clear collinearity between financial and impact targets, so that both scale together.

These should be part of a broader impact strategy pursued by the GP, and a new impact manager should spend the majority of their resources determining their impact definition, model, measurement, and management.

Customizing the Impact Carry Model

The model in Fig. 2 is intentionally simple but there are design elements that the implementing fund manager can customize. These design choices are listed below with a discussion of the pros and cons of each option based on the SME interviews.

1. Set the Share of Total Carry Tied to Achieving Impact Goals. First, recall that financial returns will be the starting point for any calculation – if a firm delivers financial results below its hurdle rate, there will be no carried interest to allocate. From there, the manager can choose the proportion of this carry pool that is paid out based on the achievement of impact goals. The authors have seen examples that range from 10% to 100%, where the main decision factor was the strength of the signal to the market and employees. The rationale for choosing a smaller portion – say 10% – is that it is sufficient to drive the desired behavior. When asked about weighting, one practitioner stated: "10% is enough for people that care about impact." Those choosing to link half the carry to impact present that argument of a dual mandate. The equal share reflects the equal importance of impact and social returns.

Some other funds chose to link 100% of the carry to impact to demonstrate strong commitment to their mission and their ability to source truly collinear investment opportunities. If investments are not collinear and 100% of the carry pool is linked to impact, there is an outside chance of creating a perverse incentive. In the case that an investment is underperforming on its impact KPIs and 100% of carry is linked to impact there is a risk that an investment team might disengage and risk delivering lower financial results. However, the chance of this seems low in an impact-first fund and if investments are truly collinear.

2. Decide How to Aggregate Investment at the Fund Level. The model described in Fig. 2 shows how Impact Linked Carry is calculated for a single investment. However, a fund makes multiple investments over its lifetime and traditionally the carry pool combines the carry of some or all those investments. Therefore, the impact-linked carry from each investment should be aggregated at the fund level, in the same way that financial carry is aggregated. Every impact target and achievement rate can be converted into a score between 0 and 1 to allow for comparison across investments.

For a single investment, let's refer to the example posed at the beginning of this section. In this example, the impact KPI is metric tons of CO_2 equivalents reduced, and the target set ex-ante is 15 metric tons by exit. If at exit, the company has achieved 12 metric tons of reduced CO_2, that would translate into a score of 0.8. This could also easily be averaged out across KPIs if there are multiple selected at entry. See Fig. 3 for an example. The weights for a single investment can be decided by the GP in collaboration with management.

The impact achievement scores would then be aggregated at the fund level through a weighted average of the different investments. The most common "weight" used in practice to determine the proportion a particular company takes at the fund level is the capital deployed in that investment. For instance, if an investor

Company X KPIs	Impact achievement	Weight in total	Weighted KPI achievement
KPI 1: Metric CO2 tonne equivalent reduction	0.8	50%	40%
KPI 2: Annual B impact Assessment in place	1	35%	35%
KPI 3: At least 3 female/ minority board members	0.67	15%	10%
Total	**0.85**	**100%**	**85%**

Figure 3. Aggregating the KPIs of an Individual Investment.

has 5 investments in a $250 million fund, and each investment is $50 million, the proportionate weight to each company would be 20%. Figure 4 below shows an example of a hypothesized impact achievement and consequent carry payment in a fictional fund.

An issue with this methodology is that it disconnects the financial returns of individual investments from their impact score. Imagine a situation where 4 of the 5 companies go bankrupt, and thus do not contribute to total carry, and do not achieve their impact goals. The last company, however, outperforms significantly, both financially and from an impact perspective. If capital deployed is used to weight the impact score, GPs would receive a large percentage of the commercial-linked carry but only a very limited portion (20%) of carry that is linked to impact.

Investment	Impact achievement	Capital deployed as share of total	Proportion impact carry paid out
Company 1	0.8	20%	16%
Company 2	1	20%	20%
Company 3	0.7	20%	14%
Company 4	0.9	20%	18%
Company 5	0	20%	0%
Total	**0.68**	**100%**	**68%**

Figure 4. Weighting the impact score by capital deployed.

The alternative solution is to use contribution to carry pool instead of capital deployed to align incentives for financial and impact return. Using contribution to carry pool as a weight is particularly relevant to earlier stage investors, who often bet big on 10 investments with the hope that 1 will "return the fund". The model becomes more aligned with financial performance and will likely have more volatile weighting. An illustrative example is in Fig. 5.

Investment	Impact achievement	Capital deployed as share of total	Contribution to carry pool as share of total
Company 1	0.8	20%	40%
Company 2	1	20%	30%
Company 3	0.7	20%	30%
Company 4	0.9	20%	0%
Company 5	0	20%	0%
Total	**0.83**	**100%**	**100%**

Figure 5. Weighting the impact score by contribution to carry pool.

3. Ensure Third-Party Verification and KPI Adjustment. All GPs that the authors spoke to who had implemented impact carry, mentioned the importance of third-party impact verification to avoid "impact washing". In other words, an independent body can help to set and verify the scope and scale of targets, quality-check chosen KPIs, and challenge the realism of goal achievement. There are two main options for this third-party: (1) an independent impact auditor chosen by LPs to assess impact goals and achievements, or (2) a committee of key LPs that need to approve suggested targets ex-ante at the time of investment.

In the case of a third-party auditor, several consultancies provide impact verification. In the interviews the authors conducted they heard examples of annual audits to track progress, and simple comparisons of pre-investment and exit audit. The involvement of the auditor can also be limited to verifing GP-calculated impact scores. In this model, the GP controls most KPIs and impact targets, with auditors verifying rather than determining.

The second option, a committee of key LPs, is more collaborative but can also drive a higher administrative burden. This is the model chosen by the European Investment Fund, where LPs play a more active role both in setting KPIs and determining absolute levels. The LP committee (usually called the AdCom) approves the impact scope and scale ex-ante for each investment and meets annually to discuss achievement for each investment. A strength of the model is that it also allows for the AdCom to approve of changes during the holding period, and there is room for discretionary agreements in very special situations such as M&A, business model pivots, and hard to measure impacts.

One interviewee argued for both – having an LP AdCom as a steering group but the actual verification of targets and achievements be conducted by a third party. In practice, smaller and newer funds may not have the resources for external audits. There are other options they can explore to reduce administrative burden, for example (and discussed later), IWEF has opted for external audit at exit and Investment Committee approval for KPIs.

4. Assess Achievement Annually or at Exit. The choice of when to evaluate impact will be influenced by the choice of KPI and the impact measurement system. The models had one of two ways to measure impact goal achievement:

1. A simple measurement at exit (i.e., 0.8 achievement of the target at exit).
2. A weighted average of impact achievement across years (e.g., a score of 0.8 in year 1, 0.9 in year 2, 1.0 in year 3 would result in an average impact score of 0.9).

Regardless of the option chosen, all GPs still measured their portfolio companies on an annual basis for the chosen KPI(s). Option 1 is the simplest but option 2 allows for changing impact goals in scope or scale over the life of an investment. For some cases, option 2 may be better. For example, in the case of climate related KPIs, it may be beneficial to achieve impact goals sooner because the impact of CO_2 on the environment compounds.

5. Set Impact Carry Thresholds. One of the more technical design choices is whether or not to include minimum and maximum thresholds for impact. In the author's base case, the portion of the impact carry distributed is directly proportional to the impact score. For example, if the impact score is 0.8 and the impact carry pool is $50 then the distributed carry is $40, and if the score is 0.4 the distributed carry is $20. A variation on this (seen at, e.g., Norrsken VC) is to introduce minimum and maximum thresholds as Fig. 6 demonstrates (Shieber 2021). If the impact score is less than 0.6 none of the impact carry is distributed, and any score above 0.8 receives full carry. Between 0.6 and 0.8 there is a linear sliding scale.

The main argument against this variation is the added complexity. Where possible, this chapter advocates for keeping the incentive system as simple as possible to reduce distraction and increase the likelihood that it will drive desired behaviors. However, there are some strong arguments in favor of this variation. First, it protects strongly against impact downside. Since half the carry will be lost if the impact score is less than 0.8, the GPs are highly motivated to put impact performance on the agenda. In the base case, there is a possibility that the investment team might focus more on the financials (which determine the overall size of the carry pool) and not push the management teams on impact goals. Second, this approach acknowledges that there is (at least for now) some imperfection in impact measurement and makes the incentive instrument a bit less granular; it rewards behavior that is broadly in the right direction but also leaves room for imperfect KPIs.

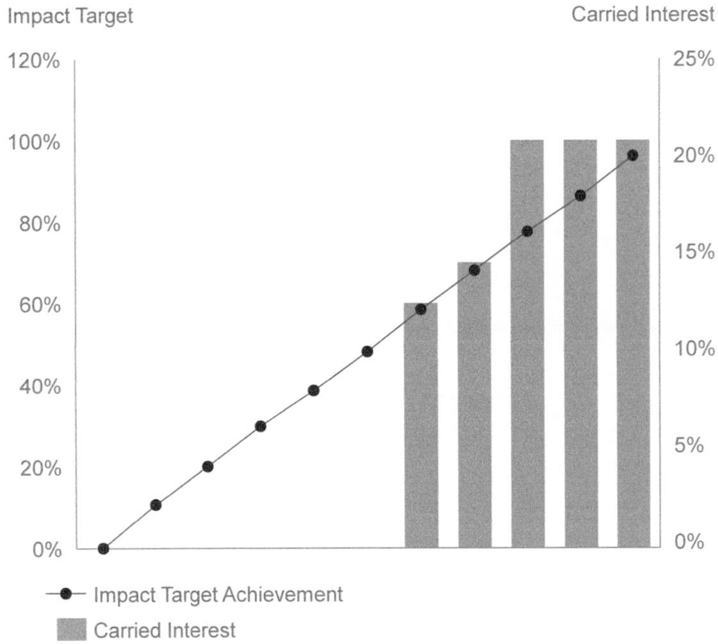

Figure 6. Introducing a threshold before impact carry is achieved.

6. Manage for Impact Upside. The proposed model only works as a 'stick' and not a "carrot," in that it punishes for underperformance in impact, but does not provide higher carry levels if impact goals are reached or exceeded. There are two compatible approaches for managing impact upside. The first (Option 1: Adjusted Stick) holds the total carry pool fixed (at, e.g., 20%) and allows a low impact score in one investment (e.g., 0.5 achievement) to be compensated with a high score on another investment (e.g., 1.3 achievement). The second (Option 2: Carrot) expands the total size of the carry pool (to, e.g., 25% or 30%) in the case of achievement of financial *and* impact outcomes. Below each option is expanded in detail.

Option 1: Adjusted Stick. The traditional carry model incentivizes outsized financial performance; better returns mean the absolute carry pool is larger. In addition to penalizing missed impact goals, fund managers may also choose to reward impact outperformance when they aggregate impact at the fund level. Whilst the absolute size of the carry pool is fixed by the financial performance, the low impact score of one investment could be offset by an excess impact score of another investment. As an example, imagine a fund with only 2 investments, both with the same initial investment and contribution to the carry pool. Investment 1 scores 0.7 at exit and investment 2 scores 1.3 (i.e., 30% greater than the target). The +0.3 overperformance could count towards the fund's total average impact score, as opposed to capping target achievement at 100%. Of the managers that did this, they all had a cap on overperformance, typically at 15–30% above target (1.15–1.3).

The approach of blending impact scores at the fund level is an approach commonly employed by development finance institutions to increase the scope

Scenario 1: No reward for outperformance

Investment	Impact achievement	Contribution to carry pool as share of total	Proportion impact carry paid out
Company 1	0.7	50%	35%
Company 2	1	50%	50%
Total	**0.85**	**100%**	**85%**

Scenario 2: Reward for outperformance

Investment	Impact achievement	Contribution to carry pool as share of total	Proportion impact carry paid out
Company 1	0.7	50%	35%
Company 2	1.3	50%	65%
Total	**1**	**100%**	**100%**

Figure 7. Comparing options for rewarding outperformance on impact.

of possible investments. In the same way, a portfolio allows for investments with different financial risks, it allows for investments with different impact risks. Figure 7 shows an example showing the potential improvement in financial outcomes for the GP when allowing for overperformance.

Option 2: Carrot. None of the funds interviewed had agreements with their LPs that would allow for a so-called "super carry", whereby the carry pool is increased if both financial and impact goals were exceeded. However, Aureos Capital's 2009 Africa Health Fund introduced a carrot that increased the size of the carry pool from 20% to 30% of fund returns, with 15% tied to financial returns only and the remaining 15% tied to the achievement of impact outcomes. The impression from both GPs and LPs was that expanding the total carry pool was unlikely to happen since "GPs are already seen as being overpaid by most LPs".

7. Donate Non-Achieved Impact Carry. It is recommended that any portion of the impact carry that is not achieved be donated to a third-party NGO with an arms-length distance from the GP. In other words, the carry is paid out in full, but parts of it go to GPs and the remaining part is distributed to a third party. GPs may choose to nominate an organization that is closely aligned with the fund's mission or an industry organization (e.g., an organization focusing on impact measurement). The rationale is that some social goal was missed and that the missed financial reward should in some way contribute to "offset" that miss. Almost all funds interviewed have or plan to adopt this approach. Most conclude that the alternative, returning the missed carry to LPs, could create skewed incentives whereby LPs want the GPs to miss their impact goals to pay less total carry.

A Proof of Concept Case Study: Indonesia Women Empowerment Fund (IWEF)

Indonesia Women Empowerment Fund (IWEF)[13] is an impact investment fund jointly managed by Moonshot Ventures and YCAB ventures. The targeted 10 m USD fund invests small checks in startups at pre-seed and seed stages. The fund aims to address the missed economic opportunities for women in Indonesia, which is the most populous country and largest digital economy in Southeast Asia. In Indonesia, 90 million women face barriers preventing them from reaching their full economic potential and the country performs lowest in Southeast Asia on several measures of women's participation in the workforce (McKinsey Global Institute 2018). If women can be brought into the economy, the gross domestic product is expected to increase by 9%, relative to business as usual. IWEF aims to capture the opportunity this challenge presents by investing in technology-driven solutions that seek to address systemic barriers to women's economic opportunities and livelihoods.

IWEF is among the first venture funds in Indonesia to adopt a Gender Lens Investing (GLI) approach. GLI is an approach that intentionally takes into account gender-based factors across the investment process to advance gender equality and better inform investment decisions (GIIN n.d.). GLI is generally practiced by investing in women-led and women-owned enterprises, investing in firms enabling diversity in the workplace and supply chains, or investing in firms building products and services that bridge gender gaps in society. For IWEF, its gender lens comes from investing in Indonesian firms that are women-led (women co-founders must be allocated a minimum of 20% of founders' shares) *and* that are building scalable solutions that aim to have a positive impact on women's ability to achieve economic security.

Furthermore, IWEF's team is gender-balanced, with over 50% being women. Rooted in the opportunity to improve economic opportunities for women, the fund's investment thesis relies on partnering with such ventures where revenue growth is achieved by providing solutions generating more or better impact for women. This collinearity allows the fund to pursue investments that are positioned to simultaneously generate superior impact and financial returns. For example, IWEF invested in Binar Academy,[14] an ed-tech startup offering accelerated vocational courses and a job-matching platform that enables employed women in Indonesia to access better paid jobs in the country's fast-growing digital economy. The firm's revenue generation (through its vocational courses and job platform) is in alignment with the impact goal of helping more women upskill and access better jobs.

Through its investments, IWEF aims to contribute to improving livelihoods for 2.5 million women, by making up to 100 pre-seed and seed investments in firms that fit its investment thesis. In the short term, IWEF's impact objectives are to increase access to better jobs (superior pay and/or flexibility) for women, improve access to essential services (like education, health, financial services, etc.), and make it easier

[13] Indonesia Women Empowerment Fund: https://www.moonshotventures.org/.

[14] Binar Academy: https://www.binaracademy.com/.

for women to launch, run, and grow their businesses. Over the long run, the fund expects to contribute toward improving women's livelihoods and nurture a startup ecosystem that is more inclusive to women. The fund aims to capture impacts in the form of an increase in income for women, the number of quality jobs created, and the number of women accessing services that improve their earning potential.

Within fund management, Impact Measurement and Management (IMM) is the emerging practice of capturing and reporting on the impact that a fund may generate via its investments and operations. At its broadest level, it includes setting fund goals, defining strategies, defining relevant metrics and targets, and reporting on those metrics to manage performance (GIIN n.d.). For IWEF, impact measurement is focused on the fund's gender goals, and is approached in an integrative manner, being embedded throughout IWEF's decision making, thus serving to plan and guide outcomes.

As such, gender metrics guide the IWEF team throughout the investment cycle. To start with, metrics are evaluated regarding the extent of women's leadership in a startup, as a minimum eligibility requirement for investment. Next, a startup is considered on its potential to help drive the fund toward its gender targets, in parallel to its financial goals. Then, after an investment has been approved, metrics are captured and considered regarding the degree of co-investment from other investors into a startup, as a way of measuring success in IWEF's goal of catalyzing other investors toward GLI. After the investment, gender metrics and targets are agreed upon with a startup's leadership team alongside periodic operational goals, which in the previous illustration of Binar Academy might include the number of women admitted and also successfully completing its courses or being hired into better paid jobs within six months of graduating.

More broadly, these goals include metrics and targets around gender equality in the workplace and a startup's supply chain, as well as related to its customers and beneficiaries in the context of improving livelihoods for women. Examples of metrics include the percentage of women in the workforce and the share of female suppliers. Follow-up investment from IWEF may be contingent upon progress toward these performance targets. The metrics are captured and assessed several times a year and reported to IWEF's capital providers.

Objectives of Adopting Impact-Linked Carry

IWEF's investment thesis is centered on the notion that investments in firms that enable women to access better economic outcomes will also generate better financial returns. The fund views impact-linked carry as a tool that incentivizes investment managers to identify, invest, and support such firms where financial returns are in sync with better economic opportunities for women. For this reason, impact-linked carry calculations are based solely on the impact that products and services of portfolio firms create for women consumers. Simply the presence of women owners or women in leadership positions is not sufficient to unlock impact-linked carry.

By linking carry to impact, the fund aims to signal to entrepreneurs, and potential capital providers to IWEF, its intentional commitment toward investing for impact.

To drive the whole team toward a shared mission, IWEF adopts the practice common among European PE/VC funds, where the carry is calculated based on the full fund's financial performance, instead of calculating carry on a deal-by-deal basis as is often done in the United States.

Challenging the Status Quo in Venture Capital

Furthermore, IWEF aims to challenge the status quo in venture capital, where fund managers focus solely on optimizing for maximum financial returns. By adopting impact carry, IWEF hopes to offer a proof of concept to other fund managers, especially those investing with a gender lens, on how they can be incentivized to achieve both superior financial and social returns at the same time.

Principles for Design Choices.

Given the fund's strong commitment to impact and its relatively small size, all the decisions involved with building the impact-linked carry model are based on three principles:

1. The design choices must drive the fund toward achieving impact outcomes.
2. The model should consider the context of seed-stage venture capital investing where IWEF is often the first institutional investor. Most ventures at this early stage have limited resources and lack robust monitoring systems.
3. Simplicity is key; the design choices are made to reduce complexity while ensuring completeness. Furthermore, the fund intends to update the model and its assumptions over the fund's duration based on its experiences and changing market conditions.

Design Choice 1: Share of Total Carry Tied to Achieving Impact Goals. IWEF has chosen to link 100% of its carry to achieving impact goals. This decision is rooted in the fund's commitment to generating both financial and impact returns. Typically, the existence of a carry pool in itself is an incentive mechanism for fund managers to pursue financial returns. If the fund does not generate returns beyond the hurdle rate, managers do not unlock any carry. In extension to this practice, IWEF views that the same level of incentive must remain for achieving impact outcomes. That is, fund managers shall not enjoy any carry if impact returns are not achieved. On the other hand, all the carry should be unlocked only when both the expected financial and impact outcomes are achieved. IWEF achieves such alignment in incentives by linking 100% of the carry to impact. This drives the team to invest in such opportunities that are poised to generate high financial returns (to generate a sizable carry pool in the first place) and high impact returns (to access the carry pool).

There remains a concern that 100% linkage to impact may create a moral hazard where the fund managers sidestep the pursuit of financial returns. IWEF's argues that its mandate is to invest in only such firms where impact and financial returns are collinear. Then, the focus on impact returns will also serve as the incentive for investors motivated by financial returns. While 100% linkage to impact is still rare, IWEF aims to test the model and serve as a blueprint to funds that aim to be held fully accountable for their impact goals.

Design Choice 2: Fund Aggregation Weighting. IWEF assesses the impact created by portfolio firms along two dimensions: breadth (measuring quantity of impact) and depth (measuring the quality of impact). Scores in each dimension, measured on a 0–1 scale, are multiplied by 50% of the fund's carry pool to calculate impact-linked carry.

Impact-linked carry = (0.5 x carry pool x breadth score) +
(0.5 x carry pool x depth score)

Breadth is a portfolio-level measure of the extent to which the fund has achieved its target of enabling 2.5 million women to improve their economic livelihoods. The aggregate number of women served by all portfolio firms is divided by the above target to calculate the breadth score. For example, if the fund has served 2 million women, it would achieve a breadth score of 0.8.

Breadth score = (No. of women served by all portfolio firms)/(2.5 million)

For each firm in the portfolio, IWEF identifies depth metrics that best reflect the extent to which the firm's products and services enable women to achieve better economic outcomes. Impact targets are set for each firm at the time of investment and are tracked regularly. For example, an EdTech startup like Binar Academy might include self-reported depth metrics for women graduates that are aligned with standard KPIs for the coding school sector, including the percentage who find full time employment related to their newly learned skills within 6 months of graduation; the increase in salary upon placement within the tech industry and salary levels compared to industry benchmarks; and Net Promoter Scores[15] related to overall satisfaction, learning experience, and expected outcomes.

When a firm exits, its aggregate impact performance is measured against the target to calculate the firm-level depth score. These scores are then aggregated at the portfolio level and weighted by each firm's contribution to the carry pool, to calculate portfolio level depth score.

Firm-level depth score = depth impact(s) achieved/depth impact target(s).

Portfolio-level depth score = Σ (*firm's contribution to carry pool/
total carry pool*) *x firm level depth score*

Design Choice 3: Third-Party Verification and KPI Adjustment. The fund will seek approval from the independent members of its investment committee while setting ex-ante impact metrics and KPIs. Given the nature of seed-stage investing, some portfolio firms may pivot and target different impact outcomes at later stages. To allow for such a dynamic nature of early-stage firms, independent members of the investment committee will also approve any changes to impact KPIs during the

[15] What is Net Promoter? https://www.netpromoter.com/know/.

holding period. As the fund intends to return unrealized carry to LPs, a sub-committee of LPs deciding on the impact targets may lead to a conflict of interest. For this reason, IWEF relies on the independent members of the investment committee, who do not receive carry and have no conflicting interests.

In addition, the fund plans to commission an independent third-party audit to verify the data integrity and calculate impact achievements against the set targets whenever a firm in the portfolio experiences an exit or liquidity event. An independent auditor will also verify and validate the impact-linked carry calculations at the end of fund life.

Design Choice 4: Achievement Assessment – Annual vs. at Exit. IWEF has opted to assess the impact achievements of each firm against its targets at the firm's exit event (M&A, IPO, management buy-out, etc.) for 2 reasons. First, it is aligned with IWEF's principle of simplicity. Second, as discussed earlier, IWEF's target ventures are typically nascent and annual assessment would add an unnecessary burden early in their growth. In the event of a failed venture, impacts are not assessed within the context of impact-linked carry because they make no contribution to the carry pool.

Design Choice 5: Impact Carry Thresholds. Abiding by the principle of keeping the model simple to use, IWEF does not set any minimum or maximum thresholds for impact carry. Carry will be unlocked on a linear scale for both breadth and depth dimensions.

Design choice 6: Managing for Impact Upside. IWEF's carry pool is fixed at 20% of the capital gains beyond the hurdle rate. The fund thus adopts an "Adjusted Stick" approach where a low impact score of depth measure in one investment will be compensated with a higher score of depth measure in another investment. The model also allows for such compensation between the two dimensions (breadth and depth) where underperformance in one can be compensated by the other until an upper limit of 15% (so, ¾ of the total 20%).

Design choice 7: Distribution of Non-Achieved Impact Carry. At the end of fund life, IWEF plans to return to LPs the portion of carry that may remain unclaimed because of not achieving set impact goals. Specifically, impact-focused LPs who provided grant or concessionary capital will receive such claims in proportion to their original commitments. If there remain any proceeds from the unclaimed impact-linked carry after this, the fund aims to donate that capital to one or more non-profit organizations selected by IWEF and approved by the external members of the investment committee. Through this decision, IWEF aims to signal its impact intentionality to impact focused LPs by offering them some sort of "money back guarantee".

Implementation Challenges and Learnings

Implementation of the ILC model at IWEF is still in the early stages, so it is not yet possible to provide in-depth learnings. However, several challenges were anticipated in designing the Impact Linked Carry model. Chief among these challenges was determining how to measure impact across a range of very different sectors and with the resource constraints of a small fund. In seeking solutions, IWEF was inspired by the Lean Data approach championed by 60 Decibels,[16] including the use of Net Promoter Score (NPS) to gauge beneficiaries' self-reported impact. Based on this, IWEF plans to capture impact metrics periodically via short market-research styled surveys that are administered by startups among a representative sampling of their customers or beneficiaries. The surveys will include NPS questions that allow comparability between startups in different sectors, as well as questions specific to a startup's product and the 'depth' of its impact in improving women's livelihoods to allow for contextualization. This approach has numerous advantages, including cost-effective data capture and analysis, while also providing valuable market insights to a startup's management team. Startups that gain value from such surveys are also less likely to view them as burdensome and will be more incentivized to ensure the quality of the data. Furthermore, costs can be controlled by engaging third party independent verification of the data and analysis only for the subset of startups that exit and otherwise contribute to the carry pool.

Another anticipated challenge was the need to balance rigor versus ease of implementation. The model needed to be rigorous enough to be effective, but also simple enough for fund teams and capital providers to understand. On one hand, it was important to carefully foresee the model's unintended consequences on the fund team's investment decisions, while also providing sufficient veracity to the intended impact goals and reported outcomes in alignment with the objectives of investors in the fund. On the other hand, the IWEF team had to consider the steep learning curve involved with implementing a model that is not yet standardized. As an early adopter with an intent that its approach should encourage wider adoption, IWEF's impact-linked carry model favors ease of use and practicability. This enables future proponents to build upon the current model and refine it further.

Conclusion

As the industry grows and concerns about impact washing continue, impact funds may choose to link their incentive structures to impact metrics to signal their commitment to impact and reinforce rigorous impact strategies. While impact measurement remains a challenging topic, several players have successfully implemented impact-linked incentives. Findings from the authors' qualitative research led to the conclusion that there is interest in impact-linked carry but a lack of clarity on the approach, and a misguided belief that it would create "mind-boggling levels" of complexity.

[16] *60 Decibels* is a consulting firm that provides services around impact measurement, which was spun-off from Acumen in 2019. It champions customer-centric and more accessible approaches to impact measurement. See: Acumen, June 05, 2019. "Acumen Launches 60 Decibels to Make Lean Data an Impact Measurement Standard for Impact Investing". https://acumen.org/blog/acumen-launches-60-decibels/.

The proposed model is designed for fund managers who are looking to implement impact-linked carry. The model is simple to implement (as an additional layer to an existing carry process) and can be customized at the GPs' discretion. There are several design choices, including the impact measurement approach, percentage of carry linked to impact, and recipient of 'unearned' carry. This chapter does not suggest that this model alone will revolutionize impact investing, but it is an important part of a robust impact strategy. A fund that incorporates impact-linked carry will need to integrate considerations of impact in every part of its investment process: from identifying key impact goals early on in sourcing, quantifying impact in due diligence, setting clear goals (and targets) when underwriting, and ensuring robust tracking throughout the holding period and at exit.

Of course, this approach is not foolproof and would likely face considerable challenges if applied in an organization that lacks intentionality and a strong commitment. But for a fund with a clear impact strategy and vigorous measurement and management, this is a powerful tool for focusing the investment team on their dual mandate and signaling the fund's commitment to impact outside parties.

Indonesia Women Empowerment Fund offers a useful case study. As a fund investing in technology-enabled solutions that improve economic opportunities for women, its investment thesis is centered on the notion that impact and financial returns are collinear. By linking 100% of its carry to impact, the fund has built a strong incentive mechanism to identify opportunities that are poised to generate top-tier financial *and* impact returns. It sends a clear signal to investors and other stakeholders about the fund's commitment to impact. The fund has also adopted a novel aggregation and weighting method to prioritize different kinds of impact. Furthermore, by committing to return potential unclaimed carry to impact-focused LPs, the fund has leveraged impact-linked carry as a mechanism to identify partners who are in full alignment with its impact objectives. This case shows how different components of the impact-linked carry model can be tailored to suit a fund's impact objectives. Note also that the design choices mentioned in this chapter offer a useful starting point, but fund managers can build their versions relevant to their context.

Academic research now recognizes that a firm's performance goes beyond financial returns to include outcomes for stakeholders. The previous chapter (Sandoval and Holladay) highlights how new ventures can go about measuring and improving "non-financial" outcomes. Several studies are being conducted to identify the determinants of a firm's social/environmental outcomes (Baldini et al. 2018), and their link with financial outcomes (Friede et al. 2015). However, research exploring compensation structures that drive startup teams or private equity and venture capital investors toward achieving such non-financial outcomes is still nascent. By introducing the conceptual framework that links investor compensation with impact, this paper provides a basis for future research studying the link between compensation structures and non-financial outcomes.

Furthermore, previous research on venture capital and private equity fund compensation agreements made efforts toward identifying equilibrium conditions for financial incentives of limited partners and general partners (Flor and Grell

2013; Litvak 2009). As impact investing is on the rise, compensation agreements should evolve to consider financial and non-financial incentives. The design choices outlined in this paper introduce several variables through which limited partners and general partners can establish an equilibrium in incentive alignment. As more fund managers adopt the impact-linked carry model, future research can study the structure of compensation agreements that best suit the evolving needs of investors. The influence of impact-linked carry on the nature of the relationship between fund managers and entrepreneurs is also an important avenue for future research.

Finally, this paper introduces impact-linked carry as a mechanism by which a fund manager is financially incentivized to achieve non-financial outcomes. However, it is not clear if financial incentives are the only solution to advance impact outcomes. Future research can explore alternative mechanisms that align investors with the act of impact creation.

In conclusion, the process of linking carried interest to impact is still not common, and several new models will emerge as fund managers find ways to incentivize their teams toward generating impact alongside financial returns. As one interviewee mentioned, "this is a new model that has not yet played out, and it remains to be seen what mistakes we have already made". The authors hope the impact-linked carry model introduced in this chapter serves as an open-source guide for firms and researchers intent on testing the mechanisms by which venture capital and private equity investments can drive more impact.

References

Baldini, M., Maso, L.D., Liberatore, G., Mazzi, F. and Terzani, S. (2018). Role of Country- and Firm-level determinants in environmental, social, and governance disclosure. *Journal of Business Ethics*, 150: 79–98. https://doi.org/10.1007/s10551-016-3139-1.

Butler, J. (2019, April 26). *Like Honey to a Bee – Why Do Millennials Like Purpose Driven Companies*. Jeff Butler. Retrieved June 1, 2022, from https://jeffjbutler.com/2019/04/26/why-do-millennials-like-purpose-driven-companies/.

Flor, C.R. and Grell, K.B. (2013). Venture capital budgeting - carry and correlation. *Journal of Corporate Finance*, 21: 216–234. https://doi.org/10.1016/j.jcorpfin.2013.02.004.

Friede, G., Busch, T. and Bassen, A. (2015). ESG and financial performance: Aggregated evidence from more than 2000 empirical studies. *Journal of Sustainable Finance & Investment*, 5(4): 210–233. 10.1080/20430795.2015.1118917.

Furlow, N.E. (2010). Greenwashing in the new millennium. *The Journal of Applied Business and Economics*, 10(6): 22.

Global Impact Investment Network. (n.d.). *Gender Lens Investing Initiative | The GIIN*. Global Impact Investing Network. Retrieved June 1, 2022, from https://thegiin.org/gender-lens-investing-initiative.

Global Impact Investment Network. (n.d.). *An Introduction to Impact Measurement and Management | IRIS+ System*. IRIS+. Retrieved June 6, 2022, from https://iris.thegiin.org/introduction/.

Grabenwarter, U. (2013, April 18). *The Gamma Model for Impact Measurement in Fund Investments*. SSRN. Retrieved June 1, 2022, from https://papers.ssrn.com/sol3/papers.cfm?abstract_id=2381129.

Hand, D., Dithrich, H., Sunderji, S. and Nova, N. (2020, June 11). *Annual Impact Investor Survey 2020* (10th ed.). GIIN. Retrieved June 1, 2022, from https://thegiin.org/research/publication/impinv-survey-2020.

Kohan, S.E. (2021, March 28). Customers Seek Purpose Driven Companies Creating A Rise In B Corps. *Forbes*. https://www.forbes.com/sites/shelleykohan/2021/03/28/customers-seek-purpose-driven-companies-creating-a-rise-in-b-corps/?sh=c6f1b306dd26.

Kreutzer, L. (2020, November 23). Bain Capital raises $800m for second impact investing fund. *Private Equity News*. https://www.penews.com/articles/bain-capital-raises-800m-for-second-impact-investing-fund-20201123.

Litvak, K. (2009). Venture Capital Limited Partnership Agreements: Understanding Compensation Arrangements. *The University of Chicago Law Review*, *76*(1), 161–218. http://www.jstor.org/stable/27654700.

McKinsey Global Institute. (2018, April). *The Power of Parity: Advancing Women's Equality in Asia Pacific*. McKinsey & Company. https://www.mckinsey.com/~/media/mckinsey/featured%20insights/gender%20equality/the%20power%20of%20parity%20advancing%20womens%20equality%20in%20asia%20pacific/mgi-the-power-of-parity-advancing-womens-equality-in-asia-pacific-full-report.pdf.

Mitchenall, T. and Stutts, J. (2021, July 27). Brookfield, TPG showcase allure of climate funds with $12bn raised. *New Private Markets*. https://www.newprivatemarkets.com/brookfield-tpg-showcase-allure-of-climate-funds-with-12bn-raised/.

Newbury, R. and Delves, D. (2020, March 27). *New Research Finds Progress on the use of ESG Incentive Metrics*. WTW. Retrieved June 1, 2022, from https://www.wtwco.com/en-US/Insights/2020/03/New-research-finds-progress-on-the-use-of-ESG-incentive-metrics.

Shieber, J. (2021, March 28). To improve accountability, Norrsken VC ties partner compensation to its portfolio's sustainable successes. *TechCrunch*. https://techcrunch.com/2021/03/28/to-improve-accountability-norrsken-vc-ties-partner-compensation-to-its-portfolios-sustainable-successes/.

U.S. Forum for Sustainable and Responsible Investment. (2020, November 16). *The US SIF Foundation's Biennial "Trends Report" Finds That Sustainable Investing Assets Reach $17.1 Trillion*. US SIF. Retrieved June 1, 2022, from https://www.ussif.org/blog_home.asp?Display=155.

Part IV

Concluding Remarks

Chapter 15

The Answer to Decreasing the Startup Failure Rate:

Human Capital

Nikki Blacksmith

Entrepreneurship is the backbone of many economies; driving the progress of civilization, addressing the world's most pressing problems, and creating new jobs (Drucker 1983). In the United States, successful entrepreneurship is largely dependent on venture-backed startups, not large corporations (Kaplan and Lerner 2010, 2016). In 2021, over 600 billion dollars were invested in startups in the United States and global VC-backed exits were over $1 Trillion (KPMG 2022). However, up to 90% of startups fail, rendering billions of dollars wasted each year (Patel 2015). Financial damage is just one consequence of the abysmal startup failure rate. When a startup fails, it inflicts harm on the founders — physically, emotionally, and financially— their employees, customers, and business stakeholders through layoffs, toxic cultures, and severed business relationships. For all of the money invested in startups, it seems irresponsible to let the entrepreneurial ecosystem continue operations as normal.

Scientists and practitioners in the entrepreneurship space should focus on the people in order to increase the startup's success rate. One promising approach to address the low success rate is through managing and optimizing human capital. Founders are typically taught to focus on product and business development when they are in the early-stage. However, this comes at an expense; people issues get pushed to the bottom of priorities. Because human capital management is not prioritized, people are often the root cause of the startup's failure (CB Insights

Blackhawke Behavior Science, American University, 1709 21st St NW #52, Washington, DC 20009, 630-888-8150.
Email: nikkiblacksmith@gmail.com

2021). Fortunately, organizational scientists have been studying people at work for over a century. Organizational research has led to evidenced-based practices for managing human capital (Kozlowski et al. 2017). It is now widely understood in the scientific community that people flourish in high-performance work systems. This volume introduces research, practice, and non-traditional measurement tools from industrial-organizational psychology that can be leveraged to increase startup success. The science, practice, and tools show that startup success is inherently multi-level. To drive startup success, we ought to focus on multiple levels of analysis: (1) the individual entrepreneur and the role entrepreneur; (2) the startup team and the social and environmental context in which entrepreneurs are embedded; and (3) the overall startup organization and the larger, entrepreneurial ecosystem. This concluding chapter is not meant to be a summary of the chapters, but rather it provides suggestions and direction for future science and practice.

1. Effective Human Capital Management Can Improve Performance

Many startups fail to grow or become profitable businesses. Despite working hard for their passions, many startups do not "perform" in the traditional sense of their startup either succeeding or failing. As the chapters in this book elucidate, performance problems in startups are often a result of human capital issues. Unfortunately, startup leaders rarely turn to human capital activities for value creation; such activities are typically viewed as a way to control costs or maintain legal requirements, taking a back seat to fundraising, developing products, and generating sales. This volume shows that startups that focus on thoughtfully managing their human capital are more likely to build high-performance work systems and gain a competitive advantage. Therefore, it is reasonable to conclude that if startups invest in effective human capital management, they can not only prevent problems from occurring but also establish another mechanism for value creation and growth, reducing their risk of failure.

2. Scientific Tools Exist to Help Startups Manage Human Capital

Each of the 14 chapters of this volume discusses a different scientific tool to help startups leverage their human capital for individual, collective, and systemic success. For example, just as a business plan provides a foundation from which one builds a business, a work analysis serves as a foundation to manage human capital strategically. A work analysis (see Chapters 1, 6, 7) is critical for human capital human management because it helps startup leaders understand the tasks that need to be completed to succeed. Work analysis also enables job crafting. Job crafting (see Chapter 7) enables founders to create meaningful work which in turn leads to lower levels of stress, higher levels of engagement, higher work performance, and overall well-being. Another tool mentioned in several of the chapters is psychometric assessments (see, in particular, Chapters 3, 6, 7, 9, 13, 14), which can be used in a variety of different ways in human capital management. Psychometric assessments are a more accurate method to measure workers' KSAOs, feelings, attitudes, and opinions than 'gut instinct' or human judgment alone. They can be used to gather many types of human capital data including team functioning, employee engagement, and organizational values. Innovative tools such as machine learning, artificial intelligence, and advanced text analyses (Chapters 3, 4, 9) unlock a new set of tools that hold a growing promise for addressing challenges of growing complexity in the future. Understanding the psychological aspects of startup workers can help leaders make decisions that will increase motivation, productivity, and job satisfaction. In addition, assessments are more predictive than traditional interviews and reference calls and, ultimately, less costly.

3. Human Capital Data is an Untapped Gold Mine

The chapters in this volume explained how these methodological and analytic techniques can use data to help startups manage human capital. A data-driven approach to human capital is a powerful tactic to calibrate and guide high-stakes decisions. Study after study has demonstrated that scientific evidence is more accurate than expert opinions and data-driven algorithms outperform the evaluations of intuitive raters. For example, standardized selection methods capture data that predicts future worker behavior and is linked to performance outcomes. Moreover, the increased adoption of technology in the workplace has opened a treasure trove of previously untapped data. Startup leaders can use this untapped data to improve their performance and gain a competitive advantage in the marketplace.

4. Identifying Diverse Talent is the Basis of High-Performance

Strategically speaking, investing in diverse human capital will set startups apart from their competitors. Startups can improve their performance by finding, selecting, and retaining diverse talent. Similarly, investors can increase their likelihood of getting a return on their investment by taking a rigorous approach to human capital due diligence (i.e., focusing on the right KSAOs). Conversely, if investors select poorly or if a startup mishires, it can lead to severe financial loss, productivity disruption, and opportunity costs. While there is no single profile that defines a successful entrepreneur, there are some attributes that set high-performing entrepreneurs and low performing entrepreneurs apart. For instance, specific knowledge, skills, abilities, and other attributes (KSAOs) impact entrepreneurship performance.

5. Startups Can Improve Performance through Upskilling

Startup workers also need to develop skills to effectively manage human capital, including leadership and management of individuals and teams. Professional development for startup workers has a positive impact on organization performance, sustainability, profitability, offers legitimacy, and reduces the likelihood of startup failure. Thus, startup leaders would be well served to create professional development plans tailored to the needs of their organization, which can contribute to higher worker satisfaction, team effectiveness, retention, and performance.

6. Fostering Effective Teamwork is Non-negotiable

Startup leaders are reliant on their capabilities to effectively work together as a founding team and manage multiple employee teams. Effective teams require multiple individuals from diverse backgrounds to combine their efforts. For a team to reach apex performance, they need time to grow and develop — just like a living organism. As teams grow over time, psychological states that facilitate teamwork (e.g., cohesion) emerge. Growing effective teams is complex work that requires unique knowledge and skill sets. For instance, team leaders need to ensure members are equipped to operate effectively through communication, coordination, mutual performance monitoring, adaptability, and conflict management.

7. Building a Positive Organizational Culture Prevents People Problems

Entrepreneurship is stressful; building a company requires constant adaptation, navigating an overload of information and distraction, working long hours, interacting with numerous personalities, and dealing with an inadequate supply of resources. Therefore, a positive work environment is integral to the long-term flourishing of a startup. For instance, a positive social environment fosters continued creativity and innovation as a startup matures. Founders can foster a social environment by establishing open communication channels, providing training, and addressing misbehavior instantaneously. Given the challenge and stress that entrepreneurs face, effectively designing their work and career is vital to their well-being and success.

8. It Takes a Village; Startup Success Depends on Its Ecosystem

Startups are more likely to succeed if they have financial capital from investors, insight from experts, and support from coaches or startup development organizations (e.g., accelerators and incubators). Therefore, investors, advisors, coaches, and startup development organizations must invest in human capital. Investors can take a data-driven approach to select prospective startups and invest in the development of their portfolio companies. Startups can seek advisors with expertise in human capital management. Startup development organizations can include team and leadership skill development components alongside the skills needed for typical business functions.

Conclusion

The startup failure rate is upsetting. Clearly, the startup growth and value creation approaches used today are not working. It is time for drastic change and a different mindset. A focus on human capital can be that change and mindset. Bodies of evidence exist to demonstrate that effective human capital management increases organizational performance and survival, removing the risk of trying something new. Moreover, a focus on people will improve outcomes in the broader entrepreneurial ecosystem, including increased well-being, economic growth, and improved social justice.

References

CB Insights. (2021). *The Top 12 Reasons Startups Fail.* CB Insights. Retrieved from: https://www.cbinsights.com/research/report/startup-failure-reasons-top/.

Drucker, P. (1983). *Innovation and Entrepreneurship. Practice and Principles.* New York, NY: Harper.

Kaplan, S.N. and Lerner, J. (2010). It ain't broke: The past, present, and future of venture capital. *Journal of Applied Corporate Finance*, 22(2): 36–47.

Kaplan, S.N. and Lerner, J. (2016). Venture capital data: Opportunities and challenges. *Measuring Entrepreneurial Businesses: Current Knowledge and Challenges*, 413–431.

Kozlowski, S.W., Chen, G. and Salas, E. (2017). One hundred years of the Journal of Applied Psychology: Background, evolution, and scientific trends. *Journal of Applied Psychology*, 102(3): 237.

KPMG. (2022). Global venture capital annual investment shatters records following another healthy quarter. Retreived from: https://kpmg.com/xx/en/home/media/press-releases/2022/01/global-venture-capital-annual-investment-shatters-records-following-another-healthy-quarter.html.

Patel, N. (2015). 90% Of Startups Fail: Here's What You Need To Know About The 10%. Retrieved from: https://www.forbes.com/sites/neilpatel/2015/01/16/90-of-startups-will-fail-heres-what-you-need-to-know-about-the-10/?sh=59352d296679.

Index

D

E

S

Y

About the Editors

Maureen E. McCusker, Ph.D.

Dr. Mo McCusker is a Co-Founder of Blackhawke Behavior Science, an adjunct faculty member at American University, and a corporate practitioner, specializing in talent analytics and predictive modeling. She has spent the last decade and a half working with and studying leaders and teams in organizations from a wide variety of sectors, including the entrepreneurial ecosystem, corporate Fortune 100, medical, legal, academic, military, and non-profit. Using non-traditional data analytical research methods, Mo enjoys studying team dynamics and composition, team and leader effectiveness as a function of the socio-environment, and informal social hierarchies. She has published over 30 different journal articles, scientific conference papers, and book chapters. She has earned several awards and grants for scholarly research from the U.S. Department of Health and Human Services, Virginia Tech, and the Department of Defense, including a post-doctoral research fellowship with the U.S. Army. She holds a Ph.D. and an M.S. in Industrial-Organizational Psychology from Virginia Tech and a B.A. in Psychology and French from Georgetown University.

Nikki Blacksmith, Ph.D.

Dr. Nikki Blacksmith is a Co-Founder and CEO of Blackhawke Behavior Science and an adjunct faculty member at American University in Washington, DC. She has nearly 15 years of experience as a scientist-practitioner of Industrial-Organizational Psychology focusing on psychometrics, selection, decision-making, and entrepreneurial performance. Her clients have included large corporations and startups from a wide range of industries including government, healthcare, MedTech, manufacturing, and venture capital. She has published over 80 different conference papers, book chapters, and journal articles. Her scientific research has been published in top-tier academic publications, including the *American Psychologist*, *Journal of Business and Psychology, Personality and Individual Differences,* and *Journal of Behavioral Decision-Making*. Her research has also featured in many media outlets, including *New York Magazine*, *Real Simple Magazine*, and *BBC.* Dr. Blacksmith received her B.A. from the University of Iowa, her M.A. from the University of North Carolina at Charlotte, and her Ph.D. from George Washington University. She was a Postdoctoral Research Fellow at the Army Research Institute for Social and Behavioral Sciences. She serves on the board of directors for Sembramos, a nonprofit dedicated to building diverse and inclusive organizations. Dr. Blacksmith also serves on the executive board of the Society for Industrial-Organizational Psychology.